Medical Intelligence Unit 24

Peptide-Based Cancer Vaccines

W. Martin Kast, Ph.D.
Professor of Microbiology, Immunology and Pharmacology
Director, Cancer Immunology Program
Cardinal Bernardin Cancer Center
Loyola University Chicago
Maywood, Illinois, U.S.A.

Landes Bioscience
Georgetown, Texas
U.S.A.

Eurekah.com
Austin, Texas
U.S.A.

Peptide-Based Cancer Vaccines

Medical Intelligence Unit

Eurekah.com
Landes Bioscience

Printed in the U.S.A.

Please address all inquiries to the Publishers:
Eurekah.com / Landes Bioscience, 810 South Church Street
Georgetown, Texas, U.S.A. 78626
Phone: 512/ 863 7762; FAX: 512/ 863 0081
www.eurekah.com
www.landesbioscience.com

ISBN: 1-58706-026-4

While the authors, editors and publisher believe that drug selection and dosage and the specifications and usage of equipment and devices, as set forth in this book, are in accord with current recommendations and practice at the time of publication, they make no warranty, expressed or implied, with respect to material described in this book. In view of the ongoing research, equipment development, changes in governmental regulations and the rapid accumulation of information relating to the biomedical sciences, the reader is urged to carefully review and evaluate the information provided herein.

Library of Congress Cataloging-in-Publication Data

Peptide-based cancer vaccines/(edited by) W. Martin Kast
p. ; cm.--(Medical Intelligence Unit)
Includes bibliographical references and index.
ISBN 1-58706-026-4 (alk. paper)
1. Cancer--Immunotherapy. 2. Vaccines. 3.Peptide drugs. I. Kast W. Martin. II. Series.
(DNLM: 1. Cancer Vaccines. 2. Peptides--therapeutic use. QZ 267 P4235 2000) RC271.I45 P462000
616.99'406--dc21

CONTENTS

EDITOR

W. Martin Kast, Ph.D.
Professor of Microbiology, Immunology and Pharmacology
Director, Cancer Immunology Program
Cardinal Bernardin Cancer Center
Loyola University Chicago
Maywood, Illinois, U.S.A.
Chapters 3, 4, 11

CONTRIBUTORS

Gosse J. Adema
Department of Tumor Immunology
University Hospital Nijmegen
Nijmegen, The Netherlands
Chapter 13

Vasso Apostolopoulos
The Austin Research Institute
Austin and Repatriation Medical Center
Heidelberg, Australia
Chapter 7

Jay A. Berzofsky
Molecular Immunogenetics
and Vaccine Research Section
Metabolism Branch
National Cancer Institute
Bethesda, Maryland, U.S.A.
Chapter 2

Remco M.P. Brandt
Cancer Immunology Program
Cardinal Bernardin Cancer Center
Loyola University Chicago
Maywood, Illinois, U.S.A.
Chapters 4, 11

David P. Carbone
Division of Oncology,
Department of Medicine
Vanderbilt Cancer Center
Vanderbilt University School
of Medicine
Nashville, Tennessee, U.S.A.
Chapter 2

Esteban Celis
Department of Immunology
Mayo Graduate School
Mayo Clinic
Rochester, Minnesota, U.S.A.
Chapter 1

Martin A. Cheever
Corixa Corporation
Seattle, Washington, U.S.A.
Chapter 9

David Contois
Molecular Immunogenetics
and Vaccine Research Section
Metabolism Branch
National Cancer Institute
Bethesda, Maryland, U.S.A.
Chapter 2

Diane M. Da Silva
Cancer Immunology Program
Cardinal Bernardin Cancer Center
Loyola University Chicago
Maywood, Illinois, U.S.A.
Chapter 4

Ian D. Davis
Ludwig Institute Oncology Unit
Austin Repat Cancer Centre
Heidelberg, Victoria, Australia
Chapter 14

Joan H. de Jong
Department of Immunohematology
and Blood Bank
Leiden University Medical Center
The Netherlands
Chapter 11

I. Jolanda M. De Vries
Department of Tumor Immunology
University Hospital Nijmegen
Nijmegen, The Netherlands
Chapter 13

Mary L. Disis
Division of Oncology
University of Washington
Seattle, Washington, U.S.A.
Chapter 9

Carl G. Figdor
Department of Tumor Immunology
University Hospital
Nijmegen, The Netherlands
Chapter 13

Dmitry I. Gabrilovich
Cardinal Bernardin Cancer Center
Loyola University Chicago
Maywood, Illinois, U.S.A.
Chapter 3

Theresa Goletz
Molecular Immunogenetics
and Vaccine Research Section
Metabolism Branch
National Cancer Institute
Bethesda, Maryland, U.S.A.
Chapter 2

Lee J. Helman
Pediatric Oncology Branch
National Cancer Institute
National Institutes of Health
Bethesda, Maryland, U.S.A.
Chapter 2

Gregory E. Holt
Cancer Immunology Program
Cardinal Bernardin Cancer Center
Loyola University Chicago
Maywood, Illinois, U.S.A.
Chapter 4

Michael J. Kelley
Medicine Branch
National Cancer Institute
Bethesda, Maryland, U.S.A.
Chapter 2

Samir N. Khleif
Medicine Branch
National Cancer Institute
Bethesda, Maryland, U.S.A.
Chapter 2

R. Kiessling
Department of Oncology
Karolinska Institute
Immune and Gene Therapy Laboratory
Stockholm, Sweden
Chapter 8

Koji Kono
Department of Surgery
Yamanashi Medical University
Yamanashi, Japan
Chapter 8

Michael T. Lotze
Division of Biologic Therapeutics
University of Pittsburgh Cancer Institute
Pittsburgh, Pennsylvania, U.S.A.
Chapter 14

Crystal Mackall
Pediatric Oncology Branch
National Cancer Institute
National Institutes of Health
Bethesda, Maryland, U.S.A.
Chapter 2

V. Ellen Maher
Molecular Immunogenetics
and Vaccine Research Section
Metabolism Branch
National Cancer Institute
Bethesda, Maryland, U.S.A.
Chapter 2

Ian FC McKenzie
The Austin Research Institute
Austin and Repatriation Medical Center
Heidelberg, Australia
Chapter 7

Cornelis J.M. Melief
Department of Immunohematology
and Blood Bank
Leiden University Medical Center
The Netherlands
Chapters 3, 11

Rienk Offringa
Dept. of Immunohematology
and Blood Bank
Leiden University Medical Center
The Netherlands
Chapters 3, 11

Egbert Oosterwijk
Department of Urology
University Hospital Nijmegen
Nijmegen, The Netherlands
Chapter 13

Willem W. Overwijk
Surgery Branch
National Cancer Institute
National Institutes of Health
Bethesda, Maryland, U.S.A.
Chapter 12

Geoffrey Pietersz
The Austin Research Institute
Austin and Repatriation Medical Center
Heidelberg, Australia
Chapter 7

Maurizio Provenzano
Cancer Immunology Program
Cardinal Bernardin Cancer Center
Loyola University Chicago
Maywood, Illinois, U.S.A.
Chapter 4

Maaike E. Ressing
Department of Immunohematology
and Blood Bank
Leiden University Medical Center
The Netherlands
Chapter 11

Nicholas P. Restifo
Surgery Branch
National Cancer Institute
National Institutes of Health
Bethesda, Maryland, U.S.A.
Chapter 12

Y. Rongcun
Department of Oncology
Karolinska Institute
Immune and Gene Therapy Laboratory
Stockholm, Sweden
Chapter 8

Michael P. Rudolf
Cancer Immunology Program
Cardinal Bernardin Cancer Center
Loyola University Chicago
Maywood, Illinois, U.S.A.
Chapter 4

Michael L. Salgaller
Pacific Northwest Cancer Foundation
and Northwest Biotherapeutics, Inc.
Seattle, Washington, U.S.A.
Chapter 10

Jeffrey Schlom
Chief, Laboratory of Tumor
Immunology and Biology
National Cancer Institute
National Institutes of Health
Bethesda, Maryland, U.S.A.
Chapter 6

Barbara Seliger
IIIrd Department of Internal Medicine
Johannes Gutenberg University
Mainz, Germany
Chapter 8

Laurie A. Small
Cancer Immunology Program
Cardinal Bernardin Cancer Center
Loyola University Chicago
Maywood, Illinois, U.S.A.
Chapter 4

M. Charles Smith
Molecular Immunogenetics
and Vaccine Research Section
Metabolism Branch
National Cancer Institute
Bethesda, Maryland, U.S.A.
Chapter 2

Michael Stipanov
Division of Oncology,
Department of Medicine
Vanderbilt Cancer Center
Vanderbilt University School
of Medicine
Nashville, Tennessee, U.S.A.
Chapter 2

Simone A.J. ter Horst
Cancer Immunology Program
Cardinal Bernardin Cancer Center
Loyola University Chicago
Maywood, Illinois, U.S.A.
Chapter 4

Leon van den Broeke
Molecular Immunogenetics
and Vaccine Research Section
Metabolism Branch
National Cancer Institute
Bethesda, Maryland, U.S.A.
Chapter 2

Markwin P. Velders
Cancer Immunology Program
Cardinal Bernardin Cancer Center
Loyola University Chicago
Maywood, Illinois, U.S.A.
Chapter 4

Michel P.M. Vierboom
Dept. of Immunohematology
and Blood Bank
Leiden University Medical Center
Leiden, The Netherlands
Chapter 3

Marten Visser
Cancer Immunology Program
Cardinal Bernardin Cancer Center
Loyola University Chicago
Maywood, Illinois, U.S.A.
Chapter 4

Joost L.M. Vissers
Department of Tumor Immunology
University Hospital Nijmegen
Nijmegen, The Netherlands
Chapter 13

Jeffrey Weber
USC/Norris Comprehensive Cancer
Center
Los Angeles, California, U.S.A.
Chapter 5

Sanne Weijzen
Cancer Immunology Program
Cardinal Bernardin Cancer Center
Loyola University Chicago
Maywood, Illinois, U.S.A.
Chapter 4

B. Scott Worley
Molecular Immunogenetics
and Vaccine Research Section
Metabolism Branch
National Cancer Institute
Bethesda, Maryland, U.S.A.
Chapter 2

PREFACE

The field of peptide based cancer vaccines has evolved tremendously in the last decade of the 20th century. The events that led to the development of this field are the crystal structure of a major histocompatibility (MHC) antigen molecule that revealed a mass in the groove formed by these MHC molecules and the later identification of the peptidic nature of that mass. Since then numerous ways have been described on how to identify such peptides and rules have been determined for the lengths and preferred amino acids at certain positions in those peptides (motifs) that allowed them to bind to the different MHC molecules. In addition, extensive knowledge was gathered on how these peptides were processed from proteins in the cell before they could bind to MHC molecules. The exploration on how to apply the peptide knowledge for vaccination purposes started when it was demonstrated that these peptides after being mixed into adjuvants actually induced T cell responses that could prevent virus infections and tumor growth in experimental animal models. However, it also became apparent that the way a peptide is delivered to a host was crucial in the immunological outcome of the vaccination, leading to a large research effort in that area. Although all the original peptide based vaccine approaches aimed at inducing cytotoxic T lymphocytes (CTL) lately an additional effort is placed on inducing T helper cells in addition to CTL. The results of animal models are currently translated into clinical applications with all of their associated difficulties and heterogeneity. This is especially true with the immunosuppressed state of end stage cancer patients in which almost all of the peptide based cancer vaccines which have been tested have not allowed a major demonstration of the strengths of peptide based cancer vaccines. Nevertheless, initial promising data do appear warranting further research in this area. This book pays tribute to key researchers in the field. They have either written chapters in this book or their work is discussed. The chapters range from basic science on how to identify peptides, through what would be good target epitopes and delivery methods to clinical trials in a variety of different cancers. Some small overlap was unavoidable but actually adds to the depth and understanding of the issue in this research area since different contributors emphasize different aspects of the overall field. Times are changing in the publishing industry. This book will be first to appear in its complete form at the publisher's website, www.eurekah.com, before it is available in print. At the website it can be updated on an ongoing basis. As such it does justice to all contributors of this book with respect to the impact of their work. I hope that this book will provide a comprehensive review for my colleagues working in this field of research and for immunologists and oncologists who want to understand the history of peptide based cancer vaccines and grasp the promise that they have.

W. Martin Kast, Ph.D.

Chapter 1

Identification and Selection of T-Cell Epitopes Derived from Tumor-Associated Antigens for the Development of Immunotherapy for Cancer

Esteban Celis

Because the immune system has the capacity to recognize and in many cases destroy tumor cells, significant efforts are being devoted to the development of immune-based therapies for cancer. Both cytotoxic T lymphocytes (CTL) and helper T lymphocytes (HTL) have been shown to react with antigens expressed by tumor cells and as a result, establish protective and therapeutic effects. Since CTL and HTL recognize antigens in the form of peptide complexes with major histocompatibility complex (MHC) surface molecules (HLA in humans), it is necessary to identify the nature of tumor-derived peptides that can elicit T-cell responses capable of inhibiting tumor-cell growth. The overall objective of our work is to identify peptides derived from sequences of several known tumor-associated antigens (TAA) that are capable of stimulating CTL and HTL against tumor cells. The amino acid sequences of TAA are screened for the presence of peptides containing MHC binding motifs. Corresponding peptides are then synthesized and tested for their capacity to elicit in vitro T-cell responses to tumor cells and corresponding TAA as a final proof that they truly represent T-cell epitopes. As a consequence of these studies, the identified tumor-reactive T-cell epitopes can be developed into therapeutic compounds to treat commonly found epithelial cancers (breast, gastrointestinal and lung). The remaining challenges are how to select the most appropriate mode of vaccination and how to evaluate the effectiveness of immunotherapy in the cancer setting.

Introduction

The incidence of many types of tumors including breast, prostatic, colorectal and lung carcinomas continues to rise in the majority of developed and underdeveloped countries. Most importantly, there is a desperate need to develop non-toxic therapies to eliminate disease, prevent tumor recurrences and inhibit metastatic dissemination, all which should prolong survival while maintaining a good quality of life. Immunotherapy must be considered as the best alternative to accomplish this goal. The purpose of this Chapter is to describe our group's approach to develop effective immune-based therapies for the treatment of commonly found types of cancer. Our belief is that T-lymphocytes are the most efficient constituents of the immune system that are capable of limiting tumor cell growth. Based on this bias, the goal of our studies

Peptide-Based Cancer Vaccines, edited by W. Martin Kast. ©2000 Eurekah.com.

has been to determine the best approach to induce strong and effective anti-tumor-specific T-cell responses as a means of developing epitope-based therapeutic vaccines for cancer.

T-Cells and Cancer

It is well accepted that the immune system has the ability to recognize and eliminate many types of tumors. As a consequence, significant efforts have been devoted in the last 20 years to the development of immune-based therapies for cancer. Cytotoxic T lymphocytes (CTL) and helper T lymphocytes (HTL) have been shown to react with antigens expressed by tumor cells and in many circumstances T-cell reactivity against tumor-derived antigens results in the induction of protective and therapeutic anti-tumor effects. While CTL can directly kill the tumor cells they recognize, antigen-specific HTL will amplify CTL responses and may also exhibit anti-tumor responses by producing lymphokines that directly inhibit tumor-cell growth. In several murine tumor model systems it has been clearly demonstrated that T-cells, and in particular CTL, are capable of eliminating established tumors. Adoptive transfer of tumor-reactive CTL,[1,2,5,6] active immunization using dendritic cells (DC) pulsed with CTL epitopes [7,8] or the use of strong co-stimulatory signals which increase CTL responses have all been reported as successful means of eliminating relatively large established tumors.

In humans, adoptive transfer of tumor-reactive T-cells (sometimes in combination with cytokines), has resulted in objective anti-tumor responses and in may cases in total tumor eradication.[5,6] Although these results have been most encouraging in limited types of tumors such as melanoma, renal cell carcinoma and B-cell lymphomas, positive responses are not observed in all patients. Furthermore, this mode of therapy has not been applicable to the most frequently encountered malignancies such as breast, lung, prostate and gastrointestinal carcinomas. With respect to active immunization, impressive therapeutic responses have been reported in melanoma patients immunized with peptides corresponding to CTL epitopes that were administered in combination with GM-CSF, IL-2 or pulsed onto DC.[11-14] However, as with adoptive therapy, not all patients responded favorably to this mode of immunotherapy and the applicability of this approach to other tumor types is limited because the appropriate CTL epitopes are yet to be defined.

The most likely explanation for these inconsistent results is that tumor cells vary significantly with respect to their antigenic composition and hence melanomas, renal carcinomas and B-cell lymphomas have been considered as "immuno-responsive", while most other tumors are regarded as "poorly immunogenic". Another possible cause for the variability observed in responses to T-cell adoptive therapy within the same tumor type is that the content of "antigen-specific" effector T-cells that are present in the cell product infused into the patients is usually not equivalent from patient to patient. Thus, the identification of relevant TAA and corresponding epitopes for tumor-reactive T-cells will certainly broaden the type of tumors suitable for immunotherapy and should increase the efficacy of therapeutic vaccines or adoptive therapy approaches by facilitating the induction of antigen-specific effector cells.

Although CTL are considered to be the main effector of anti-tumor immune responses, HTL play a pivotal role in enhancing tumor-reactive effector immune responses. Furthermore, HTL may also participate in the establishment of long-term immunity, which is essential for the prevention of tumor recurrences. It has been clearly demonstrated that restoration of long-term immunity by adoptive transfer of CTL requires the presence of antigen-specific HTL.[6,15,16] Thus, a T-cell mediated immunotherapy approach for tumors must include, in addition to the induction of CTL, the concurrent stimulation of TAA-specific T helper cells that will not only potentiate the therapeutic effect, but will also provide long-lasting immunological memory.

Antigen Recognition by T-Cells

T cells recognize antigen as small peptides bound to cell surface molecules encoded by the major histocompatibility gene complex (MHC). CTL are characterized by expression of CD8 cell surface molecules and by their capacity to induce lysis of the target cells they react with via the perforin/granzyme and/or the Fas/Fas-L pathways.[17,18] The T-cell receptors for antigen (TCR) of CTL bind to a molecular complex on the surface of the antigen-presenting cells (APC) formed by peptide epitopes usually derived from viral or tumor-associated antigens (TAA) and major histocompatibility gene complex (MHC) class I molecules. The peptides that are recognized by CTL are usually fragments 8-10 residues long that associate non-covalently with polymorphic class I MHC molecules.[19] On the other hand, HTL express the CD4 surface marker and recognize slightly larger peptides (12-20 residues) in the context of MHC class II molecules, which are only expressed in specific types APC such as B-lymphocytes, monocytes/macrophages and DC.[20]

Many normal and abnormal (e.g., oncogene products) cellular components as well as proteins derived from genes of foreign intracellular microorganisms are processed into MHC-binding peptides which are transported to the APC surface for presentation to the TCR.[19,20] After TCR engagement by appropriate MHC-peptide complexes, CTL have the ability to bind and kill target cells expressing foreign (infectious) or TAA. On the other hand, as a result of MHC-peptide recognition by HTL, these cells produce lymphokines that enhance and amplify CTL immune responses, or in some cases, HTL may also induce the lysis of the cell presenting the antigen or bystander cells via cytolytic mechanisms such as TNF and Fas ligand.

Strategies to Identify TAA and Selection of T-Cell Epitopes

Some of the changes that occur during cell transformation can produce MHC-binding peptides that are immunogenic for CTL or HTL. These TAA include: 1) oncogenic viral proteins; 2) abnormal overexpression of fetal or tissue specific proteins; and 3) mutated or overexpressed oncogene or tumor suppressor gene products.[21-26]

Over several decades various TAA such as CEA, PSA, HER2 and p53, which serve as "tumor markers" have been identified and biochemically characterized.[21,23-28] Because many of these antigens were first identified serologically or genetically, their relevance to CTL and HTL immunity was unclear. However, at the present time, there is sufficient evidence indicating that these tumor markers are capable of producing MHC-binding peptides that are recognized by CTL and HTL.

Two new approaches based on advanced molecular biology and immunology techniques has made it possible to identify several additional cellular products that can function as TAA for T lymphocytes. Both approaches rely on the availability of tumor-reactive CTL obtained from tumor bearing patients to screen gene libraries or peptide fraction isolates. T. Boon and collaborators pioneered the identification of TAA encoding genes of non-viral origin, originally in murine model systems and later in human melanomas.[22] This approach was used to identify a family of genes expressed predominantly in human melanomas (but also in a small proportion of breast, lung and colon carcinomas) and not in most normal tissues (with exception of the testes) designated MAGE.[29] Several CTL epitopes recognized by a melanoma patient's CTL were defined as 9 amino acid peptides which were presented to the TCR in association with class I MHC molecules. Various additional TAA (MART1/Melan-A; pmel-17/gp100, tyrosinase, gp75, p15, BAGE, GAGE, and others), also expressed mainly in melanomas, have been identified by several groups using the same methodology.[4,34-45] These proteins in addition to being expressed in melanomas, are also found in normal melanocytes. These observations demonstrate that under some circumstances TAA can be derived from normal cell constituents, and that immune tolerance to "self-antigens" at the CTL, and possibly at the HTL level is not necessarily always complete.[46] Another approach to identify TAA is to directly sequence

MHC-binding peptides that are eluted and purified from tumor cells.[47-49] This technique requires large numbers of tumor cells from which MHC molecules can be purified together with accurate and sensitive methods to characterize the eluted peptides and as in the previous method, TAA-reactive CTL, usually isolated from tumor patients are required to identify the active peptide fractions before they are sequenced by tandem-mass spectrography.

Identification of T-Cell Epitopes for Tumor Cells Using Reverse Immunology

MHC-Binding Peptides as Potential T-Cell Epitopes

We have developed a completely different strategy to identify peptide epitopes for CTL which can be extended to HTL. These T-cell epitopes are derived from known TAA and the approach does not require the use of patient's tumor-reactive T-cells (which have been very difficult to isolate for tumors other than melanoma and renal-cell Ca) to screen DNA libraries or peptide fractions. This method involves three critical steps: i) identification of defined MHC binding motifs for the major HLA alleles; ii) selection of peptide sequences from putative or known TAA that contain these motifs and measurement of their capacity to bind to purified MHC molecules; and iii) determination of which MHC-binding peptides can elicit in vitro CTL that are capable of killing tumor cells that express the TAA.[50-53]

An important factor to consider in the identification of TAA is whether a peptide can bind to a specific MHC allele since MHC binding is a prerequisite for immunogenicity. Peptide binding to an individual MHC molecule depends on the specific sequence of the peptide.[19,54] Analysis of the sequence patterns of peptides that bind to MHC molecules in humans and mice has revealed the presence of primary and secondary anchor residues. MHC molecules are extremely polymorphic, and theoretically each allelic type will bind different sets of peptides (different alleles of the MHC tend to vary in those residues that form part of the peptide binding pockets). The MHC binding motifs for the most frequently found class I alleles (HLA-A1, -A2, -A3, -A11, -A24, -B7) as well as those for several major class II molecules (DRB*0101, -DRB*0301, DRB*0401 and DRB*0701) have been reported.[54-59] By identifying sets of tumor-associated peptides that bear these motifs and that bind to these the various HLA molecules, one could offer coverage to the majority of the human population (>80%) for developing T-cell epitope-based immunotherapy for tumors.

Once the selected TAA have been screened for sequences that contain MHC binding motifs, synthetic peptides representing these sequences can be synthesized and tested for their capacity to bind purified HLA molecules. Numerous quantitative peptide MHC binding assays have been developed which allow to screen a large number of motif-containing peptides from TAA and determine their binding affinity to several different HLA class I and II alleles. The results presented in Figure 1.1 illustrate an example of a quantitative binding assay of peptides derived from the MAGE antigens to HLA-A1 molecules.[51]

The last and probably most difficult step in the T-cell epitope identification process is to determine whether the peptides that have been identified as MHC binders are capable of inducing anti-tumor CTL or HTL responses. Primary T-cell immunization using synthetic peptides can be done in vitro using human peripheral blood mononuclear cells (PBMC) from appropriate HLA-typed individuals. In the following section we will present several examples where our group has been successful using this strategy in identifying numerous CTL epitopes for melanoma and solid epithelial tumors. The same approach should be applicable for the identification of tumor-reactive HTL.

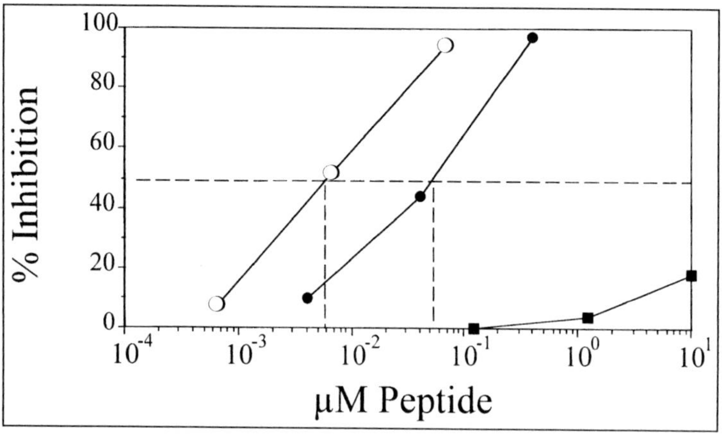

Fig. 1.1. HLA-A1 binding of synthetic peptides from MAGE-1, -2 and -3. The MAGE peptides were tested in a dose titration for the inhibition of the binding of the radiolabeled standard peptide ^{125}I-YLEPAIAKY to purified HLA-A1 molecules as described. Peptides tested were: (●), MAGE-1 peptide EADPTGHSY; (■), MAGE-2 peptide EVVPISHLY; (O), MAGE-3 peptide EVDPIGHLY. Dotted lines are used to calculate the 50% inhibitory concentration (IC_{50}) for each peptide, which inversely correlates with the binding affinity of the peptide to the HLA molecule. Reprinted with permission from: Celis E, Tsai V, Crimi C et al. Proc Natl Acad Sci (USA) 1994; 91:2105-2109. ©1994 National Academy of Sciences, USA.

In Vitro Immunization of T-Cells Using MHC-Binding Peptides from MAGE Antigens

The identification of MHC-binding peptides from known antigen sequences is not sufficient to guarantee that these peptides truly represent to T-cell epitopes. It is necessary to demonstrate that the T-cells can recognize cells that naturally process the antigen and present the corresponding peptide to the TCR as an MHC/peptide complex. In animals (mostly in mice) it is possible to immunize with peptides corresponding to putative T-cell epitopes and demonstrate that the responding T-cells can kill or proliferate to APC that process the corresponding antigen. In humans this type of approach is not feasible and one is limited to either evaluating the responses of T-cells isolated from patients or performing in vitro primary immunization of the T-cell precursors. As mentioned in previous sections, the use of patient-derived T-cell has limitations that in many types of cancer, especially the ones we wish to study here, antigen-specific T-cell lines/clones have been difficult to establish. Furthermore, as will be mentioned in more detail below, T-cells from patients will respond primarily to classical "immunodominant" epitopes whereas (in vitro) immunization studies using defined peptide epitopes should uncover both dominant and subdominant T-cell epitopes.[62]

In view of the above, we have developed an in vitro CTL immunization procedure that utilizes peptide-pulsed APC and CTL precursors from peripheral blood mononuclear cells (PBMC) of normal individuals.[61,63] Using this procedure our laboratory was the first to demonstrate the feasibility of inducing tumor-reactive CTL in normal individuals by in vitro immunization with peptide pulsed activated autologous epitopes B-cells as APC. The results presented in Figure 1.2 show that the HLA-A1-binding peptide that we identified from the MAGE-3 antigen (Fig. 1.1) was efficient in inducing peptide and tumor (melanoma) reactive CTL.

Subsequently, the technique of in vitro immunization of CTL was greatly improved by the use of tissue culture generated DC[64,65] that are used as professional APC. With DC we were able to generate primary CTL to a hepatitis B virus epitope in close to one 100% (1212)

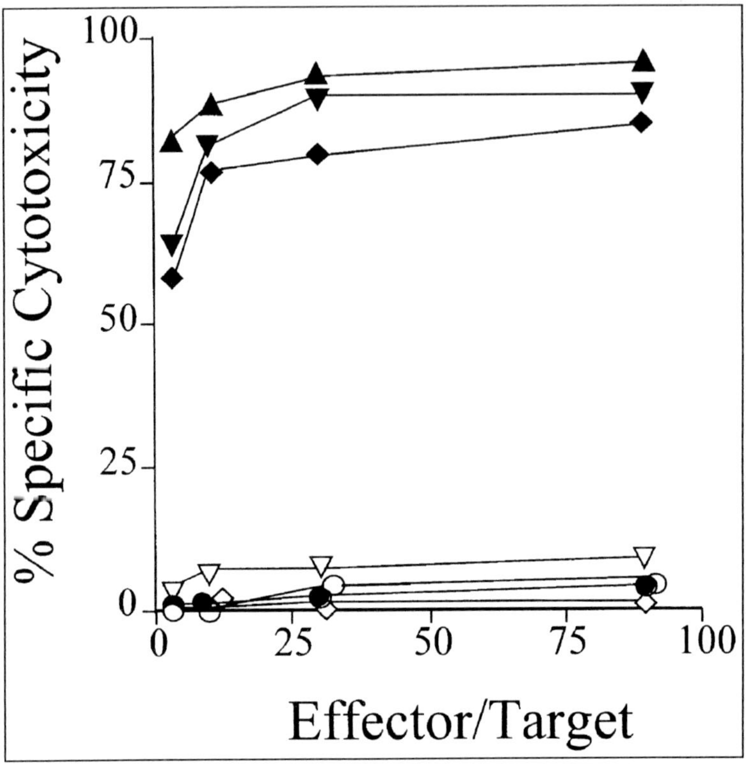

Fig. 1.2. Antigen-specificity and MHC-restriction analysis of MAGE-3 reactive CTL. Cytotoxic responses using peptide-loaded target cells and melanoma tumors: (▲), Steinlin (HLA-A1$^+$ homozygous, Epstein-Barr Virus-transformed lymphoblastoid cell line (EBV-LCL), pulsed with MAGE-3 peptide EVDPIGHLY; (●), Steinlin cells (EBV-LCL) pulsed with MAGE-1 peptide EADPTGHSY; (○), Steinlin cells with no peptide; (▼), 397mel (HLA-A1$^+$, MAGE-3$^+$); (◆), 938mel (HLA-A1$^+$, MAGE-3$^+$); (◇), 888mel (HLA-A1$^+$ MAGE-3$^-$); (∇), 526mel (HLA-A2$^+$, MAGE-3$^+$). Reprinted with permission from: Celis E, Tsai V, Crimi C et al. Proc Natl Acad Sci (USA) 1994; 91:2105-2109. ©1994 National Academy Sciences, USA.

normal individuals. Moreover, the majority of the peptide-reactive CTL (>80%) were also capable of recognizing target cells that naturally process the hepatitis B virus epitope.[61]

Using the newly optimized DC immunization protocol, we proceeded to identify additional MAGE-specific CTL epitopes, but focusing on HLA-A2, one of the most frequently found MHC alleles in humans. In addition, we wished to determine whether MAGE-2 and MAGE-3-reactive CTL could recognize and kill other tumors besides melanomas, which may express these TAA. The rationale for this experiment is based on the published reports that approximately 20-60% of breast, colon and higher numbers of gastric carcinomas express the MAGE antigens.[66-68] Although numerous reports by several groups have demonstrated that MAGE-specific CTL were effective in killing MAGE$^+$ melanoma tumors,[30-33,52,69] no one had evaluated if these CTL could also recognize and kill MAGE$^+$ tumors of other type. Using peptide-pulsed DC, we tested the capacity of two HLA-A2-binding peptides from MAGE to induce tumor-reactive CTL by in vitro immunization of T-cells, and to determine whether these effector cells were capable of recognizing epithelial (non-melanoma) tumors expressing MAGE and HLA-A2 molecules. The results in Figure 1.3 clearly demonstrate that MAGE-2

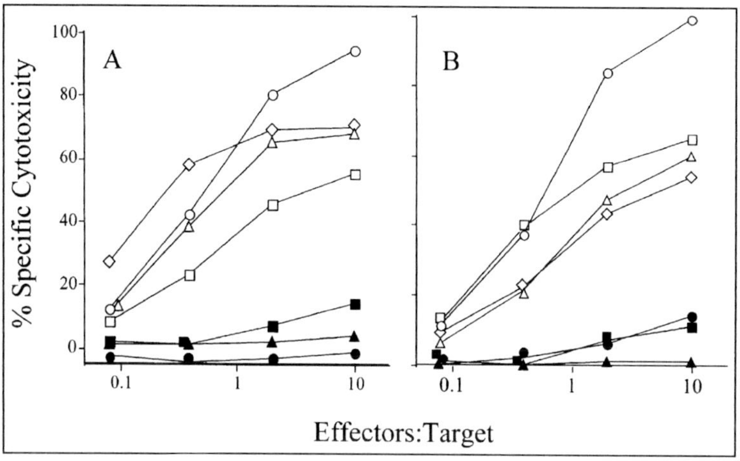

Fig. 1.3. Recognition of various tumor types by a MAGE-2 and -3 specific CTL clones. A MAGE-2 p157-166 (YLQLVFGIEV) specific CTL clone (A) and a MAGE-3 p112-120 (KVAELVHFL) specific CTL clone (B) were prepared from PBMC of a normal HLA-A2 donor using peptide-pulsed autologous DC as APC. Both clones were tested for their lytic activity against the following target cells: ○, .221A2.1(HLA-A2$^+$ homozygous, EBV-LCL) pulsed with the corresponding MAGE peptide; ●, .221A2.1 cells without peptide; △, 624mel (melanoma, HLA-A2$^+$, MAGE-2$^+$/3$^+$); □, KATO-III (gastric Ca, HLA-A2$^+$, MAGE-2$^+$/-3$^+$); ◇, SW403 (colon Ca, HLA-A2$^+$, MAGE-2$^+$/3$^+$); ■, WiDr (colon Ca, HLA-A2$^-$, MAGE-2$^-$/-3$^-$); ▲, 888mel (melanoma, HLA-A2$^-$, MAGE-2$^+$/-3$^+$). Reprinted with permission from: Kawashima I, Hudson S, Tsai V et al. Human Immunol 1998; 59:1-14. ©American Society for Histocompatibility and Immunogenetics, 1998 Elsevier Science, Inc.

and MAGE-3 specific CTL that were induced in vitro with peptide-pulsed DC, were very efficient in killing MAGE$^+$ melanoma, colon and gastric tumor cell lines.[70]

In collaboration with scientists from Takara Biotechnology, (Japan) we have identified additional new MAGE-specific CTL epitopes that are restricted by HLA-A24 which is the most common MHC class I allele in the Japanese population. The CTL elicited by the HLA-A24-binding peptides were efficient in killing both melanomas and gastric carcinomas that express the MAGE antigens.[72] These results confirm the prediction that immunotherapy using MAGE antigens may be applicable to tumors other than melanoma. However, because of the relatively low frequency of epithelial tumors that express MAGE, the use of additional TAA must be considered to offer adequate disease coverage.

In Vitro Immunization of Tumor-Reactive CTL Using MHC-Binding Peptides from Epithelial Tumor Markers

Epithelial sold tumors such as lung, gastrointestinal, breast and prostate represent the most common type of malignancies in the human population. Unfortunately, at present there is no knowledge of a single TAA that is expressed in the majority of these types of tumors which could be developed into a therapeutic vaccine. In view of this, it will become necessary to utilize several TAA to provide disease coverage for any immune-based therapeutic approach to treat epithelial-derived tumors. Our laboratory has selected in addition to MAGE, two TAA, HER2 and CEA, both which are found in a significant number of epithelial tumors. In the following we describe our efforts to identify CTL epitopes for these TAA.

A significant proportion (30-60%) of transformed breast epithelial cells overexpress and produce a proto-oncogene product known as HER2 (also known as neu, c-ErbB2, or p185^{HER2}),

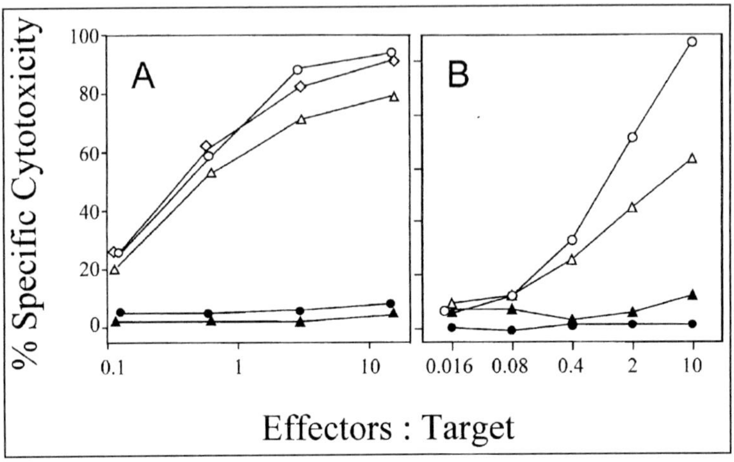

Fig. 1.4. Cytotoxic T lymphocytes induced with HLA-binding peptides from HER2 can kill tumor cells. CTL were induced with the HLA-A2-binding peptide, HER2 p435-443 (ILHNGAYSL) (**panel A**) or the HLA-A3 binding peptide HER p754-763 (VLRENTSPK) (**panel B**). Peptide specific CTL were used as effector cells to test for the lysis of following target cell lines: ○, EBV-LCL cells of the appropriate HLA type pulsed with HER p435-443 (in A) or HER p754-763 (in B); ●, EBV-LCL cells without peptide; △, SW403 cells (colon Ca, HLA-A2$^+$/-A3$^+$, HER2$^+$); ▲, colon Ca cell line, A2$^-$/A3$^-$, HER2$^+$(negative controls). Reprinted with permission from: (A) Kawashima I, Hudson S, Tsai V et al. Human Immunol 1998; 59:1-14. ©American Society for Histocompatibility and Immunogenetics, 1998 Elsevier Science, Inc. (B) Kawashima I, Tsai V, Southwood S et al. Cancer Res 1999; 59:431-435. ©1999 Amer Assoc Cancer Res, Inc.

which bears some homology to the epidermal growth factor receptor.[73] Furthermore, the amplification and overexpression of this proto-oncogene on breast tumors has been associated with aggressive disease and poor prognosis. Other tumors, mainly adenocarcinomas of the ovary, colon and lung have also been reported to overexpress HER2. Since HER2 is selectively overexpressed in malignant cells and not in normal tissues, it has been considered as a possible antigen for CTL-mediated immunotherapy.[23,75,76] Indeed, there are several reports demonstrating that some CTL from ovarian cancer patients are capable of recognizing peptides derived from the HER2 protein.[23,25,77-79] Furthermore, there is also some evidence that helper T cells are capable of reacting with MHC class II peptides from HER2.[76,80,81]

Using our T-cell epitope identification strategy we tested the capacity of MHC class I-binding peptides from HER2 to induce tumor-reactive CTL. Several HLA-A2 and a few HLA-A3 binding peptides from HER2 were analyzed for their ability to trigger primary CTL responses using DC as APC. The examples presented in Figure 1.4 demonstrate that CTL possessing a high level of cytotoxicity for HER2+ tumor cells were produced to both HLA-A2 and HLA-A3 MHC molecules.[70,82]

Another TAA which is an ideal candidate for immunotherapy is CEA (carcinoembryonic antigen). CEA is a 180-kD glycoprotein that is extensively expressed on the vast majority of colorectal, gastric, and pancreatic carcinomas, it is also found in approximately 50% of breast cancers and on 70% of non-small-cell lung cancers.[28,83-87] CEA is also present, but at usually at lower concentrations, in the normal colon epithelium and in some fetal tissues. Circulating CEA can be detected in the great majority of patients with CEA positive tumors and has been used to monitor responses to therapy and disease progression. Recently, CEA-reactive CTL responses were reported in cancer patients that were immunized with a recombinant vaccinia virus expressing CEA.[88] Following the same strategy as described above, we have identified several novel CTL epitopes restricted by commonly found MHC class I alleles such as HLA-A2, -A3 and -A24.[70,82,89] The results presented in Figure 1.5 show that as with other TAA, CEA-reactive CTL that are induced with MHC-binding peptides exhibit significant lytic activity towards CEA-expressing tumor cells.

Enhancement of T-Cell Immunogenicity by Epitope Manipulation

The main consideration of using individual T-cell epitopes instead of whole proteins for developing immune-based therapy for cancer is that while vaccinations using intact proteins representing "self antigens" rarely induce CTL and T helper responses, peptides when administered correctly, have been shown to be capable of eliciting strong T-cell responses to tumor cells.[80] Another major advantage of identifying and utilizing tumor-associated T-cell epitopes is the possibility of increasing their biological activity by the substitution of critical residues which are predicted to enhance MHC binding and/or TCR reactivity. There are several reports demonstrating that peptide analogues of CTL and HTL epitopes can be 10-1000X more efficient in stimulating antigen-specific T cells than the natural peptide sequences. For example, it was clearly shown that peptide analogues corresponding to two well known HLA-A2-restricted CTL epitopes from the melanoma-associated antigens gp100/pmel-17, p280-288 (YLEPGPVTA) and from MART-1/Melan-A, p26-35 (EAAGIGILTV), were shown to be significantly more potent in triggering CTL responses in vitro.[90,91] The analogs YLEPGPVTV and ELAGIGILTV contained substitutions in the primary HLA-A2 binding anchors (shown underlined), which increased the binding of the peptide to this MHC molecule.

In addition to the above examples, our group has demonstrated that a non-immunogenic peptide from CEA can be engineered to elicit tumor-reactive CTL in vitro.[70] In these studies we wished to address whether a low/intermediate HLA-A2-binding peptide that was not capable of inducing CTL could be rendered immunogenic by modifications designed to enhance its MHC binding. Specifically, peptide CEA p24-33 (LLTFWNPPT) is missing one of the "canonical" A2.1 anchors (a V at the carboxyl terminal end), as defined by pool sequencing analysis.[54] In order to determine whether substituting the non-canonical C-terminal anchor (T) with the canonical residue, a peptide analog from CEA p24-33 was prepared and tested for binding to purified HLA-A2.1 molecules. This analog had a "V" at the carboxyl-terminal end and an "M" at position 2 (CEA p24-33/M2V9, sequence: LMTFWNPPV), based on the finding that the presence of "M" in position 2 is frequently associated with optimal HLA-A2.1 binding capacity.[55,92] The analog peptide bound to purified HLA-A2.1 molecules with an approximately 40-fold increase CEA p24-33/M2V9 (IC_{50} 4.5 nM), as compared to the natural peptide, CEA p24-33 (IC_{50} 178.6 nM). It was notable that although peptide CEA p24-33 was not capable, at least according to our experimental protocol, of triggering CTL responses in vitro (data not shown), the peptide analog CEA p24-33/M2V9 was immunogenic in terms of CTL induction. A CTL clone induced by the analog specifically recognized and killed colon and gastric cancer cells expressing the CEA and HLA-A2 antigens (Fig. 1.6).

It has been reported that MHC class II HTL epitopes can also be engineered to increase their MHC binding affinity and T-cell immunogenicity. For example, a melanoma-reactive T helper line specific for the tyrosinase antigen and restricted by HLA-DRB*0401 was shown to recognize a peptide analog with a substitution in one of the major MHC anchors, 10-100 X better than the natural peptide sequence.[93]

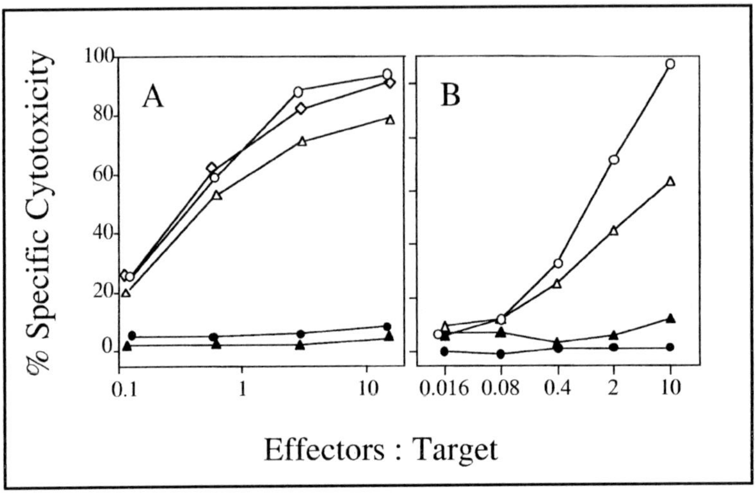

Fig. 1.5. Cytotoxic T lymphocytes induced with HLA-binding peptides from CEA can kill tumor cells. CTL were induced with the HLA-A2-binding peptide, CEA p691-700 (IMIGVLVGV) (**panel A**) or the HLA-A3 binding peptide CEA p61-70 (HLFGYSWYK) (**panel B**). Peptide specific CTL were used as effector cells to test for the lysis of following target cell lines: ○, EBV-LCL of the appropriate HLA type pulsed with CEA p691-700 (in A) or CEA p61-70 (in B); ●, EBV-LCL without peptide; △, SW403 cells (colon Ca, HLA-A2⁺/-A3⁺, CEA⁺); ◇, KATO-III cells (gastric Ca, HLA-A2⁺, CEA⁺); ▲, colon Ca cell line, HLA-A2⁻/-A3⁻, CEA⁺(negative controls). Reprinted with permission from: (A) Kawashima I, Hudson S, Tsai V et al. Human Immunol 1998; 59:1-14. ©American Society for Histocompatibility and Immunogenetics, 1998 Elsevier Science, Inc. (B) Kawashima I, Tsai V, Southwood S et al. Cancer Res 1999; 59:431-435. ©1999 Amer Assoc Cancer Res, Inc.

In summary, it is clear that T-cell epitope identification allows the unique advantage of designing more potent immunogens based on the modification of peptide sequences aimed towards increasing MHC binding affinity. Needless to say, this approach can be extremely useful when the targeted antigen represents a molecule which may be expressed in normal cells such as most TAA, where immune tolerance may need to be broken.

Conclusions

Therapeutic Approaches Using Defined T-Cell Epitopes

The ultimate goal of our research is to develop effective immunotherapies for both early (pre-metastatic) and advanced cancer patients utilizing the information derived from our T-cell epitope identification and re-engineering efforts. Because most patients suffering with advanced, metastatic disease are likely to be immunosuppressed due to their overall poor health status or as a result of chemo/radiation therapy, the logical approach to carry out anti-tumor immune intervention would be adoptive cellular therapy. TAA-specific CTL and HTL could be prepared in vitro as described above using $CD8^+$ and $CD4^+$ T-cells stimulated with autologous peptide-pulsed DC. It is foreseeable that in tissue culture (ex vivo) most of the immunosuppressive factors affecting the cancer patient can be eliminated. The in vitro-generated antigen-specific T-cells will then be expanded either by sequential stimulation with antigen or with mitogen (e.g., anti-CD3 antibody) as described by Riddell and Greenberg and later re-infused into the patients with or without the addition of systemic lymphokines (e.g., IL-2).

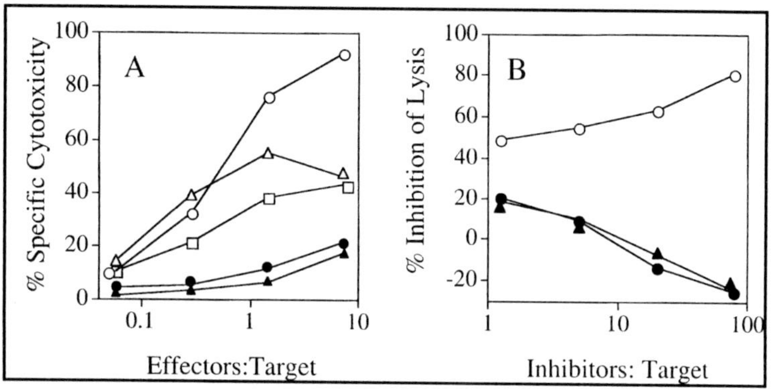

Fig. 1.6. Peptide analogs can induce tumor-reactive CTL. **Panel A:** The CEA p24-33/M2V9 (analog, sequence: LMTFWNPPV) specific CTL clone was tested for its cytolytic activity against following target cell lines: ❍, EBV-LCL cells pulsed with CEA p24-33/M2V9; ●, EBV-LCL without peptide; △, SW403 (colon Ca, HLA-A2$^+$, CEA$^+$); ❒, KATO-III (gastric Ca, HLA-A2$^+$, CEA$^+$); ▲, HT-29 (colon Ca, HLA-A2$^-$, CEA$^+$). **Panel B:** Antigen specificity demonstrated by cold target inhibition assay. Lysis of ^{51}Cr-labeled SW403 cells at an effector/target ratio of 2:1 by the CEA p24-33/M2V9 specific CTL clone was blocked at various Inhibitors / Target ratios by the following cold targets: ❍, EBV-LCL pulsed with CEA p24-33/M2V9; ▲, EBV-LCL pulsed with irrelevant HLA-A2.1 binding peptide (FLPSDYFPSV); ●, EBV-LCL without peptide. Reprinted with permission from: Kawashima I, Hudson S, Tsai V et al. Human Immunol 1998; 59:1-14. ©American Society for Histocompatibility and Immunogenetics, 1998 Elsevier Science, Inc.

Anti-tumor active immunotherapy (i.e., vaccination), on the other hand, will be more suitable for early (non-metastatic) disease, where patients are most likely to have an intact immune system. Furthermore, it becomes very attractive to utilize anti-tumor vaccines in the adjuvant setting with the aim of preventing tumor recurrences after primary conventional treatment by the elimination of micrometastases. Several potential approaches for inducing effective in vivo CTL and HTL responses to TAA can be contemplated. In many circumstances, vaccination with synthetic peptides representing the defined CTL and HTL epitopes in the absence or presence of adjuvants appears to be an easy and attractive way of inducing T-cell responses.[94-96] However, although this mode of vaccination can effectively prevent the establishment of tumors (prophylaxis) it seldom has any benefit when used in the therapeutic mode. Furthermore, it has recently been reported that vaccination with certain peptides representing CTL epitopes can result in T-cell inactivation/deletion[97,98] which would certainly be an undesirable outcome for the treatment of tumors. Notwithstanding the above concerns, it was recently reported that some melanoma patients exhibited tumor responses after receiving peptide vaccination in combination with GM-CSF or IL-2.[11-13] These results support further research in this area, where optimization of vaccination protocols could become an effective method of anti-tumor immunotherapy.

Another approach for inducing active immunity against TAA, especially when multiple T-cell epitopes are being considered, is the use of DNA-based vaccines. Plasmid vaccines containing multiple T-cell epitopes linked in tandem have been reported to induce strong T-cell responses to most of the components of the vaccine. Nevertheless, it remains to be determined whether these types of immunogens are capable of inducing sufficiently potent immune responses capable of providing benefit (extend disease-free survival) in cancer patients.

Currently, the most promising type of anti-tumor vaccine is the use of autologous peptide-pulsed DC. As mentioned previously, DC can relatively easily be prepared in tissue culture from monocytic precursors ($CD14^+$ cells) that are incubated for approximately 1 week with GM-CSF and IL-4. Experiments in mouse models have shown that peptide-pulsed DC vaccination can effectively eliminate established tumors and extend disease-free survival.[7,8] Also very encouraging is the recent observation that several melanoma patients vaccinated with autologous DC pulsed with melanoma-associated peptides demonstrated objective tumor responses.[14] Thus, it appears that antigen-presenting DC are capable of overcoming potential immune tolerance and triggering immune responses to epitopes that are expressed in some normal tissues (melanocytes).

Finally, regardless of the vaccination strategy selected to induce anti-tumor CTL and HTL responses in individuals with established tumors, it becomes critical to carry out clinical studies with realistic endpoints. Unfortunately the evaluation of anti-tumor immunotherapy has been set to the standards of conventional cancer treatments such as chemo and radiation therapy where effectiveness is the reduction or elimination of measurable tumors. Many tumor immunologists, including myself and Dr. Martin Kast feel quite strongly that is naive to expect the disappearance of large tumor masses as a result of vaccination. More realistic and desirable endpoints such as 1) disease free survival, 2) overall survival with reasonable quality of life and 3) time to recurrence should be used to evaluate immunotherapy in cancer. Unfortunately, clinical studies designed to measure these endpoint require significant number of patients, a considerable amount of time (2-5 years minimal), a high amount of resources, and most importantly patience on the part of the clinical investigators and for-profit companies involved in the testing of potential new compounds.

References

1. Melief CJ. Tumor eradication by adoptive transfer of cytotoxic T lymphocytes. Adv Cancer Res 1992; 58:143-175.
2. Greenberg PD. Adoptive T cell therapy of tumors: mechanisms operative in the recognition and elimination of tumor cells. Adv Immunol 1991; 49:281-355.
3. Rosenberg SA. Cancer vaccines based on the identification of genes encoding cancer regression antigens. Immunol Today 1997; :175-182.
4. Van Pel A, van der Bruggen P, Coulie PG et al. Genes coding for tumor antigens recognized by cytolytic T lymphocytes. Immunol Rev 1995; 145:229-250.
5. Rosenberg SA, Packard BS, Aebersold PM et al. Use of tumor-infiltrating lymphocytes and interleukin-2 in the immunotherapy of patients with metastatic melanoma. A preliminary report. N Engl J Med 1988; 319:1676-1680.
6. Heslop HE, Ng CY, Li C et al. Long-term restoration of immunity against Epstein-Barr virus infection by adoptive transfer of gene-modified virus-specific T lymphocytes. Nat Med 1996; 2:551-555.
7. Mayordomo JI, Zorina T, Storkus WJ et al. Bone marrow-derived dendritic cells pulsed with tumor peptides elicit protective and therapeutic anti-tumor immunity. Nature Medicine 1995; 1:1297-1303.
8. Mayordomo JI, Loftus DJ, Sakamoto H et al. Therapy of murine tumors with p53 wild-type and mutant sequence peptide-based vaccines. J Exp Med 1996; 183:1357-1365.
9. Chen L, McGowan P, Ashe S et al. Tumor immunogenicity determines the effect of B7 costimulation on T cell-mediated tumor immunity. J Exp Med 1994; 179:523-532.
10. Townsend SE, Allison JP. Tumor rejection after direct costimulation of CD8+ T cells by B7-transfected melanoma cells. Science 1993; 259:368-370.
11. Marchand M, Weynants P, Rankin E et al. Tumor regression responses in melanoma patients treated with a peptide encoded by gene MAGE-3. Int J Cancer 1995; 63:883-885.
12. Jager E, Ringhoffer M, Dienes HP et al. Granulocyte-macrophage-colony-stimulating factor enhances immune responses to melanoma-associated peptides in vivo. Int J Cancer 1996; 67:54-62.

13. Rosenberg SA, Yang JC, Schwartzentruber DJ et al. Immunologic and therapeutic evaluation of a synthetic peptide vaccine for the treatment of patients with metastatic melanoma. Nat Med 1998; 4:321-327.
14. Nestle FO, Alijagic S, Gilliet M et al. Vaccination of melanoma patients with peptide- or tumor lysate-pulsed dendritic cells. Nat Med 1998; 4:328-332.
15. Riddell SR, Greenberg PD. Principles for adoptive T cell therapy of human viral diseases. Annu Rev Immunol 1995; 13:545-586.
16. Riddell SR, Watanabe KS, Goodrich JM et al. Restoration of viral immunity in immunodeficient humans by the adoptive transfer of T cell clones. Science 1992; 257:238-241.
17. Nabholz M, MacDonald HR. Cytolytic T lymphocytes. Annu Rev Immunol 1983; 1:273-305.
18. Berke G. The binding and lysis of target cells by cytotoxic T lymphocytes: molecular and cellular aspects. Annu Rev Immunol 1994; 12:735-773.
19. Rammensee HG, Falk K, Rotzschke O. Peptides naturally presented by MHC class I molecules. Annu Rev Immunol 1993; 11:213-244.
20. Germain RN, Margulies DH. The biochemistry and cell biology of antigen processing and presentation. Annu Rev Immunol 1993; 11:403-450.
21. Urban JL, Schreiber H. Tumor antigens. Annu Rev Immunol 1992; 10:617-644.
22. Boon T, Cerottini JC, Van den Eynde B et al. Tumor antigens recognized by T lymphocytes. Annu Rev Immunol 1994; 12:337-365.
23. Fisk B, Blevins TL, Wharton JT et al. Identification of an immunodominant peptide of HER-2/neu protooncogene recognized by ovarian tumor-specific cytotoxic T lymphocyte lines. J Exp Med 1995; 181:2109-2117.
24. Cheever MA, Chen W, Disis ML et al. T-cell immunity to oncogenic proteins including mutated ras and chimeric bcr-abl. Ann N Y Acad Sci 1993; 690:101-112.
25. Yoshino I, Goedegebuure PS, Peoples GE et al. HER2/neu-derived peptides are shared antigens among human non-small lung cancer and ovarian cancer. Cancer Res 1994; 54:3387-3390.
26. Nijman HW, Van der Burg SH, Vierboom MP et al. p53, a potential target for tumor-directed T cells. Immunol Letters 1994; 40:171-178.
27. Keetch DW, Andriole GL. The use of tumor markers in prostate cancer. In Prostate Cancer (eds Dawson NA,Vogelzang NJ) 95-112 (Wiley-Liss, Inc., New York, 1994).
28. Shively J, Beatty J. CEA-related antigens: Molecular biological and clinical significance. CRC Crit Rev Oncol Hematol 1985; 2:355-399.
29. van der Bruggen P, Traversari C, Chomez P et al. A gene encoding an antigen recognized by cytolytic T lymphocytes on a human melanoma. Science 1991; 254:1643-1647.
30. Traversari C, van der Bruggen P, Luescher IF et al. A nonapeptide encoded by human gene MAGE-1 is recognized on HLA-A1 by cytolytic T lymphocytes directed against tumor antigen MZ2-E. J Exp Med 1992; 176:1453-1457.
31. van der Bruggen P, Szikora JP, Boel P et al. Autologous cytolytic T lymphocytes recognize a MAGE-1 nonapeptide on melanomas expressing HLA-Cw*1601. Eur J Immunol 1994; 24:2134-2140.
32. van der Bruggen P, Bastin J, Gajewski T et al. A peptide encoded by human gene MAGE-3 and presented by HLA-A2 induces cytolytic T lymphocytes that recognize tumor cells expressing MAGE-3. Eur J Immunol 1994; 24:3038-3043.
33. Gaugler B, Van den Eynde B, van der Bruggen P et al. Human gene MAGE-3 codes for an antigen recognized on a melanoma by autologous cytolytic T lymphocytes. J Exp Med 1994; 179:921-930.
34. Brichard VG, Herman J, Van Pel A et al. A tyrosinase nonapeptide presented by HLA-B44 is recognized on a human melanoma by autologous cytolytic T lymphocytes. Eur J Immunol 1996; 26:224-230.
35. Boon T, van der Bruggen P. Human tumor antigens recognized by T lymphocytes. J Exp Med 1996; 183:725-729.
36. Van den Eynde B, Peeters O, De Backer O et al. A new family of genes coding for an antigen recognized by autologous cytolytic T lymphocytes on a human melanoma. J Exp Med 1995; 182:689-698.

37. Boel P, Wildmann C, Sensi ML et al. BAGE: a new gene encoding an antigen recognized on human melanomas by cytolytic T lymphocytes. Immunity 1995; 2:167-175.
38. Wolfel T, Van Pel A, Brichard V et al. Two tyrosinase nonapeptides recognized on HLA-A2 melanomas by autologous cytolytic T lymphocytes. Eur J Immunol 1994; 24:759-764.
39. Coulie PG, Brichard V, Van Pel A et al. A new gene coding for a differentiation antigen recognized by autologous cytolytic T lymphocytes on HLA-A2 melanomas. J Exp Med 1994; 180:35-42
40. Kawakami Y, Eliyahu S, Delgado CH et al. Cloning of the gene coding for a shared human melanoma antigen recognized by autologous T cells infiltrating into tumor. Proc Natl Acad Sci (USA) 1994; 91:3515-3519.
41. Kawakami Y, Eliyahu S, Sakaguchi K et al. Identification of the immunodominant peptides of the MART-1 human melanoma antigen recognized by the majority of HLA-A2-restricted tumor infiltrating lymphocytes. J Exp Med 1994; 180:347-352.
42. Kawakami Y, Eliyahu S, Delgado CH et al. Identification of a human melanoma antigen recognized by tumor-infiltrating lymphocytes associated with in vivo tumor rejection. Proc Natl Acad Sci (USA) 1994; 91:6458-6462.
43. Robbins PF, el-Gamil M, Kawakami Y et al. Recognition of tyrosinase by tumor-infiltrating lymphocytes from a patient responding to immunotherapy. Cancer Res 1994; 54:3124-3126.
44. Kang X, Kawakami Y, el-Gamil M et al. Identification of a tyrosinase epitope recognized by HLA-A24-restricted, tumor-infiltrating lymphocytes. J Immunol 1995; 155:1343-1348.
45. Kawakami Y, Eliyahu S, Jennings C et al. Recognition of multiple epitopes in the human melanoma antigen gp100 by tumor-infiltrating lymphocytes associated with in vivo tumor regression. J Immunol 1995; 154:3961-3968.
46. Houghton AN. Cancer antigens: immune recognition to self and altered self. J Exp Med 1994; 180:1-4.
47. Castelli C, Storkus WJ, Maeurer MJ et al. Mass spectrometric identification of a naturally processed melanoma peptide recognized by CD8+ cytotoxic T lymphocytes. J Exp Med 1995; 181:363-368.
48. Cox AL, Skipper J, Chen Y et al. Identification of a peptide recognized by five melanoma-specific human cytotoxic T cell lines. Science 1994; 264:716-719.
49. Skipper JCA, Hendrickson RC, Gulden PH et al. An HLA-A2-restricted tyrosinase antigen on melanoma cells results from posttranslational modification and suggests a novel pathway for processing of membrane proteins. J Exp Med 1996; 183:527-534.
50. Celis E, Sette A, Grey HM. Epitope selection and development of peptide based vaccines to treat cancer. Sem Cancer Biol 1995; 6:329-336.
51. Celis E, Fikes J, Wentworth P et al. Identification of potential CTL epitopes of tumor-associated antigen MAGE-1 for five common HLA-A alleles. Molec Immunol 1994; 31:1423-1430.
52. Celis E, Tsai V, Crimi C et al. Induction of anti-tumor cytotoxic T lymphocytes in normal humans using primary cultures and synthetic peptide epitopes. Proc Natl Acad Sci (USA) 1994; 91:2105-2109.
53. Appella E, Loftus DJ, Sakaguchi K et al. Synthetic antigenic peptides as a new strategy for immunotherapy of cancer. Biomedical Peptides, Proteins and Nucleic Acids 1995; 1:177-184.
54. Rammensee H-G, Friede T, Stevanovic S. MHC ligands and peptide motifs: first listing. Immunogenet 1995; 41:178-228.
55. Ruppert J, Sidney J, Celis E et al. Prominent role of secondary anchor residues in peptide binding to HLA-A2.1 molecules. Cell 1993; 74:929-937.
56. Kubo RT, Sette A, Grey HM et al. Definition of specific peptide motifs for four major HLA-A alleles. J Immunol 1994; 152:3913-3924.
57. Kondo A, Sidney J, Southwood S et al. Two distinct HLA-A*0101-specific submotifs illustrate alternative peptide binding modes. Immunogenet 1997; 45:249-258.
58. Southwood S, Sidney J, Kondo A et al. Several common HLA-DR types share largely overlapping peptide binding repertoires. J Immunol 1998; 160:3363-3373.
59. Geluk A, van Meijgaarden KE, Southwood S et al. HLA-DR3 molecules can bind peptides carrying two alternative specific submotifs. J Immunol 1994; 152:5742-5748.
60. Sette A, Sidney J, del Guercio MF et al. Peptide binding to the most frequent HLA-A class I alleles measured by quantitative molecular binding assays. Molec Immunol 1994; 31:813-822.

61. Tsai V, Kawashima I, Keogh E et al. In vitro immunization and expansion of antigen-specific cytotoxic T lymphocytes for adoptive immunotherapy using peptide-pulsed dendritic cells. Crit Rev Immunol 1998; 18:65-75.
62. Sercarz EE, Lehmann PV, Ametani A et al. Dominance and crypticity of T cell antigenic determi nants. Annu Rev Immunol 1993; 11:729-766.
63. Wentworth PA, Celis E, Crimi C et al. In vitro induction of primary, antigen-specific CTL from human peripheral blood mononuclear cells stimulated with synthetic peptides. Molec Immunol 1995; 32:603-612.
64. Romani N, Gruner S, Brang D et al. Proliferating dendritic cell progenitors in human blood. J Exp Med 1994; 180:83-93.
65. Sallusto F, Lanzavecchia A. Efficient presentation of soluble antigen by cultured human dendritic cells is maintained by granulocyte/macrophage colony-stimulating factor plus interleukin 4 and downregulated by tumor necrosis factor a. J Exp Med 1994; 179:1109-1118.
66. Weynants P, Lethe B, Brasseur F et al. Expression of MAGE genes by non-small-cell lung carcinomas. Int J Cancer 1994; 56:826-829.
67. De Plaen E, Arden K, Traversari C et al. Structure, chromosomal localization, and expression of 12 genes of the MAGE family. Immunogenet 1994; 40:360-369.
68. Russo V, Traversari C, Verrecchia A et al. Expression of the MAGE gene family in primary and metastatic human breast cancer. Implications for tumor antigen-specific immunotherapy. Int J Cancer 1995; 64:216-221.
69. Fleischhauer K, Fruci D, Van Endert P et al. Characterization of antigenic peptides presented by HLA-B44 molecules on tumor cells expressing the gene MAGE-3. Int J Cancer 1996; 68:622-628.
70. Kawashima I, Hudson S, Tsai V et al. The multi-epitope approach for immunotherapy for cancer: identification of several CTL epitopes from various tumor-associated antigens expressed on solid epithelial tumors. Human Immunol 1998; 59:1-14.
71. Dong RP, Kimura A, Okubo R et al. HLA-A and DPB1 loci confer susceptibility to Graves' disease. Human Immunol 1992; 35:165-172.
72. Tanaka F, Fujie T, Tahara K et al. Induction of antitumor cytotoxic T lymphocytes with a MAGE-3-encoded synthetic peptide presented by human leukocytes antigen-A24. Cancer Res 1997; 57:4465-4468.
73. Slamon DJ, Godolphin W, Jones LA et al. Studies of the HER-2/neu proto-oncogene in human breast and ovarian cancer. Science 1989; 244:707-712.
74. Yokota J, Yamamoto T, Toyoshima K et al. Amplification of the c-erbB-2 oncogene in human adenocarcinomas in vivo. Lancet 1986; 1:765-767.
75. Ioannides CG, Fisk B, Fan D et al. Cytotoxic T cells isolated from ovarian malignant ascites recognize a peptide derived from the HER-2/neu proto-oncogene. Cell Immunol 1993; 151:225-234.
76. Cheever MA, Disis ML, Bernhard H et al. Immunity to oncogenic proteins. Immunol Rev 1995; 145:33-59.
77. Lustgarten J, Theobald M, Labadie C et al. Identification of Her-2/Neu CTL epitopes using double trangenic mice expressing HLA-A2.1 and human CD8. Human Immunol 1997; 52:109-118.
78. Disis ML, Smith JW, Murphy AE et al. In vitro generation of human cytolytic T-cells specific for peptides derived from the HER-2/neu protooncogene protein. Cancer Res 1994; 54:1071-1076.
79. Peoples GE, Goedegebuure PS, Smith R et al. Breast and ovarian cancer-specific cytotoxic T lympho cytes recognize the same HER2/neu-derived peptide. Proc Natl Acad Sci U S A 1995; 92:432-436.
80. Disis ML, Gralow JR, Bernhard H et al. Peptide-based, but not whole protein, vaccines elicit immunity to HER-2/neu, an oncogenic self-protein. J Immunol 1996; 156:3151-3158.
81. Disis ML, Cheever MA. HER-2/neu protein: a target for antigen-specific immunotherapy of human cancer. Adv Cancer Res 1997; 71:343-371.
82. Kawashima I, Tsai V, Southwood S et al. Identification of HLA-A3-restricted cytotoxic T lymphocyte epitopes from carcinoembryonic antigen and HER-2/neu by primary in vitro immunization with peptide-pulsed dendritic cells. Cancer Res 1999; 59:431-435.
83. Vincent RG, Chu TM. Carcinoembryonic antigen in patients with carcinoma of the lung. J Thorac Cardiovasc Surg 1978; 66:320-328.
84. Thompson J, Grunert F, Zimmerman W. CEA gene family: Molecular biology and clinical perspectives. J Clin Lab Anal 1991; 5:344-366.

85. Sikorska H, Shuster J, Gold P. Clinical applications of carcinoembryonic antigen. Cancer Detect Prev 1988; 12:321-355.
86. Muraro R, Wunderlich D, Thor A. Definition by monoclonal antibodies of a repertoire of epitopes on carcinoembryonic antigen differentially expressed in human colon carcinomas versus normal adult tissues. Cancer Res 1985; 45:57
87. Steward AM, Nixon D, Zamcheck N et al. Carcinoembryonic antigen in breast cancer patients: Serum levels and disease progress. Cancer 1974; 33:1246-1252.
88. Tsang KY, Zaremba S, Nieroda CA et al. Generation of human cytotoxic T cells specific for human carcinoembryonic antigen epitopes from patients immunized with recombinant vaccinia-CEA vaccine. J Natl Cancer Inst 1995; 87:982-990.
89. Nukaya I, Yasumoto M, Iwasaki T et al. Identification of HLA-A24 epitope peptides of carcinoembryonic antigen which induce tumor-reactive cytotoxic T lymphocyte. Int J Cancer 1999; 80:92-97.
90. Parkhurst MR, Salgaller ML, Southwood S et al. Improved induction of melanoma-reactive CTL with peptides from the melanoma antigen gp100 modified at HLA-A*0201-binding residues. J Immunol 1996; 157:2539-2548.
91. Valmori D, Fonteneau JF, Lizana CM et al. Enhanced generation of specific tumor-reactive CTL in vitro by selected Melan-A/MART-1 immunodominant peptide analogues. J Immunol 1998; 160:1750-1758
92. del Guercio MF, Sidney J, Hermanson G et al. Binding of a peptide antigen to multiple HLA alleles allows definition of an A2-like supertype. J Immunol 1995; 154:685-693.
93. Topalian SL, Gonzales MI, Parhurst M et al. Melanoma-specific CD4+ T cells recognize nonmutated HLA-DR-restricted tyrosinase epitopes. J Exp Med 1996; 183:1965-1971.
94. Feltkamp MC, Smits HL, Vierboom MP et al. Vaccination with cytotoxic T lymphocyte epitope-containing peptide protects against a tumor induced by human papillomavirus type 16-transformed cells. Eur J Immunol 1993; 23:2242-2249.
95. Melief CJ. Prospects of T-cell immunotherapy for cancer by peptide vaccination. Sem Hematol 1993; 30:32-33.
96. Melief CJ, Offringa R, Toes RE et al. Peptide-based cancer vaccines. Curr Opin Immunol 1996; 8:651-657.
97. Toes RE, van der Voort EI, Schoenberger SP et al. Enhancement of tumor outgrowth through CTL tolerization after peptide vaccination is avoided by peptide presentation on dendritic cells. J Immunol 1998; 160:4449-4456.
98. Toes RE, Offringa R, Blom RJ et al. Peptide vaccination can lead to enhanced tumor growth through specific T-cell tolerance induction. Proc Natl Acad Sci U S A 1996; 93:7855-7860.in vitro

Chapter 2

Mutant Oncogene and Tumor Suppressor Gene Products and Fusion Proteins Created by Chromosomal Translocations as Targets for Cancer Vaccines

V. Ellen Maher, B. Scott Worley, David Contois, M. Charles Smith, Michael J. Kelley, Michael Stipanov, Samir N. Khleif, Theresa Goletz, Leon van den Broeke, Crystal Mackall, Lee J. Helman, David P. Carbone and Jay A. Berzofsky

Introduction

Active immunotherapy harnesses the body's own immune system to fight cancer. There is strong evidence that the immune response is a potent weapon against cancer. BCG and levamisole, which act through the active participation of the immune system, have been used as therapies against cancer.[1,2] Additional examples of the efficacy of the immune system include tumors that develop following immunosuppressive therapy and respond to withdrawal of that therapy or the tumors that develop in patients with AIDS.[3,4] More recent examples of the efficacy of the immune system include vaccine therapies which have resulted in a clinical response in patients with metastatic cancer.[5-9] The number of these vaccine approaches has grown with our understanding of the immune response and a wide variety of vaccination schemes have been carried out both in animal models and in human trials. Tumor antigens for these trials have been chosen by a variety of methods. One method is through the evaluation of the specificity of tumor infiltrating lymphocytes. Another approach, and the one we have chosen, is to use our increased understanding of tumor biology to develop tumor specific antigens. Molecular changes which are unique to the cancer cells have been described. These include mutant proteins and chromosomal translocations with resulting fusion proteins which provide a unique immunologic target. With this in mind, our targets have included the mutant portion of overexpressed and mutated p53 proteins, oncogenic ras mutations, and fusion proteins created by chromosomal translocations specific to pediatric sarcomas. These targets have the additional advantage that the mutation or fusion contributes to the malignant phenotype of the cell, so that escape mutations or loss variants cannot occur without loss or reduction of the malignant phenotype.

Our understanding of the immune response to cancer has progressed rapidly over the past two decades. However, the fundamental question of why the body's immune response does not eliminate the cancer cells has still not been answered. An understanding of this question is a key element in the development of an effective immune therapy. Tumors are known to evade

Peptide-Based Cancer Vaccines, edited by W. Martin Kast. ©2000 Eurekah.com.

the body's immune system in a variety of ways. These include a decrease in class I expression on the tumor cell surface, a lack of costimulatory molecules, the development of clonal anergy, and the induction of deletional T cell tolerance.[10] Tumor cells are also able to secrete cytokines and other active molecules such as vascular endothelial growth factor which may dampen the immune response.[4,11] Finally, tumor cells have also been shown to actively eliminate the T cells active in the immune response through the fas-fas ligand pathway.[12-14]

Antigen may be presented to T cells through the class I or class II pathway. Cytotoxic T lymphocytes generated through the interaction of T cells with the class I system are thought to be the major mechanism by which cancer cells and intracellular pathogens are eliminated. Antigens presented to the class I pathway are displayed in the form of an eight to ten amino acid peptide within the groove of the class I molecule. This complex interacts with a T cell receptor specific for that complex. Signaling through the T cell receptor causes T cell proliferation, cytokine production, and the arming of the cellular machinery involved in target cell lysis. In addition to activation of the T cell, T cells are also able to recognize targets through the interaction of their T cell receptor with the peptide-class I MHC complex.[15]

Tumor cells are able to evade the immune system in part due to the presence of a decreased density (when compared to other cells in the body) of class I molecules on their cell surface.[16,17] This causes both a decrease in the initial T cell activation and in T cell recognition of the tumor cell as a target. For example, we have seen a frequent loss of HLA-B and C locus expression, usually with retention of HLA-A locus expression, on a series of hepatoma cell lines.[18] We have also seen mutations and deletions in b2-microglobulin that result in decreased class I MHC expression in human lung and colon cancer[19] as well as examples of TAP deficiency in small cell lung cancer resulting in loss of class I MHC expression.[20] Some oncogenic mutations are able to further decrease the amount of class I expressed through a decrease in the production of beta 2 microglobulin. Beta 2 microglobulin is part of the class I complex of all cells and acts to stabilize the complex on the cell surface. Ras mutations have been shown to decrease the production of beta 2 microglobulin.[21] That this lack of class I expression is, in practice, an effective evasion of the immune system is shown by the results of the upregulation of class I molecules. Cytokines such as interferon can cause an increase in class I expression on the cell surface.[10,22] When tumor cells are incubated with these cytokines they have been shown to upregulate their class I molecules and to be more effectively lysed by cytotoxic T lymphocytes (CTL).[23] While the ability of tumor cells to evade the immune system through the downregulation of class I is well established, it is unclear why this decrease in class I expression does not lead to the effective generation of a natural killer cell response. In general, natural killer cells are inhibited by binding to class I molecules.[24] Therefore, theoretically a tumor which has decreased its class I molecules would be susceptible to natural killer cell lysis. This decrease in class I does not, however, result in an effective natural killer cell response that results in tumor elimination. One potential mechanism is the selective downregulation of only certain class I loci.[18]

An additional way in which the tumor cell can act as a poor antigen presenting cell and evade the immune system is through the lack of costimulation. Costimulation is provided through at least three molecules, B7-1, B7-2, and B7-3.[25] Additional adhesion molecules such as leukocyte functional antigen 1 and intercellular adhesion molecule 1 are also involved in enhancing the T cell response to stimulation.[26] The lack of costimulation can prevent the naive T cells from responding to the antigen presented by the tumor cell. When tumor cells are transfected with costimulatory molecules, they can then serve as effective antigen presenting cells. The immune response to murine tumor in the presence and absence of such costimulatory molecules has been examined in murine models. Here, the B7 transfected tumor cells were able to present to the immune system and to activate cytotoxic T lymphocytes while the original tumor cells did not cause a CTL response. In addition, these B7 transfected tumors grew poorly and the majority of these tumors regressed while the original tumors continued to expand. Further, mice injected with these B7 bearing tumors were now able to reject both transfected tumor cells and the original tumor.[28] Presentation of antigen by MHC in the absence of costimulation can lead to T cell anergy, one mechanism of tolerance.[29]

Tumors may evade the immune system through the development several types of tolerance, including aneregy or deletional tolerance. Anergy may be reversed by presentation of antigen accompanied by optimal costimulation or through the addition of interleukin 2,[27,30-32] whereas deletional tolerance cannot be reversed. Deletional tolerance can either be central, due to negative selection of T cells specific for self antigens in the thymus, or peripheral, induced by antigens in the periphery, potentially including antigens of the tumor if present in large amounts. With regard to thymic negative selection or central tolerance, it has been shown that the threshold for T cell deletion is much lower than that for a T cell response.[33,34] Despite this, transgenic mouse models have been developed in which a T cell response to the transgenic antigen can be seen, and human cytotoxic T cell responses to self-antigens have been observed. In addition, cytotoxic T lymphocytes have been developed toward such ubiquitous proteins as albumin.[37]

Antigen-induced apoptosis in the periphery is one mechanism of peripheral tolerance by which the T cell response to the tumor may be abrogated. This mechanism may occur under several conditions. The most closely studied in the case of $CD4^+$ cells is the introduction of a second stimulation event through either the continuous presence of the antigen or the addition of IL-2. This second stimulation results in apoptosis of the formerly responding T cells.[38-40] In the case of $CD8^+$ T cells, supraoptimal peptide MHC densities induce apoptosis without a requirement for a second stimulation.[41,42] The role of this event in the failure of the body to eliminate cancer is unclear. However, the continuous presence of tumor could, theoretically, lead to T cell apoptosis, rendering the immune response less effective. We have studied T cell apoptosis following supra-optimal antigen presentation rather than re-presentation of the antigen. High and low avidity cytotoxic T cells to the I10 peptide (from HIV-1 gp160) have been developed by growing these cells in vitro with high (low avidity) and low (high avidity) concentrations of this peptide. The I10 peptide is the immunodominant epitope in the HIV envelope protein gp160.[43,44] We have found that exposure of the high avidity CTL to increased concentrations of peptide will result in their apoptosis.[41,42] We have not achieved a concentration of antigen that induces apoptosis of the low avidity CTL. Central tolerance to self antigens may also preferentially delete high avidity T cells, as suggested in the case of T cells specific for wild type p53 in normal vs. p53-knock-out mice.[45]

Preferential deletion of high avidity T cells by either central or peripheral mechanisms is particularly important since the high affinity cells may be necessary to effectively clear an infection or tumor. Our own work has shown that high avidity cells are much more effective than low avidity cells in clearing infection.[46] Low avidity cytotoxic T lymphocytes were grown with 100 μM of the I10 peptide from HIV-1 gp160 while high avidity CTL were grown with 0.0001 μM of I10. An equal number of cells were transferred into a nude mouse infected with vaccinia expressing gp160. The mice were then sacrificed and assayed for the amount of vaccinia virus still present within the mouse. Figure 2.2 illustrates that the high avidity cells were much more effective than the low avidity cells in clearing the virus.

It is also true that cancer cells can be more effectively eliminated by high avidity cytotoxic T lymphocytes as shown in two recent studies.[47,48] This suggests that tumors, which present a relatively large amount of antigen on their surface may also be able to avoid the immune system by eliminating the most effective, high avidity cells from the T cell repertoire. Further, since many tumor antigens are self antigens, the high avidity cytotoxic T lymphocytes deleted to these self antigens during development.[41,45] With this in mind, we have chosen to focus on new and unique tumor antigens encoded by mutant or translocated genes that occur following development, for which high avidity T cells should not be deleted at least centrally.

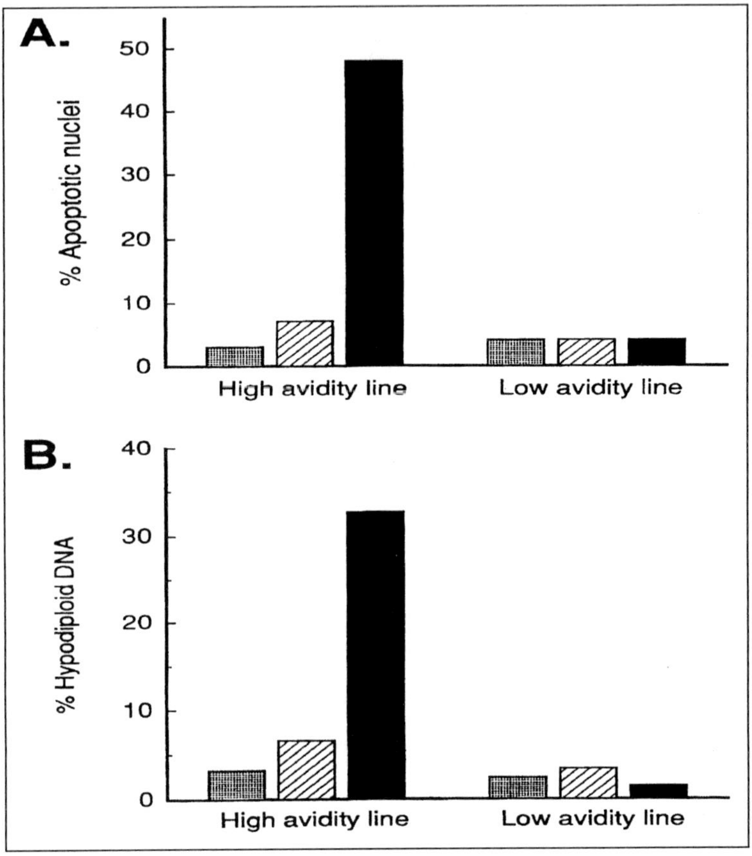

Fig. 2.1. Cytotoxic T lymphocytes were grown with high concentrations, 100 µM (low avidity) or low concentrations, 0.0001 µM (high avidity) of I10. The cells were then stimulated with spleen cells incubated with no peptide (stippled bars), 0.001 µM I10 (stripped bars), or 100 uM I10 (solid bars). High avidity cytotoxic T lymphocytes respond by undergoing apoptosis while low avidity CTL are able to tolerate high concentrations of peptide. Reproduced from ref. with permission.

We have also studied the role of frequent dendritic cell stimulation of tumor bearing mice. These experiments have shown that frequent stimulation, under conditions of optimal costimulation, do not result in a depletion of the T cell response. Theoretically, mechanisms such as anergy, tolerance, or apoptosis could dampen this T cell response. We have found, however, that frequent immunizations are more effective in holding the tumor in than immunization of dendritic cells pulsed with the same tumor antigen. Here, mice that received multiple injections showed stabilization of their tumor.

Tumors may also evade the immune system through the elaboration of various cytokines.[4] We have studied the role of vascular endothelial growth factor (VEGF) in the tumor's ability to avoid immune detection. VEGF is necessary for tumor angiogenesis and growth.[50] It provides an additional advantage to the tumor in that it inhibits the production of mature dendritic cells.[11] The tumor bearing host has been shown to produce a decreased number of mature dendritic cells which can be reversed by the growth of dendritic cells outside of the body or produced experimentally through the delivery of VEGF.[51] In Figure 2.4, we show that splenic dendritic cells from tumor bearing mice (matured in vivo), pulsed with antigen, are relatively

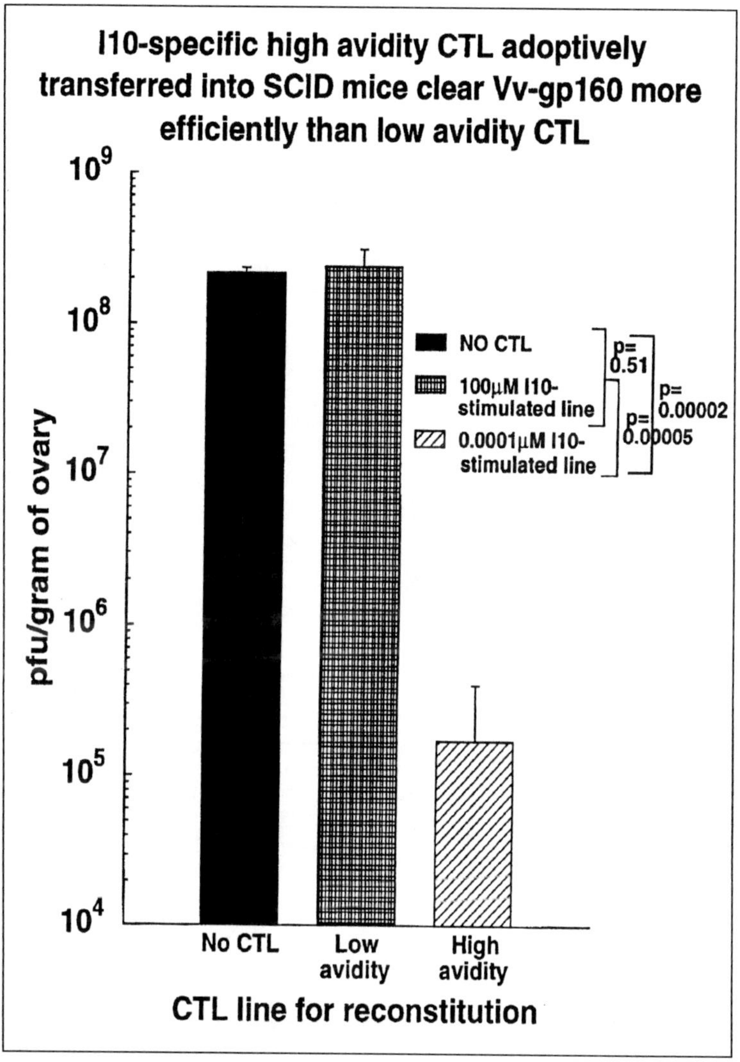

Fig. 2.2. This Figure illustrates the clearance of vaccinia virus from SCID mice that underwent adoptive transfer of high and low avidity cytotoxic T lymphocytes. Cytotoxic T lymphocytes were grown on either high concentrations, 100 μM (low avidity) or low concentrations, 0.0001 μM (high avidity) of I10. I10 is the immunodominant epitope of gp160. Either 10^7 high or low avidity cells were then transferred into each SCID mouse followed by infection with a vaccinia virus construct containing gp160. Three days later the ovarian tissues were assayed for the presence of live viral particles (measured as plaque forming units. Reproduced from ref. with permission.

ineffective in preventing tumor growth, while bone marrow cells matured into dendritic cells ex vivo are able to inhibit this growth.

We have demonstrated that this phenomenon is secondary to the ineffective dendritic cells produced by tumor bearing mice by introducing dendritic cells from syngeneic mice, by maturing dendritic cells ex vivo, by adding VEGF to maturing dendritic cells from tumor bearing mice, and by treating tumor bearing mice with anti-VEGF antibodies. Dendritic cells

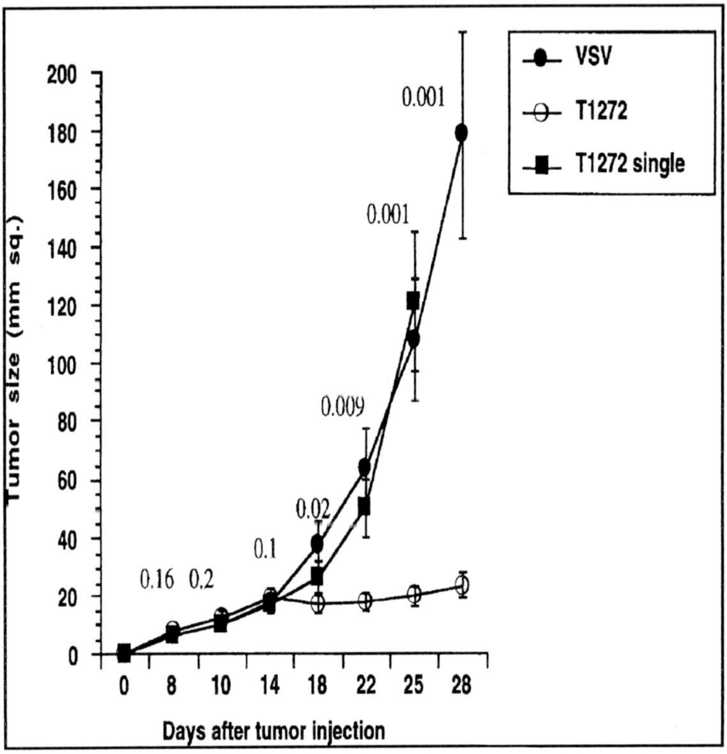

Fig. 2.3 Mice with established tumors were immunized with peptide pulsed dendritic cells either on day 8 or days 8, 14, 18, and 22. The peptide T1272 used in the immunization corresponded to the area of mutation of a mutant p53 protein endogenously processed and presented by the tumor. Mice that received a single immunization showed little if any delay in their tumor growth while mice that received the more frequent immunizations exhibited stabilization of their tumor. Cells pulsed with a VSV peptide were used as a control. Reproduced from ref. with permission.

from tumor bearing mice treated with these antibodies have been shown to be effective antigen presenting cells when compared with dendritic cells from mice which did not receive antibody.[52]

Finally, a newly elucidated mechanism by which tumor may evade the immune system is through the fas-fas ligand pathway.[12-14] Tumor cells may express both fas and fas ligand on their surface.[53] The interaction of fas ligand on the tumor cell surface with fas on the T cell may result in the death of the T cell through apoptosis, although this interaction may result in the apoptosis of the tumor cell as well.[54] However, the induction of apoptosis through fas-fas ligand interactions is often markedly reduced in cancer cells, even when fas is highly expressed.[53,54] Expression of fas ligand by tumor cells is often accompanied by loss of fas on the tumor cells, eliminating the possibility of self induced apoptosis.[13] The presence of ras mutations has been shown to alter these fas mediated interactions among tumor cells. Tumors harboring ras mutations are able to downregulate fas, making such tumor cells resistant to fas-mediated apoptosis.[55] Thus, tumor cells are able to both abrogate the immune response and to actively induce the death of responsive T cells through this mechanism.

As we have illustrated, there are multiple mechanisms by which tumor cells are able to evade the immune system. Many of these mechanisms, in animal models, can be overcome by various vaccination schemes. We have therefore chosen an approach which optimizes the

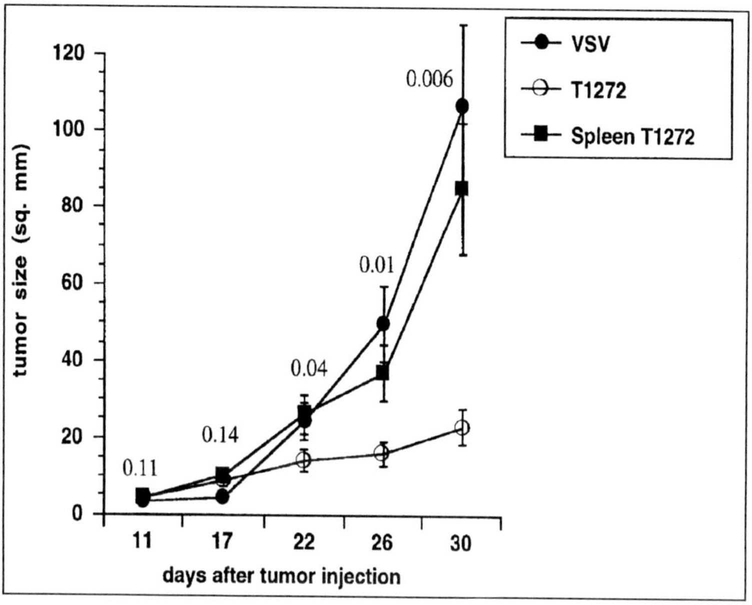

Fig. 2.4. Mice with established tumors were immunized with dendritic cells matured ex vivo from the bone marrow of tumor bearing mice (circles) or isolated from the spleens of tumor bearing mice (matured in vivo) (squares). Dendritic cells were pulsed with mutant p53 peptide T1272 or control VSV peptide. Immunizations were performed on days 11, 17, 22, 26 after tumor injection. Reproduced from ref. with permission.

response of the immune system to tumor specific antigens such as ras and p53 mutations as well as the unique peptides generated by chromosomal translocations.

Ras

While many gaps remain in our understanding of the immune response to tumor cells, this increased understanding, as well as advances in molecular biology, have lead to efforts by ourselves and others to develop effective cancer vaccines. We have chosen to use an active vaccine approach using tumor specific antigens in the development of an effective cancer vaccine. There are several advantages to the use of tumor specific antigens rather than tumor associated antigens. One is that an immune response directed against a tumor specific antigen should have very limited effects on normal tissues. A second is that the body's immune system has not developed tolerance to these unique antigens through clonal deletion during T cell development. An additional advantage is that the ras mutation is required for the malignant phenotype.[56] Therefore, theoretically, ras expression cannot be downregulated to avoid immune detection. Ras is normally an intracellular guanosine triphosphate binding protein that is important in cell signal transduction. When a signal is received by the cell, it works through the ras pathway to provide a proliferative signal to the cell nucleus. Mutated ras proteins provide a continuous proliferative signal to the cancer cell. Ras mutations occur through a single base substitution within the DNA which results in an amino acid substitution within the ras protein. These substitutions typically occur at positions 12, 13, 61 and occasionally position 59. In addition to the limited number of sites at which substitutions occur within the protein, a limited number of amino acids have been found to substitute at these sites. These properties make ras a unique candidate for specific immune therapies.

There are three ras proteins, each expressed by different tissues within the body. These are K-ras, N-ras, and H-ras. Tumors arising from the tissue in which a given protein is expressed may have a mutation in one of these proteins. Our work has predominantly focused on K-ras which is commonly mutated in lung and gastrointestinal malignancies. Mutations in N-ras are associated with hepatoma, myeloid tumors such as acute leukemia and myelodysplasia, melanoma, and other skin cancers. H-ras is the least commonly mutated ras protein and is usually associated with bladder and renal cell carcinomas.[58]

The development of an immune response to mutant ras peptides has been shown in a number of mouse model systems. Fenton et al showed that cytotoxic T lymphocytes could be developed in BALB/c mice to a mutant ras protein that had arginine substituted for glycine at position 12. These cytotoxic T lymphocytes were protective when the mice were challenged with tumors containing this specific mutation, but were not protective when mice were challenged with tumors containing other ras mutations. Here, as a tumor, BALB/c 3T3 cells were transfected with plasmids containing various ras mutations.[59] Fenton was also able to demonstrate cytotoxic T cells in C3H mice to a ras mutation (glutamine to leucine) occurring at position 61 of the ras protein.[59] Abrams et al[56,60] were also able to show in the BALB/c mouse specific lysis directed toward a second ras mutation. Here, the cytotoxic T lymphocytes were specific for ras peptides representing positions four to 12 of the ras protein with valine substituted for the usual glycine at position 12. These cytotoxic T lymphocytes were able to lyse A20 cells transfected with mutant ras containing this point mutation or cells pulsed with the peptide containing the mutation.[60] Peace et al[61] were also able to induce cytotoxic T lymphocytes that specifically recognized murine tumors transfected with the appropriate ras mutation. This group, however, used a very different method. They generated a primary in vitro response from C57BL/6 mice by exposure to a mutant ras peptide containing leucine substituted at position 61 for the normally occurring glutamine. These cytotoxic T lymphocytes were able to lyse an SV40 transformed fibroblast line transfected with H-ras containing the appropriate mutation.[61]

Skipper and Stauss[62] were able to develop cytotoxic T lymphocytes directed toward both mutant ras (lysine substituted for glutamine at position 61) and wild type ras. Here, C57BL/10 mice were vaccinated with recombinant vaccinia viruses expressing either wild type or mutant human N-ras. Although these mice were vaccinated with human N-ras, it should be noted that this is highly homologous to murine ras and is identical at positions 12, 13, and 61. The mice vaccinated with mutant ras developed cytotoxic T lymphocytes specific for the mutation at position 61 while the mice vaccinated with wild type ras developed cytotoxic T lymphocytes directed toward the wild type ras epitope at position 61. Interestingly, mice vaccinated with mutant ras had some reactivity to wild type ras (less than that of mice vaccinated with wild type ras) at position 61 while mice vaccinated with wild type ras also exhibited some cross reactivity with the mutant epitope (again less than that of mice vaccinated with mutant ras). These cytotoxic T cells were able to lyse EL4 tumor cells transfected with and hence overexpressing mutant or wild type ras. However, cytotoxic T lymphocytes from both vaccination groups failed to lyse normal murine cells. Further, mice vaccinated with both vectors developed cytotoxic T lymphocytes to a second non-mutated portion of ras at positions 152-159 with equal strength.[62] Although normal cells do not typically overexpress mutant ras, these findings may still be of concern even though these cells are artificially overexpressing ras, because of the potential risk of an autoimmune response to wild type ras. However, vaccination with mutant peptides, rather than with vectors expressing the whole protein containing the mutation, may eliminate this difficulty. Further, the ability to break tolerance may be a function of the mechanism of vaccination. In other systems, notably in mice transgenic for carcinoembryonic antigen, tolerance can be broken by vaccination with vaccinia and not by vaccination with peptide or protein in adjuvant.[63]

The development of an immune response to mutant ras peptides has been shown in the human system primarily through the development of a primary in vitro response. Van Elsas et al were able to develop HLA A2.1 restricted CTL against a mutant ras peptide containing leucine substituted for glutamine at position 61.[64] Fossum et al, were able to develop cytotoxic T lymphocytes through primary in vitro stimulation from a patient with colon carcinoma. These CTL were directed towards a substitution of asparagine for glycine at position 13 when presented by the HLA B12 molecule. Further, these cells were able to recognize a colon carcinoma that endogenously contained this ras mutation and expressed HLA B12.[23] This suggests that, in humans, epitopes from the mutant ras protein can be presented on the surface of tumor cells and more importantly, can be recognized by cytotoxic T lymphocytes.

More recently, several investigators have demonstrated a human cytotoxic T lymphocyte response in humans immunized with ras mutant peptides and have shown that these responding cytotoxic T cells are able to recognize tumor cells endogenously containing these mutations. Gjertsen et al have demonstrated the development of a cytotoxic T cell line following vaccination with peripheral blood mononuclear cells incubated with mutant ras peptides corresponding to mutations present in the patient's own tumor in two patients. These cytotoxic T lymphocytes were able to recognize mutant (substitution of valine for glycine at position 12), but not wild type ras peptides. Further, these cells were able to lyse the patient's own tumor cells which endogenously process and present the mutant ras peptide.[65] Abrams et al and Khleif et al have also produced cytotoxic T lymphocytes following vaccination with peptides containing ras mutations corresponding to that in the patient's own tumor.[66,67] They were able to demonstrate CTL from two separate patients, one vaccinated with a mutant ras peptide containing valine substituted for glycine at position 12 and the other vaccinated with a peptide containing asparagine substituted for glycine at position 12. Both cytotoxic T cell lines were able to recognize targets pulsed with the corresponding peptide. More importantly, CTL directed against the valine substitution at position 12 were able to recognize HLA matched tumor cells endogenously expressing the corresponding ras mutation.[66] These findings demonstrate a proof of principle that tumors containing mutant ras are able to process and present peptides with substitutions at position 12. Further, these results suggest that mutant ras peptides containing substitutions at position 12 can be made immunogenic, and that the resultant CTL may be able to lyse the patient's own tumor cells.

We recently completed a clinical trial in which we demonstrated a number of cytokine or cytotoxic T cell responses to vaccination with peripheral blood mononuclear cells incubated with mutant ras peptides corresponding to the mutations present in the patient's own tumor. Here, patients were not selected for HLA type or for peptide binding and presentation. Despite that, we were able to demonstrate responses. Both patients with metastatic disease and those in the high risk adjuvant setting were recruited into the trial. The trial specified four initial vaccinations, followed by additional vaccinations if the patient developed an immune response. The measurement of CTL was performed using a six to eight day in vitro stimulation with no additional in vitro restimulations. Cytokines were measured in the supernatant of these in vitro cultures at the day of harvest for the CTL assay. Cytotoxic T cell responses were considered positive if lysis of targets with mutant peptide was 10% greater than that of targets without peptide. Cytokine responses were considered positive if the stimulation index (ratio of cytokine release with peptide to that without peptide) was ≥ 2 and if the cytokine response was at least 50 pg/ml. Preliminary analysis indicates that among patients with ras mutations, six of 18 had evidence of a positive CTL response and 10 of 18 had evidence of a positive cytokine response (Carbone et al, manuscript in preparation).[68]

We have shown that a number of mutant ras peptides with substitutions at positions 12 and 13 are able to bind to the HLA A2.1 molecule, the most common human class I MHC molecule. Most substituted peptides bound with higher affinity than the wild type peptide, reducing the risk of autoimmunity. These include:

Wild Type	YKLVVVGAGGVGKSALT
PR6	YKLVVVGADGVGKSALT
PR7	YKLVVVGAAGVGKSALT
PR18	YKLVVVGACGVGKSALT
PR54	YKLVVVGAGDVGKSALT

Figure 2.5 illustrates the binding of the peptides listed above, and the corresponding 10-mers (with the suffix V10) homologous to the sequence KLVVVGAGGV in the wild type peptide, shown to be the minimal length for binding to HLA A2.1. The control peptides tested are the influenza matrix peptide FMP, and a peptide from HIV-1 IIIB gp120, P18-I10. Results showed that the ability to bind, as measured by enhanced expression of HLA-A2 on the surface of T2 cells, was dependent on the length of the peptide and the effect of the mutation on antigen processing. Some of the peptides were more effective as the minimal 10-mers, because the mutation reduced the rate of extracellular processing of the longer peptide, whereas other sequences were more effective as the longer peptides, because the minimal 10-mers were degraded too rapidly.[69] Thus, the different mutations can affect the processing of the antigen at points several residues distant from the site of the mutation.

One of our current clinical trials focuses on the ability of patients with HLA type A2.1 to present these peptides to their immune system. To increase the immunogenicity of these peptides, we have chosen to use dendritic cells matured with interleukin-4 and granulocyte-macrophage colony stimulating factor. We hope that this approach will lead to a further increase in the number of immune responses and clinical responses achieved in our future trials.

p53

We have also extended our approach of vaccination with tumor specific antigens to the p53 gene. Wild type p53 ordinarily acts as a tumor suppresser gene. However, once p53 becomes mutated, it is no longer able to act as a tumor suppresser gene. Thus if additional mutations occur, p53 is no longer able to block the full transformation into a malignant cancer cell. p53 mutations are very common, occurring in approximately half of human cancers. These mutations do not occur at a limited number of sites as they do in ras, but do tend to cluster at exons five through eight. Importantly, most p53 mutations result in overexpression of the protein, potentially making mutant p53 a highly expressed tumor antigen.[70]

Confirming the ability of p53 to act as a tumor antigen, we have demonstrated the ability of BALB/c mice to develop cytotoxic T lymphocytes upon vaccination with mutant p53 peptide pulsed spleen cells.[71] The genetic mutation used in these studies was originally found in a human lung carcinoma and corresponds to the substitution of cysteine for tyrosine at position 135. In our model system, we generated a synthetic peptide corresponding to this mutation. This substitution of cysteine for tyrosine created a new antigenic determinant by introducing an anchor residue for binding H-2 K^d. These CTL were able to kill tumors transfected with the mutant p53 that expressed this protein endogenously. Thus, endogenous mutant p53 can be processed and presented by class I MHC molecules to CTL.[71] More importantly, in established tumor models, vaccination with dendritic cells pulsed with these mutant p53 peptides was able to retard the growth of established tumor (see Figs. 2.3 and 2.4 above).[49] The addition of interleukin 12 resulted in a further inhibition of tumor growth. This is illustrated in Fig. 2.6.[72]

Other investigators have also demonstrated the development of cytotoxic T lymphocytes to mutant p53 in the BALB/c model. CTL were generated against a p53 mutation found in a murine carcinogen induced tumor. These CTL were protective against tumor challenge and slowed or prevented the growth of established tumors. More interestingly, in this model system, CTL were also induced against wild type p53 peptides. These CTL, directed against non-

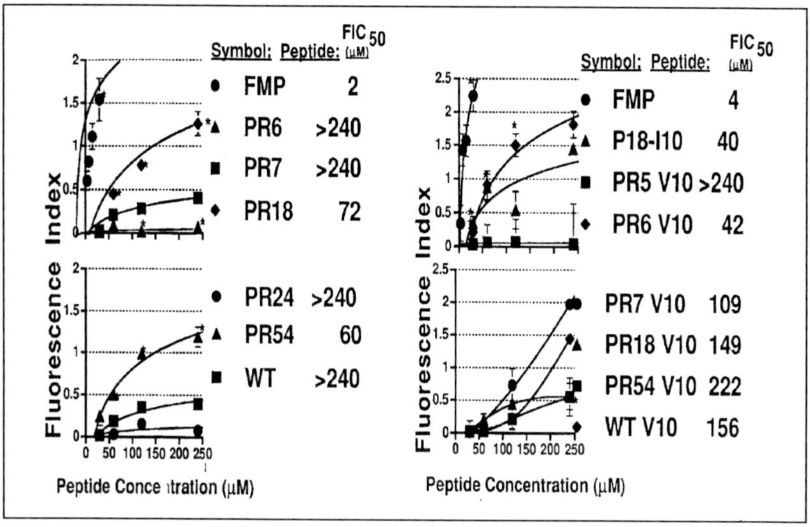

Fig. 2.5. The binding of peptides carrying various ras mutations to HLA A2.1 is shown and is compared to that of wild type peptide. Minimal peptides PR6V10 and PR7V10 are shown to bind better than the corresponding 17mers. However, PR18 and PR54 both exhibit enhanced binding as 17mers when compared to their minimal epitopes. This observation is thought to be secondary to extracellular antigen processing. All mutant peptides are compared to wild type ras and to an optimal binding peptide, flu matrix peptide. Modified from ref. with permission.

mutant sites within p53, were protective against challenge with tumors containing overexpressed, mutant p53 and were without apparent toxicity to the normal murine cells. Theobald et al looked at processing of mutant p53 epitopes. They have shown, in HLA A2Kb transgenic mice, that the mice were unable to develop CTL to epitopes near p53 mutations, but were able to develop CTL to epitopes at distant non-mutant sites. These cytotoxic T lymphocytes recognized these distant epitopes as endogenously processed and presented by tumors expressing mutant p53. Again, they were unable to recognize cells expressing normal levels of p53.[74] McCarty et al further illustrated the ability of cytotoxic T lymphocytes to provide protection in a model analogous to the human system. Here, CTL from HLA A2.1 positive human donors and HLA A2.1 transgenic mice were able to retard the growth of HLA matched human tumors within SCID mice.[75] Human CTL have also been developed from normal donors in vitro against wild type p53.[76,77] These CTL were able to lyse, in vitro, tumor cells containing p53 mutations. Peptide specificity of the CTL lysing the tumors could be confirmed by cold-target blocking with peptide-pulsed targets. These CTL recognized this wild type epitope presented by tumors containing mutant p53 since this epitope was distinct from the mutations present in these tumors. Interestingly, the ability of these CTL to lyse a given tumor was not related to the degree of p53 overexpression. However, a relatively small number of tumors were studied.[76,77]

Theobald et al were also able to compare the ability to develop cytotoxic T cells directed against wild type p53 in normal versus p53 knockout mice. Against certain epitopes, this group has been unable to develop CTL in the normal mice which expressed p53, but has been able to demonstrate CTL to these same epitopes in the p53 knockout mice, suggesting clonal deletion in the normal mice. They were also, however, able to show cytotoxic T lymphocytes to some epitopes in both normal, p53 expressing mice and in the p53 knockout mice. However, the CTL generated in the normal mice were lower affinity than those generated in the knockout mice[45] Ropke et al looked at the frequency of CTL precursors in normal human donors to

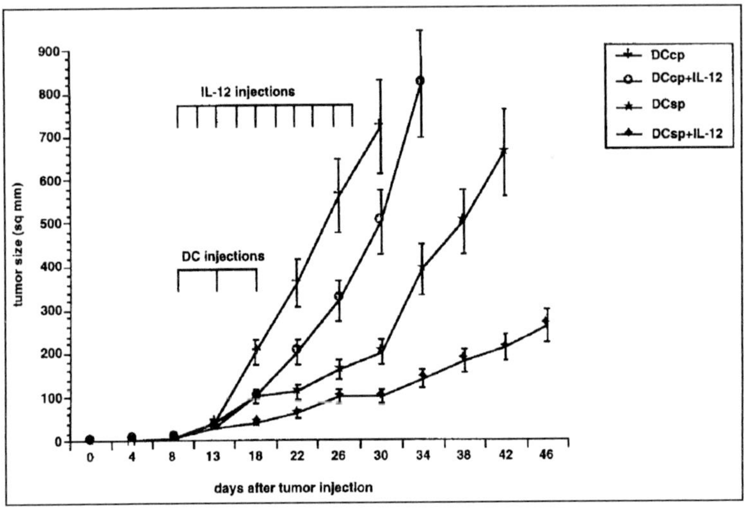

Fig. 2.6. Mice with established tumors were vaccinated with dendritic cells incubated with an irrelevant peptide (cp) or a specific peptide (sp). Dendritic cells are further given with or without IL-12. IL-12 is able to provide a further enhancement in tumor stabilization despite the use of dendritic cells as the antigen presenting cell. The specific peptide used in these studies corresponds to the mutant portion of p53. The tumors used in these studies are transfected with the p53 gene encoding the corresponding mutation. The tumor cells are able to endogenously process this mutant protein and present the epitope corresponding to our peptide on their surface. Modified from ref. with permission.

both mutant and wild type p53. Interestingly, these healthy donors had detectable precursors to both mutant and wild type p53 suggesting again that CTL reactive against wild type p53 are not deleted during T cell development.[78]

In addition to our animal model demonstrating CTL directed against mutant p53, we have also conducted human trials focusing on mutant p53. We have seen a number of immune responses in patients vaccinated with mutant p53 peptides incubated with and delivered by peripheral blood mononuclear cells. These peptides are 17mers spanning the area of mutation within the protein. This length of peptide was chosen so that every nonamer containing the point mutation (in the middle of the 17mer) would be included in the peptide.[69] Both patients with metastatic disease and those in the high risk adjuvant setting were treated on this study. The measurement of CTL was done using a single six to eight day in vitro stimulation and no additional in vitro restimulations were performed. Cytokines were measured in the supernatant of these in vitro cultures at the day of harvest for the CTL assay. Cytotoxic T cell responses were considered positive if lysis of targets with mutant peptide was 10% greater than that of targets without peptide. Cytokine responses were considered positive if the stimulation index (ratio of cytokine release with peptide to that without peptide) was ≥ 2 and if the cytokine response was at least 50 pg/ml (Carbone et al manuscript in preparation). Preliminary analysis indicates that in the patients with evident disease, 2/10 showed a CTL response, and 5/10 a cytokine response, whereas in the adjuvant group of patients, 1/5 showed a CTL response and 5/5 showed a cytokine response. In all, 3/15 had a new CTL response and 10/15 a cytokine response after immunization.[68]

Where tested, it was also notable that the T cell response was specific for the mutant peptide and did not crossreact with the wild-type peptide homologous to the mutant one (Carbone et al , manuscript in preparation). Thus, focusing the vaccine response on the site of a point mutation may minimize the risk of inducing an autoimmune T cell response.

Fusion Proteins from Sarcoma-Associated Chromosomal Translocations

Tumorigenesis results from the transformation of a single cell, which subsequently gains a proliferative advantage over normal neighboring cells. The transformation arises in part from either the inactivation of tumor suppressor genes or the activation of proto-oncogenes.The latter event can occur as a result of different genetic mechanisms, including gene amplification, point mutations, or chromosomal abnormalities such as translocations or inversions.This section will focus on several types of chromosomal translocations involved in sarcoma development as well as ongoing research efforts toward related immunologic therapy.

Mechanistic theory of chromosomal translocations has been reviewed extensively.[79-83] In one possible scenario, chromosomes are broken adjacent to two different genes, which are left intact and brought in close proximity to one another. This occurs when a T-cell receptor or immunoglobulin gene is inserted near a proto-oncogene, thereby activating it. An example of this type of alteration, which occurs primarily in hematological malignancies, is the t(8;14)(q24;q32) translocation associated with most cases of Burkitt's lymphoma.[84] The translocation juxtaposes c-*MYC* and immunoglobulin heavy-chain genes. A second scenario involves a break within the coding sequence of each constituent gene, which generates a functional chimeric gene. This occurs primarily in sarcomas and leukemias, including those reviewed here, and the affected genes often encode transcription factors, suggesting a role in the transformation process. The fusion product retains the DNA binding specificity of one parent gene while inappropriately activating or repressing transcription through the transactivation domain of the other gene. This phenomenon is exclusive to the tumor cell and thus provides a tumor-specific marker.

The t(9;22)(q34;q11) translocation characteristic of chronic myelogenous leukemia was the first documented example of this type of tumor-associated chromosomal abnormality.[85] It results in the Philadelphia chromosome that fuses *c-BCR* with c-*ABL*. [86,87] *BCR/ABL* contributes to cell transformation by activating the tyrosine kinase activity of the c-*ABL* product. This discovery paved the way for studies of mechanistically similar events in other tumors. Another example of a translocation-induced leukemia is acute lymphoblastic leukemia (ALL), and 25% of childhood cases are marked by the t(12;21)(p13;q22) translocation. This produces a novel transcription factor by fusing the 5' region of *TEL*, which contains a helix-loop-helix motif, with the 3' region of *AML1*, which contains a DNA-binding domain.[88,89]

Among the solid tumors, Ewing's sarcoma (ES), a member of the poorly differentiated ES/PNET family of neuroectodermal tumors, has been extensively studied.[90] Approximately 85% of *ES/PNET* cases are marked by t(11;22)(q24;q12),[91-93] which fuses the N-terminal region of *EWS* with the C-terminal region of *FLI1*.[94-96] *EWS* encodes a 656-amino acid protein with two major functional domains: the N-terminal region is rich in Gln-Ser-Tyr repeats, resembling the transactivation domains of transcription factors, while the C-terminal region contains a putative RNA-binding domain.[97] *FLI1* is a member of the ETS family of transcription factors and contains a conserved DNA-binding domain in the C-terminus.[98,99] DNA binding by the *FLI1* portion of the fusion protein and transcriptional activation by the *EWS* region are critical for tumorigenesis.[95] A majority of the remaining 15% of ES cases are linked to the fusion of *EWS* with another ETS family member, ERG, which shares a 98% sequence identity with *FLI1* within the ETS domain. [100-103] The two fusion proteins probably function through similar oncogenic pathways.[81]

Rhabdomyosarcoma (RMS) is the most common pediatric soft-tissue sarcoma, and approximately 20% of the cases are alveolar in morphology(ARMS).[104] About 80% of ARMS cases involve t(2;13)(q35;q14), which fuses the N-terminal region of *PAX3* with the C-terminal region of *FKHR*.[105-109] The PAX family of transcription factors plays an important role in embryonic development (reviewed in[110-113]). These proteins contain a paired-box (PB) DNA-binding domain and often contain a homeobox (HB) DNA-binding domain. Overexpression of these genes has been shown to cause oncogenic transformation.[114,115] *FKHR* is a member of the forkhead family of transcription factors, which are involved in embryogenesis and contain a winged helix DNA-binding domain (reviewed in[116]). The *PAX3-FKHR* fusion protein retains the PB and HB domains from *PAX3*, most likely conferring DNA-binding specificity, and the transactivation domain from *FKHR*.[117] In 10-20% of ARMS cases, a variant translocation, t(1;13)(p36;q14), has been described that fuses the N-terminal portion of *PAX7* with the C-terminal portion of *FKHR* . The N-terminal regions of *PAX3* and *PAX7* encode similar DNA-binding domains, and thus, the two translocations are thought to create similar transcription factors that alter expression of common target genes.[107,108,117-119] Synovial sarcoma (SS), an aggressive malignancy that occurs primarily in the extremities of adolescents and young adults, accounts for 5 to 10% of soft-tissue sarcomas. PCR analysis has shown that over 90% of SS cases contain a characteristic t(X;18)(p11.2;q11.2) translocation.[120] This translocation event fuses the N-terminal region of *SYT* with the C-terminal region of *SSX1* or *SSX2*.[120-122] The function of *SYT* remains unknown, although the N-terminal region contains a putative transactivation domain.[123] *SSX1* and *SSX2* share 80% homology,[122] with both genes encoding 188-aa proteins containing an N-terminal Kruppel-associated box (KRAB) that is thought to be involved in transcriptional repression.[124-126] In the chimeric fusion protein, the repression domain from *SSX* is replaced by the transactivation domain of *SYT*, and this presumably activates unknown target genes normally repressed by *SSX1* or *SSX2*.[127]

Malignant melanoma of soft parts (MMSP), also termed clear cell sarcoma (CCS), is a rare aggressive soft-tissue sarcoma of neuroectodermal origin,[128] affecting muscle tendons and aponeuroses typically in the extremities of young adults.[129] Over 70% of these tumors contain the t(12;22)(q13;q12) translocation, fusing the 5' region of *EWS* with the 3' region of ATF1.[130,131] ATF1 is a member of the CREB/transcription factor family which have bZIP domains for DNA-binding and protein-protein interaction.[132] ATF1 expression is induced by cAMP and is activated by cAMP-dependent protein kinase A (PKA). This phosphorylation site is lost in the fusion protein,[133] and thus *EWS*/ATF1 should not be cAMP inducible. It is not precisely known how the translocation contributes to cellular transformation, although a number of mechanisms have been postulated,[81] including constitutive activation of ATF1 target genes, repression of growth control genes, and activation of other *CREB/ATF* or non-*CREB/ATF* target genes.

Desmoplastic small round cell tumor (DSRCT), which affects mainly young males, is an aggressive malignancy occurring predominantly in abdominal serosal surfaces.[131] Almost 100% of these cases contain a characteristic t(11;22)(p13;q12) translocation which fuses the N-terminal region of *EWS* with the C-terminal region of *WT1*, a tumor suppressor gene involved in a subset of Wilms' tumors.[134-138] *WT1* contains 3 zinc-fingers in the C-terminal region responsible for DNA binding, and it has been postulated that the loss of the proximal zinc-fingers in the chimeric fusion protein converts *WT1* from a transcriptional repressor to an activator.[81,139] Resulting abnormalities in *WT1* target gene expression, including genes related to the early growth response (EGR) family, could be causative in cellular transformation.[81]

It has been shown that chimeric fusion proteins resulting from translocations can be necessary for the persistence of the tumor,[140,141] thus marking the tumor cell as a target for immunotherapy. Furthermore, the fact that these fusion proteins are unique to the disease state aids in diagnosis and serves as a prognostic indicator. Thus, with increased understanding of the role of translocations in cellular transformation, more efficient therapies can be developed. The idea of immunotherapy for translocation-associated tumors has been reviewed elsewhere[82] and is briefly presented here. The fusion proteins produced from gene translocations likely contain tumor-associated antigens (TAA), as the fusion junction region contains elements from both wild-type genes. Peptides generated from proteolytic processing of this region would thus have a "non-self" sequence of amino acids, which could mark the cell for attack by the T-cell arm of the immune system. Normal cells, on the other hand, would not contain the breakpoint-spanning sequence, thus distinguishing specific tumor cells and avoiding an autoimmune response.

The process of antigen processing and presentation has been extensively reviewed elsewhere.[145] T cells recognize peptide antigens bound to the major histocompatibility complex (MHC) on the surface of cells.[145,146] In the case of $CD8^+$ cytotoxic T lymphocytes (CTL), T cell receptors (TCR) recognize peptides 8-10 amino acids in length bound to MHC class I molecules. These peptides are derived primarily from endogenous proteins which are processed by cytosolic proteosomes. This is the case with tumor cells, where a protein responsible for cell transformation, such as a translocation fusion protein, can be processed and presented by MHC class I to $CD8^+$ CTL. Although tumor-specific fusion proteins often function as nuclear transcription factors, they are still subject to proteolytic processing, and there is evidence that tumor-associated nuclear proteins can induce CTL responses.[70,71,73,82,147-149] Even in virally infected cells, nuclear proteins are often the source of key CTL epitopes. Thus, the design of peptides based upon the sequences surrounding the breakpoints of fusion proteins has promise in immunotherapy of translocation-associated tumors, since they likely represent tumor-specific neoantigens which may be used to induce an immune response.

In order to determine the immunogenicity of fusion sequences associated with ES, ARMS, SS, MMSP and DSRCT, respective peptides were designed which contain all possible breakpoint-spanning minimal epitopes. These potentially antigenic peptides can be useful for novel immunotherapies only if they bind HLA molecules. Therefore, 2 different binding assays were utilized to assess specific peptide binding to various HLA molecules. One assay, described by Stuber et al[150] and Nijman et al,[151] utilizes the T2 cell line, a somatic cell hybrid of human B and T lymphoblastoid lines, which is deficient in TAP1 and TAP2 transporter genes.[152,153] T2 cells lack the ability to properly load cytosolic peptides onto newly synthesized class I MHC molecules in the ER, and thus empty molecules only occasionally and transiently appear at the cell surface. A peptide which specifically binds to the MHC, however, in conjunction with b2-microglobulin, stabilizes the MHC on the T2 cell surface, and this can be detected by immunofluorescence. T2 cells naturally express low levels of HLA-A2.1 and undetectable levels of HLA-B5 and can be used to study peptide binding to any HLA molecule following transfection of the gene of interest. The second binding assay is a peptide-MHC reconstitution assay, described by Zeh et al,[154] in which a human cell line is briefly acid-treated, causing surface HLA denaturation and dissociation of peptides and β2-microglobulin. Acid treated cells maintain viability as well as antigenicity of class II MHC and other non-MHC antigens, as described by Sugawara et al.[155] Subsequently, incubation with a specific HLA-binding peptide, along with b2-microglobulin and the specific anti-HLA antibody of interest, enables proper surface HLA renaturation, which can be detected by immunofluorescence.

Using the above methods, a number of peptides derived from the fusion junctions of ARMS, SS, MMSP and DSRCT were found to specifically bind to several HLA class I molecules (Table 1.1) (Worley et al, manuscript in preparation). The site of gene fusion in each case is indicated with a slash in the sequence. These results provide proof of principle that sequences in the fusion region of sarcoma-associated chimeras can bind HLA molecules and thus potentially serve as neoantigens. It was similarly demonstrated by Yotnda et al[156] that an ALL breakpoint-associated peptide binds to HLA-A2.1. In the same study, it was also shown that this peptide induces a specific CTL response in peripheral blood lymphocytes from healthy $HLA\text{-}A2^+$ donors, and that specific CTL were present in the bone marrow of a $HLA\text{-}A2^+$ ALL patient. In addition, Gambacorti-Passerini et al[157] predicted and tested a number of HLA binding motifs among fusion proteins, including ES- and MMSP-associated sequences, and two different ES-derived peptides bound specifically to HLA-Cw0702. However, despite this specific interaction, a primary in vitro cellular immune response could not be obtained from HLA-Cw0702$^+$ donor PBMC samples stimulated with the same peptides,[158] thereby asserting that HLA binding is necessary, but not always sufficient, to generate a cellular response.

To determine if the peptides in Table 2.1 can serve as neoantigens capable of eliciting a T-cell response, murine models were used for immunization studies (Goletz et al and Worley et al, manuscripts in preparation). In the case of ES, a peptide corresponding to the breakpoint region was pulsed onto dendritic cells and used to immunize mice. A peptide-specific, class I-restricted, CD8$^+$ T cell response was generated, as seen by in vitro CTL activity analysis using standard ^{51}Cr release assays. Specific CTL responses were likewise generated in mice using the SS1 peptide from SS and the P/F peptide from ARMS. The P/F-induced CTL lysed murine tumor cell targets transfected with the full-length *PAX3-FKHR* cDNA, demonstrating that the fusion protein can be endogenously processed in tumor cells and presented by class I MHC on the cell surface.[82,159] Immunization of mice with P/F conferred protection against tumor challenge, both in immunized mice challenged with tumors expressing the full-length *PAX3/FKHR* protein and in unimmunized mice with existing tumor that underwent adoptive transfer with spleen cells from immunized mice.[82,159] These studies show that chimera breakpoint-associated antigens in sarcomas are capable of eliciting T cell responses in mice. Since the peptides listed in Table 2.1 bind HLA molecules, there are promising opportunities for therapeutic studies in HLA-specific patients with the indicated sarcomas. A clinical study has recently been undertaken to determine whether immunization of ES or ARMS patients with respective peptide-pulsed dendritic cells can induce tumor-specific CTL activity.

In conclusion, there are numerous cases of sarcomas marked by distinct chromosomal translocations, only a few of which have been covered here, and the resulting chimeras may contain neoantigens within the fusion regions. To assess this, novel peptides which mimic the breakpoint-spanning regions of the fusion proteins can be designed. The antigenicity of these peptides can be studied through HLA binding, murine models and ultimately, clinical trials. HLA binding and CTL generation by sarcoma-associated peptides, such as P/F from ARMS and SS1 from SS, suggest their promising role in novel immunotherapies.

Summary

Our work has largely focused on the response to unique tumor specific antigens including mutant ras and p53 as well as the unique proteins generated by chromosomal translocations in pediatric sarcomas. We have been able to show, both in murine models and in human trials, the ability to generate an antigenic response to these unique determinants. Because these molecules are essential for the malignant state of the tumor, they cannot be lost under immuno-

Table 2.1.

Peptide	Sequence	Associated Sarcoma
P/F	TIGNGSPQ/NSIRHNLSL	ARMS
SS1	PQQRPYGYDQ/IMPKKPA	SS
SS2	RPYGYDQ/IMPKKPAE	SS
EA1	RGGGRGGMG/KILKDLSS	MMSP
EW1	SQQSSSYGQQ/SEKPY	DSRCT
EW2	SSSYGQQ/SEKPYQCDFK	DSRCT

logic presssure to produce escape variants. Thus, if we can generate a sufficient T cell response to these tumor-specific antigens, it may be possible to reject the tumor without substantial risk of autoimmune side effects. We have also developed an increased understanding of immune function, including those factors permitting the tumor to avoid immunosurveillance. These findings may in turn impact our ability to manipulate the immune system and produce an effective cytotoxic T cell response both in animals and in humans to treat or prevent cancer.

References

1. MoerTEL CG, Fleming TR, Macdonald JS et al Florouracil plus levamisole as effective adjuvant therapy after resection of stage III colon carcinoma: a final report. Ann Intern Med 1995; 122:321-326.
2. Penn I. Overview of the problem of cancer in organ transplant recipients. Ann Transp 1997; 2:5-6.
3. Greten TF, Jaffee EM. Cancer vaccines. J Clin Oncol 1999; 17:1047-1060.
4. Wojtowicz-Praga S. Reversal of tumor-induced immunosuppression: A new approach to cancer therapy. J Immunotherapy 1997; 20:165-177.
5. Hsu FJ, Caspar CB, Czerwinski D et al. Tumor -specific idiotype vaccines in the treatment of patients with B-cell lymphoma - long-term results of a clinical trial. Blood 1997; 89:3129-3135.
6. Nestle FO, Alijagic S, Gilliet M et al. Vaccination of melanoma patients with peptide- or tumor lysate-pulsed dendritic cells. Nature Medicine 1998; 4:328-332.
7. RosenbERG SA, Yang JC, Schwartzentruber DJ et al. Immunologic and therapeutic evaluation of a synthetic peptide vaccine for the treatment of patients with metastatic melanoma. Nature Medicine 1998; 4:321-327.
8. Chan AD, Morton DL. Active immunotherapy with allogeneic tumor cell vaccines: present status. Semin Oncolo 1998; 25:611-622.
9. Hsueh EC, Gupta RK, Qi K et al. Correlation of specific immune responses with survival in melanoma patients with distant metastases receiving polyvalent melanoma cell vaccine. J Clin Oncol 1998; 16:2913-2920.
10. Howard MC, Miyajima A, Coffman R. T-cell derived cytokines and their receptors. 3rd ed. New York: Raven Press, Ltd., 1993:763-765.
11. Gabrilovich DI, Chen HL, Girgis KR et al. Production of vascular endothelial growth factor by human tumors inhibits the functional maturation of dendritic cells [published erratum appears in Nat Med 1996 Nov;2(11):1267]. Nat Med 1996; 2:1096-1103.
12. Hahne M, Rimoldi D, Schroter M et al. Melanoma cell expression of Fas (Apo-1/CD95) ligand: implications for tumor immune escape. Science 1996; 274:1363-1366.
13. Strand S, Hofmann WJ, Muller M et al. Lymphocyte apoptosis induced by CD95 (APO-1/Fas) ligand-expressing tumor cells—a mechanism of immune evasion? Nature Medicine 1996; 2:1361-1366.
14. O'Connell J, O'Sullivan GC, Collins JK et al. The Fas counterattack: Fas-mediated T cell killing by colon cancer cells expressing Fas ligand. J Exp Med 1996; 184:1075-1082.
15. Germain RN. Antigen processing and presentation. 3rd ed. New York: Raven Press, Ltd., 1993:629-676.
16. Restifo NP, Kawakami Y, Marincola F et al. Molecular mechanisms used by tumors to escape immune recognition: immunogenetherapy and the cell biology of major histocompatibility complex class I. J Immunotherapy 1993; 14:182-190.
17. Eisenbach L, Kushtai G, Plaksin D et al. MHC genes and oncogenes controlling the metastatic phenotype of tumor cells. Cancer Rev 1986; 5:1-18.
18. Kurokohchi K, Carrington M, Mann DL et al. Expression of HLA class I molecules and the transporter associated with antigen processing (TAP) in hepatocellular carcinoma. Hepatology 1996; 23:1181-1188.
19. Chen HL, Gabrilovich D, Virmani A et al. Structural and functional analysis of beta2 microglobulin abnormalities in human lung and breast cancer. Int J Cancer 1996; 67:756-763.
20. Chen HL, Gabrilovich D, Tampe R et al. A functionally defective allele of TAP1 results in loss of MHC class I antigen presentation in a human lung cancer [see comments]. Nat Genet 1996; 13:210-213.
21. Lohmann S, Wollscheid U, Huber C et al. Multiple levels of MHC class I down-regulation by ras oncogenes. Scand J Immunol 1996; 43:537-544.

22. Durum SK, Oppenheim JJ. Proinflammatory cytokines and immunity. 3rd ed. New York: Raven Press, Ltd., 1993:801-817.
23. Fossum B, Olsen AC, Thorsby E et al. CD8+ T cells from a patient with colon carcinoma, specific for a mutant p21-Ras derived peptide (GLY13 ASP), are cytotoxic towards a carcinoma cell line harbouring the same mutation. Cancer Immunol Immunother 1995; 40:165-172.
24. Hoglund P, Sundback J, Olsson-Alheim MY et al. Host MHC class I gene control of NK-cell specificity in the mouse. Immunol Rev 1997; 155:11-28.
25. Mondino A, Jenkins MK. Surface proteins involved in T cell costimulation. J Leukoc Biol 1994; 55:805-815.
26. Shevach EM. Accessory molecules. In: Paul WE, ed. Fundamental Immunology. 3rd ed. New York: Raven Press, Ltd., 1993:531-575.
27. Harding FA, McArthur JG, Gross JA et al. CD28-mediated signalling co-stimulates murine T cells and prevents induction of anergy in T-cell clones. Nature 1992; 356:607-609.
28. Townsend SE, Allison JP. Tumor rejection after direct costimulation of CD8+ T cells by B7-transfected melanoma cells. Science 1993; 259:368-370.
29. Schwartz RH. Immunological tolerance. In: Paul WE, ed. Fundamental Immunology. 3rd ed. New York: Raven Press, 1993:677-731.
30. Powell JD, Ragheb JA, Kitagawa-Sakakida S et al. Molecular regulation of interleukin-2 expression by CD28 co-stimulation and anERGy. Immunol Rev 1998; 165:287-300.
31. Schultze J, Nadler LM, Gribben JG. B7-mediated costimulation and the immune response. Blood Rev 1996; 10:111-127.
32. Essery G, Feldmann M, Lamb JR. Interleukin-2 can prevent and reverse antigen-induced unresponsiveness in cloned human T lymphocytes. Immunology 1988; 64:413-417.
33. Pircher H, Rohrer UH, Moskophidis D et al. Lower receptor avidity required for thymic clonal deletion than for effector T-cell function. Nature 1991; 351:482-485.
34. Takahama Y, Kosugi A, Singer A. Phenotype, ontogeny, and repertoire of CD4-CD8- T cell receptor alpha beta + thymocytes. Variable influence of self-antigens on T cell receptor V beta usage. J Immunol 1991; 15:1134-1141.
35. Suzue K, Zhou X, Eisen HN et al. Heat shock fusion proteins as vehicles for antigen delivery into the major histompatibility complex class I presentation pathway. Proc Natl Acad Sci 1997; 94:13146-13151.
36. Gendler S, Taylor-Papadimitriou J, Duhig T et al. A highly immunogenic region of a human polymorphic epithelial mucin expressed by carcinomas is made up of tandem repeats. J Biol Chem 1988; 263:12820-12823.
37. Shimizu Y, Guidotti LG, Fowler P et al. Dendritic cell immunization breaks cytotoxic T lymphocyte tolerance in hepatitis B virus transgenic mice. J Immunol 1998; 161:4520-4529.
38. Lenardo MJ. Interleukin-2 programs mouse $a\beta$ T lymphocytes for apoptosis. Nature 1991; 353:858-861.
39. Critchfield JM, Zúñiga-Pflücker JC, Lenardo MJ. Parameters controlling the programmed death of mature mouse T lymphocytes in high-dose suppression. Cell Immunol 1995; 160:71-78.
40. Rafaeli Y, van Parijs L, London CA et al. Biochemical mechanisms of IL-2-regulated Fas-mediated T cell apoptosis. Immunity 1998; 8:615-623.
41. Alexander-Miller MA, Leggatt GR, Sarin A et al. Role of antigen, CD8, and Cytotoxic T Lymphocyte avidity in high dose antigen induction of apoptosis of effector CTL. J Exp Med 1996; 184:485-492.
42. Alexander-Miller MA, Derby MA, Sarin A et al. Supra-optimal peptide/MHC causes a decrease in Bcl-2 and allows TNF-a receptor II-mediated apoptosis of CTL. J Exp Med 1998; 188 (8):1391-1399.
43. Takahashi H, Cohen J, Hosmalin A et al. An immunodominant epitope of the HIV gp160 envelope glycoprotein recognized by class I MHC molecule-restricted murine cytotoxic T lymphocytes. Proc Natl Acad Sci USA 1988; 85:3105-3109.
44. Takeshita T, Takahashi H, Kozlowski S et al. Molecular analysis of the same HIV peptide functionally binding to both a class I and a class II MHC molecule. J Immunol 1995; 154:1973-1986.
45. Theobald M, Biggs J, Hernández J et al. Tolerance to p53 by A2.1-restricted cytotoxic T lymphocytes. J Exp Med 1997; 185:833-841.

46. Alexander-Miller MA, Leggatt GR, Berzofsky JA. Selective expansion of high or low avidity cytotoxic T lymphocytes and efficacy for adoptive immunotherapy. Proc Natl Acad Sci U S A 1996; 93:4102-4107.
47. Zeh III HJ, Perry-Lalley D, Dudley ME et al. High avidity CTLs for two self-antigens demonstrate superior in vitro and in vivo antitumor efficacy. J Immunol 1999; 162:989-994.
48. Yee C, Savage PA, Lee PP et al. Isolation of high avidity melanoma-reactive CTL from heterogeneous populations using peptide-MHC tetramers. J Immunol 1999; 162:2227-2234.
49. Gabrilovich DI, Nadaf S, Corak J et al. Dendritic cells in anti-tumor immune reponses. II. Dendritic cells grown from bone marrow precursors, but not mature DC from tumor-bearing mice are effective antigen carriers in the therapy of established tumors. Cellular Immunol 1996; 170:111-119.
50. Leung DW, Cachianes G, Kuang WJ et al. Vascular endothelial growth factor is a secreted angiogenic mitogen. Science 1989; 246:1306-1309.
51. Gabrilovich D, Ishida T, Oyama T et al. Vascular endothelial growth factor inhibits the development of dendritic cells and dramatically affects the differentiation of multiple hematopoietic lineages in vivo. Blood 1998; 92:4150-4166.
52. Ishida T, Oyama T, Carbone DP et al. Defective function of Langerhans cells in tumor-bearing animals is the result of defective maturation from hemopoietic progenitors. J Immunol 1998; 161:4842-4851.
53. Shima Y, Nishimoto N, Ogata A et al. Myeloma cells express Fas antigen/APO-1 (DC95) but only some are sensitive to anti-Fas antibody resulting in apoptosis. Blood 1995; 85:757-764.
54. Rokhlin OW, Bishop GA, Hostager BS et al. Fas-mediated apoptosis in human prostatic carcinoma cell lines. Cancer Res 1997; 57:1758-1768.
55. Fenton RG, Hixon JA, Wright PW et al. Inhibition of fas (CD95) expression and fas-mediated apoptosis by oncogenic *ras*. Cancer Res 1998; 58:3391-3400.
56. Abrams SI, Hand PH, Tsang KY et al. Mutant ras epitopes as targets for cancer vaccines. Semin Oncolo 1996; 23:118-134.
57. Pulendran B, Smith JL, Caspary G et al. Distinct dendritic cell subsets differentially regulate the class of immune response in vivo. Proc Natl Acad Sci 1999; 96:1036-1041.
58. Bos JL. *ras* oncogenes in human cancer: A review. Cancer Res 1989; 49:4682-4689.
59. Fenton RG, Keller CJ, Hanna N et al. Induction of T-cell immunity against *Ras* oncoproteins by soluble protein or Ras-expressing *Escherichia coli*. J Natl Cancer Inst 1995; 87:1853-1861.
60. Abrams SI, Stanziale SF, Lunin SD et al. Identification of overlapping epitopes in mutant *ras* oncogene peptides that activate $CD4^+$ and $CD8^+$ T cell responses. Eur J Immunol 1996; 26:435-443.
61. Peace DJ, Smith JW, Chen W et al. Lysis of Ras oncogene-transformed cells by specific cytotoxic T lymphocytes elicited by primary in vitro immunization with mutated Ras peptide. J Exp Med 1994; 179:473-479.
62. Skipper J, Stauss HJ. Identification of two cytotoxic T lymphocyte-recognized epitopes in the ras protein. J Exp Med 1993; 177:1493-1498.
63. Kass E, Schlom J, Thompson J et al. Induction of protective host immunity to carcinoembryonic antigen (CEA), a self-antigen in CEA transgenic mice by immunizing with a recombinant vaccinia-CEA virus. Cancer Res 1999; 59:676-683.
64. Elsas AV, Nijman HW, Van der minne CE et al. Induction and characterization of cytotoxic T-lymphocytes recognizing a mutated p21RAS peptide presented by HLA-A*0201. Int J Cancer 1995; 61:389-396.
65. Gjertsen MK, Bjorheim J, Saeterdal I et al. Cytotoxic $CD4^+$ and $CD8^+$ T lymphocytes, generated by mutant p21-*ras* (12VAL) peptide vaccination of a patient, recognize 12VAL-dependent nested epitopes present within the vaccine peptide and kill autologous tumour cells carrying this mutation. Int J Cancer 1997; 72:784-790.
66. Abrams SI, Khleif SN, Bergmann-Leitner ES et al. Generation of stable $CD4^+$ and $CD8^+$ T cell lines from patients immunized with ras onocogene-derived peptides reflecting codon 12 mutations. Cell Immunol 1997; 182:137-151.
67. Khleif SN, Abrams SI, Hamilton JM et al. A phase I vaccine trial with peptides reflecting ras oncogene mutations of solid tumors. J Immunotherapy 1999.

68. Carbone D, Kelley M, Smith MC et al. Results of NCI T93-0148: detection of immunologic responses and vaccination against tumor specific mutant ras and p53 peptides in patients with cancer. Proc ASCO 1999; 18:438a.
69. Smith MC, Pendleton CD, Maher VE et al. Oncogenic mutations in *ras* create HLA-A2.1 binding peptides but affect their extracellular processing. Internat Immunol 1997; 9:1085-1093.
70. Carbone DP, Ciernik IF, Yanuck M et al. Mutant p53 and ras proteins as immunotherapeutic targets. Annals of Oncology 1994; 5:117.
71. Yanuck M, Carbone DP, Pendleton CD et al. A mutant p53 tumor suppressor protein is a target for peptide-induced CD8+ cytotoxic T cells. Cancer Res 1993; 53:3257-3261.
72. Gabrilovich DI, Cunningham HT, Carbone DP. IL-12 and mutant P53 peptide-pulsed dendritic cells for the specific immunotherapy of cancer. J Immunother Emphasis Tumor Immunol 1996; 19:414-418.
73. Mayordomo JI, Loftus DJ, Sakamoto H et al. Therapy of murine tumors with p53 wild-type and mutant sequence peptide-based vaccines. J Exp Med 1996; 183:1357-1365.
74. Theobald M, Ruppert T, Kuckelkorn U et al. The sequence alteration associated with mutational hotspot in p53 protects cells from lysis by cytotoxic T lymphocytes specific for a flanking peptide epitope. J Exp Med 1998; 188:1017-1028.
75. McCarty TM, Liu X, Sun JY et al. Targeting p53 for adoptive T-cell immunotherapy. Cancer Res 1998; 58:2601-2605.
76. Ropke M, Hald J, Guldberg P et al. Spontaneous human squamous cell carcinomas are killed by a human cytotoxic T lymphocyte clone reognizing a wild-type p53-derived peptide. Proc Nat Acad Sci USA 1996; 93:14704-14707.
77. Gjnatic S, Cai Z, Viguier M et al. Accumulation of the p53 protein allows recognition by human CTL of a wild-type p53 epitope presented by breast carcinomas and melanomas. J Immunol 1998; 160:328-333.
78. Ropke M, Regner M, Claesson MH. T cell-mediated cytotoxicity against p53-protein derived peptides in bulk and limiting dilution cultures of healthy donors. Scand J Immunol 1995; 42:98-103.
79. Rowley JD. Recurring chromosome abnormalities in leukemia and lymphoma. Seminars in Hematology 1990; 27:122-136.
80. Rabbitts TH. Chromosomal translocations in human cancer. Nature 1994; 372:143-149.
81. Sorensen PHB, Triche TJ. Gene fusions encoding chimaeric transcription factors in solid tumors. Seminars Cancer Biol 1996; 7:3-14.
82. Goletz TJ, Mackall CL, Berzofsky JA et al. Molecular alterations in pediatric sarcomas: potential targets for immunotherapy. Sarcoma 1998; 2:77-87.
83. Cobaleda C, Perez-Losada J, Sanchez-Garcia I. Chromosomal abnormalities and tumor development: from genes to therapeutic mechanisms. BioEssays 1998; 20:922-930.
84. Zech I, Haglund U, Nilsson K et al. Characteristic chromosomal abnormalities in biopsies and lymphoid-cell lines from patients with Burkitt and non-Burkitt lymphomas. Int J Cancer 1976; 17:47-56.
85. Nowell PC, Hungerford DA. A minute chromosome in human chronic granulocytic leukemia. Science 1960; 132:1497.
86. De Klein A, Hagemeijer A, Bartram CR et al. *bcr* Rearrangement and translocation of the *c-abl* oncogene in philadelphia positive acute lymphoblastic leukemia. Blood 1986; 68 (6):1369-1375.
87. Bartram CR, De Klein A, Hagemeijer A et al. Translocation of *c-abl* oncogene correlates with the presence of a philadelphia chromosome in chronic myelocytic leukaemia. Nature 1983; 306:277-280.
88. Shurtleff S, Buijs A, Behm F et al. *TEL/AML1* fusion resulting from a cryptic t(12;21) is the most common genetic lesion in pediatric ALL and defines a subgroup of patients with and excellent prognosis. Leukemia 1995; 9:1985-1989.
89. Raynaud S, Cave M, Baens C et al. The 12;21 translocation involving *TEL* and deletion of the other *TEL* allele: Two frequently associated alterations found in childhood acute lymphoblastic leukemia. Blood 1996; 87:2891-2897.
90. Cavazzana AO, Miser JS, Jefferson J et al. Experimental evidence for a neural origin of Ewing's sarcoma of bone. Amer J Path 1987; 127:507-518.
91. Turc-Carel C, Aurias A, Mugneret F et al. Chromosomes in Ewing's sarcoma: An evaluation of 85 cases and remarkable consistency of t(11;22)(q24;q12). Cancer Genet Cytogenet 1988; 32:229-238.

92. Turc-Carel C, Philip I, Berger M et al. Chromosomal translocations in Ewing's sarcoma. New Engl J Med 1983; 309:497-498.
93. Aurias A, Rimbaut C, Buffe D et al. Chromosomal translocations in Ewing's sarcoma. New Engl J Med 1983; 309:496-497.
94. Delattre O, Zucman J, Plougastel B et al. Gene fusion with an ETS DNA-binding domain caused by chromosome translocation in human tumors. Nature 1992; 359:162-165.
95. May WA, Gishizky ML, Lessnick SL et al. Ewing's sarcoma 11;22 translocation produces a chimeric transcription factor that requires the DNA-binding domain encoded by *FLI1* for transformation. Proc Natl Acad Sci USA 1993; 90:5752-5756.
96. Zucman J, Delattre O, Desmaze C et al. Cloning and characterization of the ewing's sarcoma and peripheral neuroepithelioma t(11;22) translocation breakpoints. Genes, Chromosomes & Cancer 1992; 5:271-277.
97. Plougastel B, Zucman J, Peter M et al. Genomic structure of the *EWS* gene and its relationship to *EWSR1*, a site of tumor-associated chromosome translocation. Genomics 1993; 18:609-615.
98. Ben-David Y, Giddens EB, Letwin K et al. Erythroleukemia induction by Friend murine leukemia virus: Insertional activation of a new member of the ETS family, Fli 1. Genes Dev 1991; 5:908-918.
99. Wasylyk B, Hahn SL, Giovane A. The ETS family of transcription factors. Eur J Biochem 1993; 211:7-18.
100. Zucman J, Melot T, Desmaze C et al. Combinatorial generation of variable fusion proteins in the Ewing family of tumors. EMBO J 1993; 12:4481-4487.
101. Sorensen PHB, Lessnick SL, Lopez-Terrada D et al. A second Ewing's sarcoma translocation, t(21;22), fuses the *EWS* gene to another ETS-family transcription factor, ERG. Nature Genetics 1994; 6:146-151.
102. Giovannini M, Biegel JA, Serra M et al. *EWS-ERG* and *EWS-FLI1* fusion transcripts in Ewing's sarcoma and primitive neuroectodermal tumors with variant translocations. J Clin Invest 1994; 94:489-496.
103. Prasad DDK, Rao VN, Reddy ES. Structure and expression of human *fli-1* gene. Cancer Res 1992; 52:5833-5837.
104. Triche TJ. Pathology of pediatric malignancies. Philadelphia: Lippincott, 1993:110-152.
105. Turc-Carel C, Lizard-Nacol S, Justrabo E et al. Consistent chromosomal translocation in alveolar Rhabdomyosarcoma. Cancer Genet Cytogenet 1986; 19:361-362.
106. Barr FG, Galili N, Holick J et al. Rearrangement of the *PAX3* paired box gene in the paediatric solid tumor alveolar rhabdomyosarcoma. Nature Genetics 1993; 3:113-117.
107. Galili N, Davis RJ, Fredericks WJ et al. Fusion of a fork head domain gene to *PAX3* in the solid tumour alveolar rhabdomyosarcoma. Nature Genetics 1993; 5:230-235.
108. Shapiro DN, Sublett JE, Li B et al. Fusion of *PAX3* to a member of the forkhead family of transcription factors in human alveolar rhabdomyosarcoma. Cancer Res 1993; 53:5108-5112.
109. Biegel JA, Nycum LM, Valentine V et al. Detection of the t(2;13)(q35;q14) and *PAX3-FKHR* fusion in alveolar rhabdomyosarcoma by fluorescence in situ hybridization. Genes, Chromosones & Cancer 1995; 12:186-192.
110. Tremblay P, Gruss P. *Pax*: Genes for mice and men. Pharmac Ther 1994; 61:205-226.
111. Strachan T, Read AP. *PAX* genes. Current Opinion in Genetics and Development 1994; 4:427-438.
112. Read AP. *Pax* genes-Paired feet in three camps. Nature Genetics 1995; 9:333-334.
113. Mansouri A, Hallonet M, Gruss P. *Pax* genes and their roles in cell differentiation and development. Curr Opin Cell Biol 1996; 8:851-857.
114. Maulbecker CC, Gruss P. The oncogenic potential of *Pax* genes. EMBO J 1993; 12:2361-2367.
115. Scheidler S, Fredericks WJ, Rauscher FJIII et al. The hybrid *PAX3-FKHR* fusion protein of alveolar rhabdomyosarcoma transforms fibroblasts in culture. Proc Natl Acad Sci USA 1996; 93:9805-9809.
116. Brennan RG. The winged helix DNA-binding motif: another helix-turn-helix takeoff. Cell 1993; 74:773-776.
117. Davis RJ, D'Cruz CM, Lovell MA et al. Fusion of *PAX7* to *FKHR* by the variant t(1;13)(p36;q14) translocation in alveolar rhabdomyosarcoma. Cancer Res 1994; 54:2869-2872.

118. Davis RJ, Barr FG. Fusion genes resulting from alternative chromosomal translocations are overexpressed by gene-specific mechanisms in alveolar rhabdomyosarcoma. Proc Nat Acad Sci USA 1997; 94:8047-8051.
119. Bernasconi M, Remppis A, Fredericks WJ et al. Induction of apoptosis in rhabdomyosarcoma cells through down-regulation of PAX proteins. Proc Natl Acad Sci USA 1996; 93:13164-13169.
120. Crew AJ, Clark J, Fisher C et al. Fusion of *SYT* to two genes, *SSX1* and *SSX2*, encoding proteins with homology to the Kruppel-associated box in human synovial sarcoma. EMBO J 1995; 14:2333-2340.
121. de Bruijn DRH, Baats E, Zechner U et al. Isolation and characterization of the mouse homolog of *SYT*, a gene implicated in the development of human synovial sarcomas. Oncogene 1996; 13:643-648.
122. Clark J, Rocques PJ, Crew AJ et al. Identification of novel genes, *SYT* and *SSX*, involved in the t(X;18)(p11.2;q11.2) translocation found in human synovial sarcoma. Nature Genet 1994; 7:502-508.
123. Brett D, Whitehouse S, Antonson P et al. The *SYT* protein involved in the t(X;18) synovial sarcoma translocation is a transcriptional activator localised in nuclear bodies. Hum Mol Genet 1997; 6(9):1559-1564.
124. Licht JD, Ro M, English MA et al. Selective repression of transcriptional activators at a distance by the *Drosophila Kruppel* protein. Proc Natl Acad Sci USA 1993; 90:11361-11365.
125. Margolin JF, Friedman JR, Meyer WK et al. *Kruppel*-associated boxes are potent transcriptional repression domains. Proc Natl Acad Sci U S A 1994; 91:4509-4513.
126. Lim FL, Soulez M, Koczan D et al. A KRAB-related domain and a novel transcription repression domain in proteins encoded by *SSX* genes that are disrupted in human sarcomas. Oncogene 1998; 17:2013-2018.
127. Kawai A, Woodruff J, Healey JH et al. *SYT-SSX* Gene Fusion As A Determinant of Morphology and Prognosis In Synovial Sarcoma. N Engl J Med 1998; 338:153-160.
128. Chung EB, Enzinger FM. Malignant melanoma of soft parts. Am J Surg Pathol 1983; 7:405-413.
129. Hicks MJ, Saldivar VA, Chintagumpala MM et al. Malignant melanoma of soft parts involving the head and neck region: review of literature and case report. Ultrastruct Pathol 1995; 19:395-400.
130. Bridge JA, Sreekantaiah C, Neff JR et al. Cytogenetic findings in clear cell sarcoma of tendons and aponeuroses. Cancer Genet Cytogenet 1991; 52:101-106.
131. Stenman G, Kindblom L, Angervall L. Reciprocal translocation t(12;22)(q13;q13) in clear cell sarcoma of tendons and aponeuroses. Genes, Chromosones & Cancer 1992; 4:122-127.
132. Hai T, Liu F, Coukos WJ et al. Transcription factor ATF cDNA clones: An extensive family of leucine zipper proteins able to selectively form DNA-binding heterodimers. Genes Dev 1989; 3:2083-2090.
133. Zucman J, Delattre O, Desmaze C et al. *EWS* and *ATF-1* gene fusion induced by t(12;22) translocation in malignant melanoma of soft parts. Nature Genetics 1993; 4:341-345.
134. Sawyer JR, Tryka AF, Lewis JM. A novel reciprocal chromosome translocation t(11;22)(p13;q12) in an intraabdominal desmoplastic small round-cell tumor. Am J Surg Pathol 1992; 16:411-416.
135. Shen WP, Towne B, Zadeh TM. Cytogenetic abnormalities in an intra-abdominal desmoplastic small cell tumor. Cancer Genet Cytogenet 1992; 64:189-191.
136. Ladanyi M, Gerald W. Fusion of the *EWS* and *WT1* genes in the desmoplastic small round cell tumor. Cancer Res 1994; 54:2837-2840.
137. Gerald WL, Rosai J, Ladanyi M. Characterization of the genomic breakpoint and chimeric transcripts in the *EWS-WT1* gene fusion of desmoplastic small round cell tumor. Proc Natl Acad Sci USA 1995; 92:1028-1032.
138. Benjamin LE, Fredericks WJ, Barr FG et al. Fusion of the *EWS*1 and *WT1* genes as a result of the t(11;22)(p13;q12) translocation in desmolplastic small round cell tumors. Med Pediatr Oncol 1996; 27:434-439.
139. Karnieli E, Werner H, Rauscher FJ, III et al. the IGF-I receptor gene promoter is a molecular target for the ewing's sarcoma-wilms' tumor 1 fusion protein. J Biol Chem 1996; 271:19304-19309.
140. Sanchez-Garcia L, Grutz G. Tumorigenic activity of the *BCR-ABL* oncogenes is mediated by *BCL2*. Proc Natl Acad Sci USA 1995; 92:5287-5291.
141. Yi H, Fujimura Y, Ouchida M et al. Inhibition of apoptosis by normal and aberrant Fli-1 and ERG proteins involved in human solid tumors and leukemias. Oncogene 1997; 14:1259-1268.
142. Delattre O, Zucman J, Melot T et al. The Ewing family of tumors- a subgroup of small-round-cell tumors defined by specific chimeric transcripts. N Engl J Med 1994; 331:294-299.

143. Downing JR, Head DR, Parham DM et al. Detection of the (11;22)(q24;q12) translocation of Ewing's sarcoma and peripheral neuroectodermal tumor by reverse transcription polymerased chain reaction. Amer J Path 1993; 143:1294-1300.
144. Downing JR, Khandekar A, Shurtleff SA et al. Multiplex RT-PCR assay for the differential diagnosis of alveolar rhabdomyosarcoma and Ewing's sarcoma. Amer J Path 1995; 146:626-634.
145. Germain RN, Margulies DH. The biochemistry and cell biology of antigen processing and presentation. Annu Rev Immunol 1993; 11:403-450.
146. Townsend A, Bodmer H. Antigen recognition by class I-restricted T lymphocytes. Annu Rev Immunol 1989; 7:601-624.
147. Houbiers JGA, Nijman HW, van der Burg SH et al. In vitro induction of human cytotoxic T lymphocyte responses against peptides of mutant and wild-type p53. Eur J Immunol 1993; 23:2072-2077.
148. Theobald M, Biggs J, Dittmer D et al. Targeting p53 as a general tumor antigen. Proc Natl Acad Sci U S A 1995; 92:11993-11997.
149. Noguchi Y, Richards EC, Chen Y-T et al. Influence of interleukin 12 on p53 peptide vaccination against established Meth A sarcoma. Proc Natl Acad Sci USA 1995; 92:2219-2223.
150. Stuber G, Modrow S, Hoglund P et al. Assessment of major histocompatibility complex class I interaction with Epstein-Barr virus and human immunodeficiency virus by elevation of membrane H-2 and HLA in peptide loading-deficient cells. Eur J Immunol 1992; 22:2697-2703.
151. Nijman HW, Houbiers JGA, Vierboom MPM et al. Identification of peptide sequences that potentially trigger HLA-A2.1-restricted cytotoxic T lymphocytes. Eur J Immunol 1993; 23:1215-1219.
152. Salter RD, Howell DN, Cresswell P. Genes regulating HLA class I antigen expression in T-B lymphoblast hybrids. Immunogenetics 1985; 21:235-246.
153. Spies T, DeMars R. Restored expression of major histocompatibility class I molecules by gene transfer of a putative peptide transporter. Nature 1991; 351:323-324.
154. Zeh III HJ, Leder GH, Lotze MT et al. Flow-cytometric determination of peptide-class I complex formation identification of p53 peptides that bind to HLA-A2. Hum Immunol 1994; 39:79-86.
155. Sugawara S, Abo T, Kumagai K. A simple method to eliminate the antigenicity of surface class I MHC molecules from the membrane of viable cells by acid treatment at pH 3. J Immunol Methods 1987; 100:83-90.
156. Yotnda P, Garcia F, Peuchmaur M et al. Cytotoxic T cell response against the chimeric ETV6-*AML1* protein in childhood acute lymphoblastic leukemia. J Clin Invest 1998; 102:455-462.
157. Gambacorti-Passerini C, Bertazzoli C, Dermime S, Scardino A, Schendel D, Parmiani G. Mapping of HLA class I binding motifs in forty-four fusion proteins involved in human cancers. Clin Cancer Res 1997; 3:675-683.
158. Bertazzoli C, Marchesi E, Dermime S et al. HLA binding characteristics and generation of cytotoxic lymphocytes against peptides derived from oncogenic proteins. Tumori 1997; 83:847-855.
159. Goletz T, Zhan S, Pendleton C et al. Cytotoxic T cell responses against the *EWS*/FLI-1 Ewing's sarcoma fusion protein and the *PAX3*/*FKHR* alveolar rhabdomyosarcoma fusion protein. Proc AACR 1996; 3243.

CHAPTER 3

p53: A Target for T-Cell Mediated Immunotherapy

Michel P.M. Vierboom, Dmitry I. Gabrilovich, Rienk Offringa, W. Martin Kast and Cornelis J.M. Mclief

Introduction

Burnet's theory of immunosurveillance postulates that malignant transformation causes the expression of neoantigens. The theory states tumor specific antigens are recognized by the immune system, which then destroys the malignant cells.[1] A role for the immune system, especially T cells, in controlling cancer, is supported by the observation of a higher incidence of virus-associated cancer in immunocompromised patients compared to healthy individuals.[2,3] Virally induced tumors only make up a minority (15%-20%) of all tumors.[4-7] Although immunosurveillance is operational against virus-induced tumors this is still a matter of debate for non-virus induced tumors. However, for immunotherapy to have a wider application other targets besides viral targets need to be identified. Over the past decade new tumor antigens have been identified. Many of these new antigens are encoded by cellular genes representing unique tumor antigens, shared tumor-specific antigens and tissue specific differentiation antigens.[8,9] The occurrence of natural T cell responses to these cellular antigens in cancer patients demonstrates the presence of a T cell repertoire which can react against tumor associated self antigens. Although promising targets for immunotherapy, these antigens are limited in their expression to certain types of tumors, particularly in the case of tissue specific differentiation antigens

Next to these antigens targets which are expressed in a wide variety of tumors are desirable. An additional property of choice would be direct involvement of target antigens in malignant transformation, thereby reducing the possibility of immune escape. The tumor suppressor protein p53 meets these requirements.

p53 and Cancer

Mutation of the p53 gene is the most frequent event in human oncogenesis. P53 is aberrantly expressed in approximately 50% of all human malignancies.[10] There seems to be a common sequence of events, related to p53, leading to the formation of tumors. Mutation in one p53 allele and deletion of the second allele are the most frequent causes of p53 inactivation.[10,11] More than 95% of these alterations are missense mutations clustered in the central conserved DNA binding domain of the protein. As a consequence, the sequence specific binding of p53

Peptide-Based Cancer Vaccines, edited by W. Martin Kast. ©2000 Eurekah.com.

to dsDNA is abolished and p53 can no longer regulate the cell cycle. Furthermore most of these mutations dramatically increase the stability of the protein.[11]

Wild Type p53: Function and Regulation

Wild type p53 (wtp53) is a phosphoprotein which is located in the nucleus, where it performs its function as a transcriptional activator of genes regulating the cell cycle, thereby guarding the integrity of the genome as well as limiting cell proliferation. When the integrity of the genome is breached by DNA damaging agents, p53 expression is upregulated through stabilization followed by translocation to the nucleus.[12] In the nucleus p53 acts as a transcriptional activator of genes inhibiting DNA replication[13] and blocking the cell cycle at the G1/S phase.[14] No single consensus DNA binding sequence for p53 has been described. P53 has been found to bind to several different motifs. Interaction of p53 with multiple target sequences might explain the diversity of the genes regulated by this transcription factor. The growth arrest is dependent on the transcriptional induction of the protein p21 waf/CIP1.[15,16] This cell-cycle arrest allows DNA to be repaired. However, if the DNA is beyond repair, p53 activates genes causing apoptosis.[17-19] Recently, a three-step process of p53 mediated apoptosis has been suggested. It includes induction of redox related genes, the formation of reactive oxygen species and the oxidative degradation of mitochondrial components, culminating in cell death.[20]

P53 acts as a genuine tumor suppressor protein since mice lacking this protein are not hampered in their development but are highly susceptible to developing malignancies at an early age.[21] This was also demonstrated for patients suffering from Li-Fraumeni syndrome.[22,23] These patients have inherited mutated p53, which confers a greatly elevated risk of cancer at an early age.[24,25] Secondly, when wtp53 is introduced into certain tumor cells by DNA transfection it can induce growth arrest in cells as well as programmed cell death (apoptosis).[26,27] The cellular protein mdm2 is an important protein in the regulation of p53. Mdm-2 binds to p53 and targets the protein for the ubiquitin degradation pathway.[28,29] Since mdm-2 is a transcriptional target of p53, it forms an autoregulatory negative feedback loop. Other proteins like the protease calpain and p53 itself have recently been implicated in the regulation of p53.[30]

Analysis of p53 mRNA suggests that the gene is expressed in all tissues of the body throughout development.[31] However, this ubiquitous expression cannot be detected at the protein level using immunohistochemical methods.[12] A reason for this incongruity might be the very short half live of wtp53 in normal tissues.[11,32] Mutations in p53 markedly increase the protein stability and as a consequence the intracellular p53 concentration.[11] Viral proteins have also been shown to influence the function of p53. SV40 large tumor antigen[33] and the adenovirus E1b 55K protein[34] inactivate p53 by binding and blocking the DNA binding domain of the protein. The stable complex that is formed leads to overexpression of p53. In contrast, the binding of p53 with the HPV16 E6 oncoprotein leads to an increased ubiquitin-dependent proteasome-mediated degradation of p53.[35,36] Although mutations in p53 are considered to be the main cause of overexpression of p53 in tumors, intervention in the regulation of p53 stability either by interaction with viral proteins or cellular proteins can also influence the overall expression of p53 fragments in the context of MHC class I molecules. We propose that the overexpression of p53 seen in many tumors can lead to altered processing and presentation by MHC class I molecules of p53 derived peptides either in quality or in quantity, which in turn can allow the selective display of p53-derived peptides at the tumor cell surface.

Immunity Against p53

Antibodies

The first demonstration that p53 can be a target for the immune system was the presence of antibodies against p53 in mice with a p53 overexpressing tumor.[37,38] Subsequently, antibodies against human p53 have been detected in the sera of patients with colon cancer, [39,40] breast cancer,[41-45] lung cancer[46,47] hepatocellular carcinoma,[48] childhood lymphoma,[49] prostate, thyroid, bladder, pancreas cancer[50] and ovarian cancer.[39,51,52].Antibodies against p53 are virtually absent in sera from healthy donors.[53,54] The presence of anti-p53 antibodies is proportional to the occurrence of p53 mutations[53] and is an indicator of poor prognosis for colorectal cancer,[54] breast cancer[44] and lung cancer.[47] More importantly, in the case of lung cancer[47] and in patients suffering from angiosarcoma of the liver[55] anti-p53 antibodies were detected in the sera before the clinical manifestation of the disease. In these cases anti-p53 antibodies can be used as a marker of p53 alterations and early detection of these cancers. Isotyping of the antibodies against p53 revealed that they were mainly of the IgG1 and IgG2 subclasses,[53]which points at the involvement of CD4$^+$ T helper cells that are necessary for isotype class switching of antibodies.[56,57] Although much effort has been devoted to the identification of the B cell epitopes recognized in these studies, the role of CD4$^+$ T helper cells in orchestrating the immune response against p53 expressing tumors has received much less attention. One report by Tilkin et al[58] has shown proliferative responses against wild type p53 of T cells from breast cancer patients, correlating with the presence of anti-p53 antibodies in the serum of these patients. However, they did not identify the peptide that was recognized by these T helper cells. A second report by Fujita et al[59] has shown the induction of wtp53 peptide-specific T cell responses also recognizing processed wild type and mutant p53 protein. This further demonstrates that the tolerance against wild type p53 is far from complete at the level of CD4$^+$ T cells.

CTL Mediated Immunotherapy Against p53

The potency of specific CTL immunotherapy was greatly facilitated by several findings in the second half of the 1980's and the early 90's. Townsend et al.[60] discovered that CTL recognize short fragments of protein (peptides) presented by molecules of the major histocompatibility complex (MHC) class I. A year later the group of Wiley[61] elucidated the crystal structure of the human MHC class I molecule, HLA-A*0201. The MHC class I molecule contained a peptide in the groove of the molecule. These two findings explained the phenomenon of MHC restriction of T cell Receptor (TcR) mediated recognition by CTL as discovered by Zinkernagel and Doherty.[62] This, together with the discovery by the group of Rammensee that these peptides in the MHC Class I binding groove fit specific binding motifs[63] made it possible to predict and identify target peptides for T cell mediated immunotherapy. In the meantime the tremendous power of CTL against defined epitopes in tumor eradication had been shown in CTL adoptive transfer systems in mice (reviewed in ref. 64). Shortly after p53 was recognized as a potential target for tumorimmunotherapy several techniques were used in order to establish actual binding of, computer predicted,[65] p53 derived MHC Class I binding peptides. Various techniques employing either purified MHC molecules[66-68] or intact cells expressing defined MHC molecules[69] were developed to assess actual binding of the peptides.

To date, 29 wt p53 and 21 mutant p53 derived peptides have been found to bind to human MHC class I molecules (Table 3.1 and 3.2). Out of the 29 human wt p53 derived peptides 6 were able to induce peptide specific responses. The presentation by human tumor cells has been indirectly confirmed for 3 of those 6 wtp53 peptides, through recognition by CTL generated against these epitopes.[70-73] The immunogenicity of these peptides was confirmed both in man and in HLA-A*0201 transgenic mice. Presentation of the peptides

Table 3.1. Human wtp53 HLA class I binding peptides

restriction element	aa pos.[a]	sequence	bind.[b]	im. af.	[c]tum.[d] rec.	reference
A*0201	24-32	KLLPENNVL	+	nd.	nd.	68,73,127-129
	25-35	LLPENNVLSPL	++	+[70]	nd.	70,73,127,128
	43-52	LMLSPDDIEQ	++*	nd.	nd	130
	65-74	RMPEAAPPVA	~/++%	nd.	nd.	127,128
	65-73	RMPEAAPPV	++	+[70]	nd	68,70,73,127,128,130
	129-137	ALNKMFCQL	~/++	+[1]	nd.	68,73,127,129,130[1]
	136-144	QLAKTCPVQ	++*	nd.	nd.	130
	139-147	KTCPVQLWV	+	nd.	nd	128
	149-157	STPPPGTRV	~	+	+[70]	70,128
	186-196	DGLAPPQHLIR	+/++*	nd.	nd	68,130
	187-195	GLAPPQHLI	++	nd.	nd.	73,130
	187-197	GLAPPQHLIRV	++	+	+[71]	71,73,127,128
	193-201	HLIRVEGNL	+/+*	nd.	nd	68,130
	193-203	HLIRVEGNLRV	~/++%	nd.	nd.	127,128
	245-253	GMNRRPILT	+*	nd.	nd.	130
	256-265	TLEDSSGNLL	~/+*	nd.	nd.	127,128,130
	263-272	NLLGRNSFEV	+/++	nd.	nd.	73,127.130
	264-274	LLGRNSFEVRV	+	nd.	nd.	127
	264-272	LLGRNSFEV	+/++	+	+[70/72]	68,7073,103,127,130
	322-330	PLDGEYFTL	++	nd.	nd	68,130,134
	339-407	EMFRELNEA	+	nd.	nd.	127
A*01	196-205	RVEGNLRVEY	+	nd.	nd.	73
	226-234	GSDCTTIHY	++	nd.	nd.	7
B*07	26-35	LPENNVLSPL	++	nd.	nd.	73
	63-73 (R72)	APRMPEAAPRV	++	nd.	nd.	73
	189-197	APPQHLIRV	++	nd.	nd.	73
	189-197$_{(L194R)}$	APPQHRIRV	++	nd.	nd.	73
	321-330	KPLDGEYFTL	++	nd.	nd.	73
B*08	210-218$_{(R213Q)}$	NTFQHSVVV	++	nd.	nd.	73

[a] amino acid position; [b] binding affinity (in arbitrary scale); [c] immunogenicity; [d] tumor recognition; nd = not determined. * = only reconstitution assay; % = competition assay; 134) M.Schirle (personal communication) Peptides of low affinity and tested by only one group are not included in this list. In bold immunogenicity is tested with human PBL. [128]

STPPPGTRV (aa$_{149-157}$) and GLAPPQHLIRV (aa$_{187-197}$) on human tumor cells was confirmed by the recognition of these cells by murine CTL obtained from HLA-A*0201 transgenic mice.[70,71] The third human wtp53 derived peptide LLGRNSFEV (aa$_{264-272}$) is the only peptide which is recognized by human CTL that also recognize endogenously processed peptide in

Table 3.2. Human mutant p53 HLA class I binding peptides

restriction element	aa pos.[a]	sequence	bind[b] af.	im.[c]	tum.[d] rec.	reference
A*0201	129-137	ALNKMLCQL	~	nd.	nd.	127
	129-137	ALNKMFYQL	~/+	nd.	nd.	127
	132-140	NMFCQLAKT	+	nd.	nd.	127
	132-140	KLFCQLAKT	~	nd.	nd.	127
	132-140	KMFYQLAKT	~	nd.	nd.	127
	168-176	HMTEVVRHC	~	+	nd.	127
	187-195	GLAPPQHEI	++	nd.	nd.	130
	187-197	GLAPPQHFIRV	~	nd.	nd.	127
	193-201	HLIKVEGNL	+*	nd.	nd	130
	245-253	GMNCRPILT	+*	nd.	nd.	130
	245-253	GMNKRPILT	+*	nd.	nd.	130
	245-253	GMNERPILT	+*	nd.	nd	130
	245-253	GMNRHPILT	+*	nd.	nd.	130
	264-272	LLGRNSFEM	++	nd.	nd.	130
	264-272	LLGRNSEEM	+*	nd.	nd.	130
	264-274	LLGRNSFEVCV	~	nd.	nd.	127
	264-274	LLGRNSFEVC	~	nd.	nd.	127

[a] amino acid position; [b] binding affinity (in arbitrary scale); [c] immunogenicity; [d] tumor recognition; nd = not detemined ; *= only reconstitution assay.

human tumors.[72,73] Although the induction of wtp53 specific responses with cross-reactivity on endogenous p53-expressing targets is an important observation, the in vivo relevance of these responses remains to be established. Recently, a fourth human wtp53 peptide was identified by the group of Rammensee (M.Schirle personal communication). They eluted the peptide PLDGEYFTL ($aa_{322-330}$) from HLA-A*0201 molecules. The immunogenicity of this peptide remains to be determined.

Tables 3.3 and 3.4 provide an overview of the murine p53-derived peptides binding to MHC Class I and II molecules. Information concerning these peptides will be discussed in the next sections.

Mutant p53: A Target for CTL Immunotherapy

Due to the ubiquitous expression of p53 in normal tissue and particularly the thymus[31], tolerance against wt p53 might lead to the deletion of high affinity wtp53 specific CTL. Therefore, mutant p53, having a "foreign" quality not leading to tolerance, was initially targeted for CTL mediated immunotherapy against cancer. This foreign quality should not only enhance its immunogenicity but also provide great selectivity of the immunotherapy since it will only be expressed by the tumor.

Yanuck et al[74] were the first to show the feasibility of mutant p53 as a target for T cells. They vaccinated BALB/c mice with a 21-mer peptide corresponding to the sequence of mutant p53 found in a human non-small lung cell carcinoma (TYSPALNKMFYQLAKT CPVQL: aa $_{125-145}$). The mutation in this peptide (underlined) generated an H-$2K^d$ binding peptide that is

Table 3.3. Murine p53 MHC class I binding peptides

restriction element	aa pos.[a]	sequence	bind.[b] af.	im.[c]	tum.[d] rec.	reference
H-2K^d	125-145	TYSPALNKMFY QLAKTCPVQL	+	+	74	
	134-143	FYQLAKTCPV	+	+	74	
	232-240	KYMCNSSCM	+	+	+	104
	232-240	KYICNSSCM	+	+	+	82,83,104
	122-130	TYSPPLNKL	+	+	nd	131
	122-130	TYSPPLNEL	+	+	nd	131
H-2K^b	18-25	SQETFSGL	+	-	nd.	132
	119-127	VMCTYSPPL	+	+[#]	nd.	102 [#]
	122-130	TYSPPLNKL	+	+[#]	nd.	102 [#]
	123-131	YSPPLNKLF	+	nd	nd	102
	127-134	LNKLFCQL	+	+[#]	nd.	132, 25[#]
	158-166	AIYKKSQHM	+	-	+	102
	222-229	AGSEYTTI	+	-	nd.	102,132
	227-234	TTIHYKYM	+	+	nd.	102,132
	320-327	LDGEYFTL	+	-	nd.	132
	334-341	RFEMFREL	+	-	nd.	132
H-2D^b	232-240	KYMCNSSCM	+	+	+[133]	132,133
	240-248	MGGMNRRPI	+	+	nd.	132,133

[a] amino acid position; [b] binding affinity (in arbitrary scale); [c] immunogenicity; [d] tumor recognition, nd = not determined. [#] unpublished observation

Table 3.4. Murine p53 MHC class II binding peptides

restriction element	aa pos.[a]	sequence	bind.[b]	im.[c] af.	tum.[d]	reference rec
?	232-240	KYICNSSCM		+	nd.	83
?	232-240	KYMCNSSCM		+	nd	83

[a] amino acid position; [b] binding affinity (in arbitrary scale); [c] immunogenicity; [d] tumor recognition, nd = not determined

recognized by $CD8^+$ T cells (FYQLAKTCPV: $aa_{134-143}$; Table III). More importantly these CTL recognized BALB/c 3T3 fibroblasts transfected with the human mutant p53 gene[74], P815 cells transfected with the human mutant p53 gene[75] and a human lung cancer cell line, transfected with the human mutant p53 gene and the restriction element H-2K^d[76]. Furthermore vaccination with DNA containing this single mutant p53 epitope, induced a CTL response which could control the outgrowth of a P815 tumor expressing this epitope.[77] The 21-mer peptide contained, outside the 10-mer fragment that is recognized by the CTL, 2 amino acids (in bold) non homologous to the murine wtp53 sequence. These differences in sequence might add to the "foreign" nature of this peptide and account for its immunogenicity in mice.

This peptide was subsequently used as a model peptide to evaluate the immune response inducing potential of dendritic cells (DC) from tumor bearing animals and the role of T cells in this response.[78,79] It was shown that peptide vaccination 5 days after tumor challenge could not induce a CTL response. Even in pre-immunized mice no peptide specific CTL could be detected 11 days after tumor injection, indicating a general tumor induced impairment of the immune system. It was demonstrated that mature DC's, isolated from the spleen of tumor bearing animals, cocultured with supernatant of the tumor were not affected in their ability to stimulate allogenic responses in a mixed lymphocyte culture (MLR) assay. However, the presence of supernatant from tumor cultures during culture of maturing bone marrow derived DC's severely impaired the immune stimulating potential.[79] The important message from these data is that not only the effector arm of the immune system can be hampered by the presence of a tumor[80] but also that the tumor can secrete factors that affect the function of DC as priming APC.

In this model the efficacy of vaccination by peptide pulsed DC was significantly enhanced by adding IL-12.[81] This effect of IL-12 had previously been shown in another model with a murine mutant p53 peptide by Noguchi et al.[82] They demonstrated that BALB/c mice immunized with a mutant peptide (KYICNSSCM: $aa_{232-240}$) in QS-21 adjuvant in combination with IL-12 were able to protect against a subsequent challenge with a Meth A sarcoma expressing this mutant peptide. In addition, they were able to eradicate established tumors with this combination therapy. These effects were mutant specific because vaccination with the wild type counterpart of this mutant peptide (KYMCNSSCM: $aa_{232-240}$) using the same protocols did not retard progressive tumor growth. The mutant p53 H-2K^d presented peptide had been identified by this group after analysing a methylcholantrene-induced sarcoma containing 3 mutations in p53.[83] This peptide, containing a mutation of aa_{234} (M- >I), was shown to induce proliferative responses and CTL responses after peptide immunization in IFA. Immunization with this peptide heightened the resistance to a subsequent challenge with Meth A sarcoma.

Mutant peptides do not differ from viral peptides in their ability to induce responses with respect to tolerance, since CTL directed against mutant peptides are not subject to thymic deletion. So these findings, though interesting with regard to general immunotherapeutic parameters, could have been obtained in tumor models with exogenous viral antigens and do not add to the potential utility of autologous p53 sequences as a target for tumorimmunotherapy.

Although targeting mutations in p53 may be logical in view of tolerance against wild type p53, it limits the use of p53 as a broad spectrum target for cancer immunotherapy for several reasons.

Mutations in p53 are very heterogeneous which would necessitate analysis of the mutation prior to therapy for every new patient. This is further complicated by the fact that mutations found in the primary tumor can differ from the mutations found in metastases.[84] However, more importantly, mutations in p53 are rarely contained in MHC class I binding peptides. This is supported by a study by Wiedenfeld et al,[85] which shows for 6 HLA-A*0201 expressing human lung cell carcinomas all p53 mutations are outside the putative HLA-A*0201 binding peptides. They suggest that during the development of the tumor selection occurs against tumors expressing HLA binding and processed mutated p53 products underlining even more the need for targeting wild type p53 versus mutated sequences.

Wild Type p53: A Target for CTL Immunotherapy

Several observations preceeded the recognition of wild type p53 as a target for tumorimmunotherapy.

Hu et al[86] demonstrated in a Friend leukemia virus envelope (FLVenv) transgenic mouse model that adoptively transferred FLVenv-autoreactive T cells can eradicate FLVenv expressing tumor, leaving normal tissue expressing this artificial self antigen undamaged. It has been appreciated that every individual contains a large repertoire of self directed T cells which have evaded thymic deletion.[87] This is supported by the identification of target antigens recognized by CTL on non-virally-induced tumors, notably in patients with melanoma.[8,88-101] Following these observations the idea of using wild type p53 as target for T cell mediated immunotherapy gained momentum.

Two important issues need to be addressed if wild type p53 is to be used as a target for T cell mediated immunotherapy against cancer. Tolerance against p53 can seriously complicate the ability to induce responses against wtp53 derived sequences. This was elegantly shown by Theobald et al[70,71] They used HLA-A*0201 transgenic C57BL/6 mice in order to generate responses against human wt p53 peptides. Due to the sequence differences between human and murine p53 they were able to induce human wt p53 specific CTL against 2 epitopes (STPPPGTRV: $aa_{149-157}$; LLGRNSFEV: $aa_{264-272}$). These CTL also recognized endogenously processed wtp53 on HLA-A*0201 positive human cancer cell lines.[70] Both peptides are not homologous to the murine counterpart and murine CTL against those peptides are therefore not subject to self tolerance induction.[70] They subsequently showed that the peptide GLAPPQHLIRV ($aa_{187-197}$), which is homolgous to the murine wtp53 sequence could only generate a response in p53 gene deficient (p53 -/-) HLA-A*0201 transgenic C57BL/6 mice.[71] The peptide LLGRNSFEV ($aa_{264-272}$), previously identified by Houbiers et al Houbiers, 1993 #3468], was also shown to be presents on a human squamous cell carcinoma and recognized by human peptide induced CTL.[72] The data presented by Theobald et al, also suggest that a gap exists in the T cell repertoire independent of p53 expression, since not all non-homologous human p53 peptides with high HLA-A*0201 binding affinity were able to induce a response in P53 +/+ HLA-A*0201 transgenic mice. Also the inability of certain human wtp53 peptides, homologous to murine p53, to induce a response in p53 -/- HLA-A*0201 transgenic mice points to a gap in the T cell repertoire.

Next to the issue of tolerance there is the possible danger of autoimmunity if a response can be induced against wtp53. The p53 protein is ubiquitously expressed in normal tissues throughout the body albeit at low levels.[31] For the evaluation of autoimmunity high affinity CTL recognizing a murine wild-type p53 peptide AIYKKSQHM ($aa_{158-166}$) presented by the MHC class I molecule H-2K^b were generated by immunizing p53 -/- C57BL/6 mice with irradiated syngeneic p53-overexpressing tumor cells. Adoptive transfer of these wt p53 specific CTL into tumor bearing p53 +/+ nude mice led to the eradication of large p53 overexpressing tumors without demonstrable damage to normal tissue.[102] This indicated that the overexpression of p53 in the tumor and the low expression of p53 in normal tissue creates a therapeutic window. It was recently suggested by Gnjatic et al[73] that the overexpression of p53 in human tumors, as analysed with cytospin preparation, is required to allow recognition of tumor cells by wtp53 specific CTL. However, our murine wtp53 specific CTL were shown to recognize various tumor cells with no detectable p53 overexpression.[102] Vice versa, these wtp53 specific CTL showed poor recognition of Ad5 transformed tumor cells with highly stabilized p53 (manuscript in preparation. Vierboom et al). It has to be noted however that our murine wt p53 specific CTL were generated in vivo by tumor cell vaccination,[102] a process which is likely to involve crosspriming. In contrast, Gnjatic et al[103] generated human wtp53 specific CTL through in vitro stimulation by exogenous peptide loaded stimulator cells. These peptide-generated CTL are usually of lower affinity which may necessitate overexpression of p53 for recognition.

Mayordomo et al[104] were the first to demonstrate in BALB/c mice the feasability of vaccination against wtp53. They evaluated the efficacy of peptide pulsed DC with a previously identified mutant peptide 234CM[83] and showed effective prophylactic and therapeutic vaccination with mutant p53 peptide pulsed dendritic cells against the methylcholantrene-induced Meth A sarcoma. They subsequently vaccinated BALB/c mice with DC loaded with the wt p53 counterpart 234CW. Wtp53 peptide specific responses were induced which also recognized P815 and a panel of chemically induced p53 overexpressing sarcomas among which was CMS4. More importantly, vaccination with wtp53 peptide (234CW) pulsed DC's protected against a subsequent challenge with CMS4 and could eradicate established CMS4 tumors.

The important message is that in normal p53 +/+ mice a repertoire of wtp53 specific CTL is present which can be activated to perform tumoricidal activity after proper vaccination. Even in tumor bearing animals which is important for the clinical application.

Prospects of Peptide Based Vaccines Against p53

Several clinical trials of p53 directed immunotherapy are underway, but crucial questions are still to be answered. The tolerance at the level of $CD4^+$ T cells against wt p53 is incomplete, as demonstrated by the literature on antibodies and proliferative responses against wtp53. However the tolerance of wtp53 at the level of $CD8^+$ T cells, believed to be the main effector in tumor eradication, is subject to tight control. The CTL tolerance of wtp53 poses an important problem that has to be resolved for effective use of wtp53 as a target for CTL mediated immunotherapy in the clinic.Tolerance against wtp53 was demonstrated by Theobald et al.[70,71] However, the HLA- A*0201 transgenic mice they used to evaluate tolerance against wtp53 can also be used as a way to circumvent tolerance. The high affinity CTL generated against wt p53 by vaccinating HLA-A*0201 transgenic mice with human wtp53 derived peptides (STPPPGTRV ($aa_{149-157}$); LLGRNSFEV ($aa_{264-272}$) and GLAPPQHLIRV ($aa_{187-197}$)) could potentially be used for adoptive transfer in HLA-A*0201 positive patients carrying a p53 overexpressing tumor. McCarty et al.[105] have indeed shown that CTL directed against STPPPGTRV ($aa_{149-157}$) were able to control the growth of a human pancreatic carcinoma in severe combined immunodeficient (SCID) mice. However, an obvious problem limiting this approach will be the anti-murine CTL response induced in humans next to problems related to CTL/host species barriers. Alternatively wtp53 specificity can be conferred to human T cells by transducing human T cells with the high affinity, wtp53 specific T cell receptor (TcR) obtained from murine CTL generated in HLA transgenic mice.Another way to overcome tolerance has recently been suggested by Sadovnikova et al.[106,107] The idea is to generate from donor PBL, CTL specific for wtp53-derived peptides in the context of allogeneic MHC Class I molecules, specifically recognizing the tumor of the patient (recipient). The patient then receives an allogeneic bone marrow transplantation from the same donor, followed by the adoptive transfer of these so called allorestricted wtp53 specific CTL. We have demonstrated selective eradication of a p53 overexpressing tumor by wt p53 specific CTL without autoimmune damage to normal tissue.

The role of $CD4^+$ T cells in breaking tolerance and inducing p53 specific responses is underexposed. Although the presence of anti-p53 IgG antibodies, which indirectly points to the presence of $CD4^+$ T cells,[56,57] was the initial argument that p53 can induce a T cell mediated immune response, research to evaluate p53 as a target for immunotherapy has been mainly concerned with the identification of MHC Class I presented peptides recognized by $CD8^+$ CTL (Table x.1 and x.2). Recent literature underscores the importance of $CD4^+$ T cells in orchestrating an immunologic attack mediated by multiple effector arms of the immune system.[108,109] $CD4^+$ T cells play a pivotal role in the induction and maintenance of CTL responses by activating DC's through CD40-CD40-ligand interaction.[110] Moreover, $CD4^+$ T cells are important in the eradication of MHC Class II negative tumors by recruiting MHC unrestricted effector cells with tumoricidal activities such as eosinophils and macrophages.[108,111] Recent findings underscore the specific nature of T cell help in the induction of CTL responses[112] also against MHC Class II negative tumors.[113] Furthermore $CD4^+$ T cells play a crucial role in the induction of autoimmune disease.[114] Since therapy against tumor associated self antigens deals with induction of a limited beneficial autoimmune response directed against the tumor, the role of $CD4^+$ T cells in inducing responses against p53 needs to be evaluated. These findings even further necessitate the identification of p53 derived T helper epitopes. We have recently identified 4 T helper epitopes in C57BL6 p53 -/- mice, one of which recognized endogenously processed p53 protein (manuscript in preparation; Vierboom et al). The use of peptide vaccines containing only CTL epitopes has some limitations. MHC Class I restricted presentation of antigens constrains the use of a particular peptide in a genetically diverse population. The efficacy of binding of such peptides cannot be predicted in advance. Besides that, peptide based immunization does not take into account the ability of the peptides to bind to MHC Class II, playing a critical role in the induction of anti tumor immune response through the activation of $CD4^+$ T cells. The alternative approach to p53 specific immunization could be the use of whole p53 molecules with the aim to allow for expression of a variety of different tumor-associated epitopes. One strategy exploiting whole protein vaccination is the use of canarypox virus (ALVAC) containing coding DNA of wtp53.[115,116] A clinical trial using ALVAC wtp53 as immunogen recently started in our department in collaboration with Pasteur Merieux Connaught. Another strategy, as was shown previously in model experiments, using minimal epitopes linked to a single fusion protein allows separate presentation on the cell surface and recognition by specific CTLs.[117] Therefore, one or several of these antigens could be effective in the induction of anti-tumor immune response in a given patient. This approach has recently been tested using a recombinant adenovirus-wtp53 construct. Mouse dendritic cells (DC) were transduced with a human wtp53 containing recombinant adenovirus (Ad-p53). Mice immunized twice with Ad-p53 infected DC developed substantial CTL responses against tumor cells expressing either wild type or different mutant human and murine p53 genes. Only weak CTL responses were observed against target cells infected with control adenovirus (Ad-c). Immunization with Ad-p53 provided complete tumor protection in 85% of mice challenged with tumor cells expressing human mutant p53 and in 73% of mice challenged with tumor cells containing murine mutant p53.[118] This study has demonstrated that this approach might be potentially effective. More studies in syngeneic systems are needed to prove the possibility of breaking tolerance to p53 protein using this approach. A mechanism of immune escape of tumors, previously described by Ossendorp et al,[119] showed that a single residue change within a viral CTL epitope introduced a cleavage site in the epitope thereby destroying the peptide and preventing its presentation. A related mechanism, influencing the presentation of wtp53 derived epitopes, was recently described by Theobald et al[120] They have demonstrated that the replacement of an arginine by a histidine, through a hotspot mutation in a flanking sequence (aa_{273} R->H) of the human epitope LLGRNSFEV ($aa_{264-272}$) could prevent the processing and presentation of this epitope, while the recognition of another epitope (STPPPGTRV; $aa_{149-157}$) was left intact. This observation underlines the importance of targeting multiple epitopes of wt p53.Vaccines based on p53 specific immune response induction in combination with biological modifiers like anti-CTLA-4[121] and/or chemotherapeutic agents, like cyclophosphamide (a potent inducer of autoimmune disease,)[122,123] which may alleviate $CD4^+$ T cell mediated suppression or reverse tolerance,[124-126] should also be considered.

Acknowledgment

We would like to thank Frank Verreck and Sjoerd van der Burg for critically reading this manuscript.

References

1. Burnet FM. Immunological surveillance in neoplasia. Transplant Rev 1971; 7:3-25.
2. Beral V and Newton R. Overview of the epidemiology of immunodeficiency-associated cancers. Monogr Natl Cancer Inst 1998; 23: 1-6.
3. Kieff E. Current perspectives on the molecular pathogenesis of virus-induced cancers in human immunodeficiency virus infection and acquired immunodeficiency syndrome. Monogr Natl Cancer Inst 1998; 23: 7-14.
4. Zur Hausen H. Viruses in human cancer. Science 1991; 254: 1167-1172.
5. Masucci MG. Viral immunopathology of human tumors. Curr. Opin. Immunol. 1993; 5: 693-700.
6. Offermann MK. HHV-8: a new herpes virus associated with Kaposi's sarcoma. Trends Microbiol. 1996; 4: 383-386.
7. Carbone M, Rizzo P and Pass HI. Simian virus 40, poliovaccines and human tumors: a review of recent developments. Oncogene 1997; 15: 1877-1888.
8. Boon T, Cerottini J-C, Van den Eynde B et al. Tumor antigens recognized by T lymphocytes. Annu. Rev. Immunol. 1994; 12: 337-365.
9. Pardoll DM. Cancer vaccines. Nat Med 1998; 4: 525-531.
10. Hollstein M, Sidransky D, Vogelstein B et al. p53 Mutation in human cancers. Science 1991; 253: 49-53.
11. Zambetti GP and Levine AJ. A comparison of the biological activities of wild-type and mutant p53. FASEB J. 1993; 7: 855-865.
12. Lane DP. p53 and human cancer. Brit.Med.Bull. 1994; 50: 582-599.
13. Dutta A, Ruppert JM, Aster JC et al. Inhibition of DNA replication factor RPA by p53. Nature 1993; 365: 79-82.
14. Martinez J, Georgoff I and Levine AJ. Cellular localization and cell cycle regulation by a temperature sensitive p53 protein. Genes Dev 1991; 5: 151-159.
15. Waldman T, Kinzler KW and Vogelstein B. p21 Is necessary for the p53-mediated G_1 arrest in human cancer cells. Canc. Res. 1995; 55: 5187-5190.
16. Wilson JW, Pritchard DM, Hickman JA et al. Radiation-induced p53 and $p21^{WAF-1/CIP1}$ expression in the murine intestinal epithelium. Am J Pathol 1998; 153: 899-909.
17. Lowe SW, Schmitt EM, S.W. S et al. p53 is required for radiation-induced apoptosis in mouse thymocytes. Nature 1993; 362: 847-849.
18. Shaw P, Bovey R, Tardy S et al. Induction of apoptosis by wild-type p53 in a human colon tumor-derived cell line. Proc. Nat. Acad. Sci. 1992; 89: 4495-4499.
19. Yonish-Rouach E, D. R, Lotem J et al. Wild-type p53 induces apoptosis of myeloid leukaemic cells that is inhibited by interleukin-6. Nature 1991; 353: 345-347.
20. Polyak K, Xia Y, Zweier JL et al. A model for p53-induced apoptosis. Nature 1997; 389: 300-305.
21. Donehower LA, Harvey M, Slagle BL et al. Mice deficient for p53 developmentally normal but susceptible to spontaneous tumors. Nature 1992; 356: 215-221.
22. Frebourg T and Friend SH. Cancer risks from germline p53 mutations. J.Clin.Invest. 1992; 90: 1673-1678.
23. Malkin D. p53 and the Li-Fraumeni syndrome. Cancer Genet Cytogenet 1993; 66: 83- 92.
24. Malkin D, Li FP, Strong LC et al. Germline p53 mutations in a familial syndrome of breast cancer, sarcomas and other neoplasms. Science 1990; 250: 1233-1238.
25. Malkin D, Jolly KW, Barbier N et al. Germline mutations of the p53 tumor-suppressor gene in children and young adults with second malignant neoplasms. N Engl J Med 1992; 326: 1309-1315.
26. Finlay CA, Hinds PW, Levine AJ. The p53 proto-oncogene can act as a suppressor of transformation. Cell 1989; 57: 1083-1093.
27. Baker SJ, Markowitz S, Fearon ER et al. Suppression of human colorectal carcinoma cell growth by wild-type p53. Science 1990; 249: 912-915.

28. Momand M, Zambetti GP, Olson DC et al. The mdm-2 oncogene product forms a complex with the p53 protein and inhibits p53- mediated transactivation. Cell 1992; 69: 1237-1245.
29. Haupt Y, Maya R, Kazaz A et al. Mdm-2 promotes the rapid degradation of p53. Nature 1997; 387: 296-299.
30. Kubbutat MHG and Vousden KH. Keeping an old friend under control: Regulation of p53 stability. Molecular Medicine Today 1998; June: 250-256.
31. Rogel A, Popliker M, Webb CG et al. p53 cellular tumor antigen: Analysis of mRNA levels in normal adult tissues, embryo, and tumors. Molecular and Cellular Biology 1985; 5: 2851-2855.
32. Reich N and Levine AJ. Growth regulation of a cellular tumour antigen, p53, in nontransformed cells. Nature 1984; 308: 199-201.
33. Ludlow JW. Interactions between SV 40 large-tumor antigen and the growth suppressor proteins pRB and p53. FASEB J 1993; 7: 866-871.
34. Moran E. Interaction of adenoviral proteins with Rb and p53. FASEB J. 1993; 7: 880- 885.
35. Scheffner M, Werness B, Huibregtse JM et al. The E6 oncoprotein encoded by HPV16 and 18 promotes degradation of p53. Cell 1990; 63: 1129-1136.
36. Scheffner M, Huibregtse JM, Vierstra RD et al. The HPV16 E6 and E6-AP complex functions as a ubiquitin-protein ligase in the ubiquitination of p53. Cell 1993; 75: 495-505.
37. DeLeo AB, Jay G, Apella E et al. Detection of transformation-related antigen in chemically induced sarcomas and other transformed cells of the mouse. Proc Nat Acad Sci USA 1979; 76: 2420-2424.
38. Lane DP and Crawford LV. T antigen is bound to a host protein in SV40-transformed cells. Nature 1979; 278: 261-263.
39. Angelopoulou K and Diamandis EP. Autoantibodies against the p53 tumor suppressor gene product quantified in cancer patient serum with time-resolved immunofluorometry. The Cancer J 1993; 6: 315-319.
40. Crawford LV, Pim DC and Lamb P. The cellular protein p53 in human tumors. Mol Biol Med 1984; 2: 261-272.
41. Crawford LV, Pim DC and Bulbrook RD. Detection of antibodies against the cellular protein p53 in sera from patients with breast cancer. Int J Cancer 1982; 30: 403-408.
42. Davidoff AM, Iglehart JD, Marks JR. Immune response to p53 is dependent upon p53/hsp70 complexes in breast cancer. Proc Natl Acad Sci USA 1992; 89: 3439-3442.
43. Schlichtholz B, Legros Y, Gillet D et al. The immune response to p53 in breast cancer patients is directed against immunodominant epitopes unrelated to the mutational hot spot. Cancer Res 1992; 52: 6380-6384.
44. Barnes DM, Dublin EA, Fisher CJ et al. Immunohistochemical detection of p53 protein in mammary carcinoma: An important new independent indicator of prognosis. Hum Pathol 1993; 24: 469-476.
45. Peyrat J-P, Bonneterre J, Lubin R et al. Prognostic significance of cicrculating p53 antibodies in patients undergoing surgery for locoregional breast cancer. Lancet 1995; 345: 621-622.
46. Winter SF, Minna JD, Johnson BE et al. Development of antibodies against p53 in lung cancer patients appears to be dependent on the type of p53 mutation. Cancer Res 1992; 52: 4168-4174.
47. Lubin R, Zalcman G, Bouchet L et al. Serum p53 antibodies as early markers of lung cancer. Nature Med 1995; 1: 701-702.
48. Volkmann M, Müller M, Hofmann WJ et al. The humoral immune response to p53 in patients with hepatocellular carcinoma is specific for malignancy and independent of the α-fetoprotein status. Hepatology 1993; 18: 559-565.
49. Caron de Fromentel C, May-Levin F, Mouriesse H et al. Presence of circulating antibodies against cellular protein p53 in a notable proportion of children with B-cell lymphoma. Int.J.Cancer 1987; 39: 185-189.
50. Lubin R, Schlichtholz D, Bengoufa D et al. Analysis of p53 antibodies in patients with various cancers define B-cell epitopes of human p53: distribution on primary structure and exposure on protein surface. Cancer Res. 1993; 53: 5872-5876.
51. Labrecque S, Naor N, Thomson D et al. Analysis of the anti-p53 antibody response in cancer patients. Cancer Res. 1993; 53: 3468-3471.

52. Vennegoor CJGM, Nijman HW, Drijfhout JW et al. Auto-antibodies to p53 in ovarian cancer patients and healthy women: a comparison between the whole p53 protein and 18-mer peptides for screening purposes Cancer Lett 1997; 116: 93-101.
53. Lubin R, Schlichtholz B, Teillaud JL et al. p53 Antibodies in patients with various types of cancer: assay, identification and characterization. Clin.Canc.Res. 1995; 1: 1463-1469.
54. Houbiers JGA, van der Burg SH, van de Watering LMG et al. Antibodies against p53 are associated with poor prognosis of colorectal cancer. Br.J.Cancer 1995; 72: 637-641.
55. Trivers GE, Cawley HL, DeBenedetti VMG et al. Anti-p53 antibodies in sera of workers occupationally exposed to vinyl chloride. J Natl Cancer Inst 1995; 87: 1400-1407.
56. Bergstedt-Lindqvist S, Sideras P, MacDonald HR et al. Regulation Ig class secretion by soluble products of certain T-cell lines. Immunol. Rev. 1984; 78: 25-50.
57. Coffman RL, Seymour BW, Lebman DA et al. The role of helper T cell products in mouse B cell differentiation and isotype regulation. Immunol.Rev. 1988; 102: 5-28.
58. Tilkin A-F, Lubin R, Soussi T et al. Primary proliferative T cell response to wild-type p53 protein in patients with breast cancer. Eur.J.Immunol. 1995; 25: 1765-1769.
59. Fujita H, Senju S, Yokomizo H et al. Evidence that HLA class II-restricted human CD4+ T cells specific to p53 self peptides respond to p53 proteins of both wild and mutant forms. Eur.J.Immunol. 1998; 28: 305-316.
60. Townsend ARM, Rothbard J, Gotch FM et al. The epitopes of influenza nucleoprotein recognized by cytotoxic T lymphocytes can be defined with short synthetic peptides. Cell 1986; 44: 959-968.
61. Bjorkman PJ, Saper MA, Samraoui B et al. Structure of the human class I histocompatibility antigen, HLA-A2. Nature 1987; 329: 506-512.
62. Zinkernagel RM and Doherty PC. Restriction of in vitro T cell mediated cytotoxicity in lymphocytic choriomeningitis within a syngeneic or allogeneic system. Nature 1974; 284: 701-702.
63. Falk K, Rötschke O, Stevanovic S et al. Allele-specific motifs revealed by sequencing of self-peptides eluted from MHC molecules. Nature 1991; 351: 290-296.
64. Melief CJM. Tumor eradication by adoptive transfer of cytotoxic T lymphocytes. Adv. Cancer Res. 1992; 58: 143-175.
65. D'Amaro J, Houbiers JG, Drijfhout JW et al. A computer program for predicting possible cytotoxic T lymphocyte epitopes based on HLA class I peptide-binding motifs. Hum Immunol 1995; 43: 13-18.
66. Choppin J, Martinon F, Gomard E et al. Analysis of physical interactions between peptides and HLA molecules and application to the detection of Human Immunodeficiency Virus 1 antigenic peptides. J.Exp.Med. 1990; 172: 889-899.
67. Storkus WJ, Zeh III HJ, Salter HD et al. Identification of T cell epitopes: Rapid isolation of class I presented peptides from viable cells by mild acid elution. J. Immunother. 1993; 14: 94-105.
68. Zeh III HJ, Leder GH, Lotze MT et al. Flow-cytometric determination of peptide-Class I complex formation. Hum. Immunol. 1994; 39: 79-86.
69. Nijman HW, Houbiers JGA, Vierboom MPM et al. Identification of peptide sequences that potentially trigger HLA-A2.1-restricted cytotoxic T lymphocytes. Eur J Immunol 1993; 23: 1215-1219.
70. Theobald M, Biggs J, Dittmer D et al. Targeting p53 as a general tumor antigen. Proc. Natl. Acad. Sci. USA 1995; 92: 11993-11997.
71. Theobald M, Biggs J, Hernandez J et al. Tolerance to p53 by A2.1-restricted cytotoxic T lymphocytes. J.Exp.Med. 1997; 185: 833-841.
72. Röpke M, Hald J, Guldenberg P et al. Spontaneous human squamous cell carcinomas are killed by a human cytotoxic T lymphocyte clone recognizing a wild-type p53-derived peptide. Proc. Natl. Acad. Sci. USA 1996; 93: 14704-14707.
73. Gnjatic S, Bressac-de Paillerets B, Guillet J-G et al. Mapping and ranking of potential cytotoxic T epitopes in the p53 protein: effect of mutations and polymorphism on peptide binding to purified and refolded HLA molecules. Eur.J.Immunol 1995; 25.: 1638-1642.
74. Yanuck M, Carbone DP, Pendleton D et al. A mutant p53 tumor suppressor protein is target for peptide-induced CD8+ cytotoxic T cells. Cancer Res. 1993; 53: 3257-3261.

75. Ciernik IF, Berzofsky JA and Carbone DP. Mutant oncopeptide immunization induces CTL specifically lysing tumor cells endogenously expressing the corresponding intact mutant p53. Hybridoma 1995; 14: 139-142.
76. Ciernik IF, Berzofsky JA and Carbone DP. Human lung cancer cells endogenously expressing mutant p53 process and present the mutant epitope and are lysed by mutant-specific cytotoxic T lymphocytes. Clin Cancer Res 1996; 2: 877-882.
77. Ciernik IF, Berzofsky JA and Carbone DP. Induction of cytotoxic T lymphocytes and antitumor immunity with DNA vaccines expressing single T cell epitope. J.Immunol 1996; 156: 2369-2375.
78. Gabrilovich DI, Ciernik IF and Carbone DP. Dendritic cells in antitumor immune responses. I. Defective antigen presentation in tumor-bearing hosts. Cell Immunol 1996; 170: 101-110.
79. Gabrilovich DI, Nadaf S, Corak J et al. Dendritic cells in antitumor immune responses. II. Dendritic cells grown from bone marrow precursors, but not mature DC from tumor-bearing mice, are effective antigen carriers in the therapy of established tumors. Cell Immunol 1996; 170: 111-119.
80. Ochoa AG and Longo DL. Alteration of signal transduction in T cells from cancer patients. In: Important advances in Oncology 1995; DeVita VT, Hellman A, and Rosenberg SA, eds: 43-54.
81. Gabrilovich DI, Cunningham HT and Carbone DP. IL-12 and mutant p53 peptide-pulsed dendritic cells for the specific immunotherapy of cancer. J Immunother 1997; 19: 414-418.
82. Noguchi Y, Richards EC, Chen Y-T et al. Influence of interleukin 12 on p53 peptide vaccination against established Meth A sarcoma. Proc Natl Acad Sci USA 1995; 92: 2219-2223.
83. Noguchi Y, Chen Y-T and Old LJ. A mouse mutant p53 product recognized by CD4+ and CD8+ T cells. Proc. Natl. Acad. Sci. USA 1994; 91: 3171-3175.
84. Chung KY, Mukhopadhyay T, Kim J et al. Discordant p53 gene mutations in primary head and neck Cancers and corresponding second primary cancers of the upper aerodigestive tract. Cancer Res 1993; 53: 1676-1683.
85. Wiedenfeld EA, Fernandez-Viña M, Berzofsky JA et al. Evidence for selection against human lung cancers bearing p53 missense mutations which occur within the HLA A*0201 peptide consensus motif. Cancer Res 1994; 54: 1175-1177.
86. Hu J, Kindsvogel W, Busby S et al. An evaluation of the potential to use tumor-associated antigens as targets for antitumor T cell therapy using transgenic mice expressing a retroviral tumor antigen in normal lymphoid tissues. J Exp Med 1993; 177: 1681-1690.
87. Schoenberger SP and Sercarz EE. Harnessing self-reactivity in cancer immunotherapy. Sem in Immunol 1996; 8: 303-309.
88. Van der Bruggen P, Traversari C, Chomez P et al. A gene encoding an antigen recognized by cytolytic T lymphocytes on a human melanoma. Science 1991; 254: 1643-1647.
89. Van der Bruggen P, Szikora J-P, Boël P et al. Autologous cytolytic T lymphocytes recognize a MAGE-1 nonapeptide on melanomas expressing HLA-Cw*1601. Eur J Immunol 1994; 24: 2134-2140.
90. Van den Eynde B, Peeters O, de Backer O et al. A new family of genes coding for an antigen recognized by autologous cytolytic T lymphocytes on a human melanoma. J Exp Med 1995; 182:689-698.
91. Van den Eynde B, Lethé B, van Pel A et al. The gene coding for a major tumor rejection antigen of tumor P815 is identical to the normal gene of syngeneic DBA/2 mice. J Exp Med 1991; 173: 1373-1384.
92. Van Pel A, Van der Bruggen P, Coulie PG et al. Genes coding for tumor antigens recognized by cytolytic T lymphocytes. Immunol Rev 1995; 145: 229-250.
93. Traversari C, Van der Bruggen P, Lüscher IF et al. A nonapeptide encoded by human gene MAGE-1 is recognized on HLA-A1 by cytolytic T lymphocytes directed against tumor antigen MZ2-E. J Exp Med 1992; 176: 1453-1457.
94. Brichard V, Van Pel A, Wolfel T et al. The tyrosinase gene codes for an antigen recognized by autologous cytotoxic T lymphocytes on HLA-A2 melanomas. J Exp Med 1993; 178: 489-495.
95. Visseren MCW, Van Elsas A, Van der Voort EIH et al. CTL specific for the tyrosinase autoantigen can be induced from healthy donor blood to lyse melanoma cells. J Immunol 1995; 154: 3991-3998.
96. Gaugler B, van den Eynde B, van der Bruggen P et al. Human gene MAGE-3 codes for an antigen recognized on a melanoma by autologous cytolytic T lymphocytes. J Exp Med 1994; 179: 921-930.

97. De Plaen E, Arden K, Traversari C et al. Structure, chromosomal localization, and expression of 12 genes of the MAGE family. Immunogenetics 1994; 40: 360-369.
98. Kawakami Y, Eliyahu S, Delgado CH et al. Cloning of the gene coding for a shared human melanoma antigen recognized by autologous T cells infiltrating into tumor. Proc Natl Acad Sci USA 1994; 91: 3515-3519.
99. Kawakami Y, Eliyahu S, Sakaguchi K et al. Identification of the immunodominant peptides of the MART-1 human melanoma antigen recognized by the majority of HLA-A2-restricted tumor infiltrating lymphocytes. J Exp Med 1994; 180: 347-352.
100. Kawakami Y, Eliyahu S, Delgado CH et al. Identification of a human melanoma antigen recognized by tumor-infiltrating lymphocytes associated with in vivo tumor rejection. Proc Natl Acad Sci USA 1994; 91: 6458-6462.
101. Bakker ABH, Schreurs MWJ, De Boer AJ et al. Melanocyte lineage-specific antigen gp100 is recognized by melanoma-derived tumor-infiltrating lymphocytes. J Exp Med 1994; 179: 1005-1009.
102. Vierboom MPM, Nijman HW, Offringa R et al. Tumor eradication by wild-type p53-specific cytotoxoc T lymphocytes. J Exp Med 1997; 186: 695-704.
103. Gnjatic S, Cai Z, Viguier M et al. Accumulation of the p53 protein allows recognition by human CTL of a wild-type p53 epitope presented by breast carcinomas and melanomas. J Immunol 1998; 160: 328-333.
104. Mayordomo JI, Loftus DJ, Sakamoto H et al. Therapy of murine tumors with p53 wild-type and mutant sequence peptide based vaccines. J Exp Med 1996; 183: 1357-1365.
105. McCarty TM, Liu X, Sun J-Y et al. Targeting p53 for adoptive T-cell immunotherapy. Canc Res 1998; 58: 2601-2605.
106. Sadovnikova E, Stauss HJ. Peptide specific cytotoxic T lymphocytes restricted by non-self major histocompatibility complex class I molecules: Reagents for tumor immunotherapy. Proc Natl Acad Sci 1996; 93:13114-13118.
107. Sadovnikova E, Jopling LA, Soo KS et al. Generation of human tumor-reactive cytotoxic T cells against peptides presented by non-self HLA class I molecules. Eur J Immunol 1998; 28: 193-200.
108. Hung K, Hayashi R, Lafond-Walker A et al. The central role of $CD4^+$ T cells in the antitumor immune response. J Exp Med 1998; 188:2357-2368.
109. Toes REM, Ossendorp F, Offringa R et al. $CD4^+$ T cells and their role in antitumor immune repsonses. J Exp Med 1999; 189: 753-756.
110. Schoenberger SP, Toes REM, van der Voort EIH et al. T-cell help for cytotoxic T lymphocytes is mediated by CD40-CD40L interactions. Nature 1998; 393: 480-483.
111. Greenberg PD, Kern DE, Cheever MA. Therapy of disseminated murine leukemia with cyclophosphamide and immune Lyt-1^+,2^- T cells. Tumor eradication does not require participation of cytotoxic T cells. J Exp Med 1985; 161: 1122-1134.
112. Bennett SRM, Carbone FR, Karamalis F et al. Induction of a $CD8^+$ cytotoxic T lymphocyte response by cross-priming requires cognate $CD4^+$ T cell help. J Exp Med 1997; 186: 65-70.
113. Ossendorp F, Mengedé E, Camps M et al. Specific T helper cell requirement for optimal induction of cytoxic T lymphocytes against major histocompatibility complex class II negative tumors. J Exp Med 1998; 187: 693-702.
114. O'Garra A, Steinman L, Gijbels K. $CD4^+$ T-cell subsets in autoimmunity. Curr Opinion in Immunol 1997; 9: 872-883.
115. Roth J, Dittmer D, Rea D et al. p53 as a target for cancer vaccines: Recombinant canarypox virus vectors expressing p53 protect mice against a lethal tumor cell challenge. Proc Natl Acad Sci USA 1996; 93: 4781-4786.
116. Hurpin C, Rotarioa C, Bisceglia H et al. The mode of presentation and route of administration are critical for the induction of immune responses to p53 and antitumor immunity. Vaccine 1998; 16: 208-215.
117. Thomson SA, Khanna R, Gardner J et al. Minimal epitopes expressed in a recombinant polyepitope protein are processed and presented to $CD8^+$ cytotoxic T cells: Implications for vaccine design. Proc Natl Acad Sci USA 1995; 92: 5845-5849.
118. Ishida T, Chada S, Stipanov M et al. Dendritic cells transduced with wild type p53 gene elicit antitumor immune responses. Clin Exp Immunol 1999; in press.

119. Ossendorp F, Eggers M, Neisig A et al. A single residue exchange with a viral CTL epitope alters proteosome-mediated degradation resulting in lack of antigen presentation. Immunity 1996; 5: 115-124.
120. Theobald M, Rupert T, Kuckelkorn U et al. The sequence alteration associated with a mutational hotspot in p53 protects cells from lysis by cytotoxic T lymphocytes specific for a flanking peptide epitope. J Exp Med 1998; 188: 1017-1028.
121. Leach DR, Krummel MF and Allison JP. Enhancement of antitumor immunity by CTLA-4 blockade. Science 1996; 271: 1734-1736.
122. Harada M and Makino S. Promotion of spontaneous diabetes in non-obese diabetes-prone mice by cyclophosphamide. Diabetologia 1984; 27: 604-606.
123. Lando Z, Teitelbaum D and Arnon R. Induction of experimental allergic encephalomyelitis in genetically resistant strains of mice. Nature 1980; 287: 551-552.
124. Charlton B, Bacelj A, Slattery RM et al. Cyclophosphamide-induced diabetes in NOD/Wehi mice. Evidence for suppression in spontaneous autoimmune diabetes mellitus. Diabetes 1989; 38: 441-447.
125. L'Age-Stehr J and Diamantstein T. Induction of autoreactive T lymphocytes and their suppressor cells by cyclophosphamide. Nature 1978; 271: 663-665.
126. Yoshida S, Nomoto K, Himeno K et al. Immune response to syngeneic or autologous testicular cells in mice. I. Augmented delayed footpad reaction in cyclophosphamide-treated mice. Clin Exp Immunol 1979; 38: 211-217.
127. Houbiers JGA, Nijman HW, Van der Burg SH et al. In vitro induction of human cytotoxic T lymphocyte responses against peptides of mutant and wild-type p53. Eur J Immunol 1993; 23: 2072-2077.
128. Nijman HW, Van den Burg SH, Vierboom MPM et al. p53, a potential target for tumor-directed T cells. Immunol. Letters 1994; 40: 171-178.
129. Röpke M, Regner M and Cleasson MH. T cell-mediated cytotoxicity against p53-protein derived peptides in bulk and limiting dilution cultures of healthy donors. Scand J Immunol 1995; 42: 98-103.
130. Stuber G, Leder GH, Storkus WJ et al. Identification of wild-type and mutant p53 peptides binding to HLA-A2 assessed by a peptide loading-deficient cell line assay and a novel major histocompatability complex class I peptide binding assay. Eur J Immunol 1994; 24: 765-768.
131. Bertholet S, Iggo R and Corradin G. Cytotoxic T lymphocytes responses to wild-type and mutant mouse p53 peptides. Eur J Immunol 1997; 27: 798-801.
132. Dahl AM, Beverley PCL and Stauss HJ. A synthetic peptide derived from the tumor-associated protein mdm-2 can stimulate autoreactive, high avidity cytotoxic T lymphocytes that recognize naturally processed protein. J Immunol 1996; 157: 239-246.
133. Lacabanne V, Viguier M, Guillet J-G et al. A wild-type p53 cytotoxic T cell epitope is presented by mouse hepatocarcinoma cells. Eur J Immunol 1996; 26: 2635-2639.

Chapter 4

Critical Dependence of the Peptide Delivery Method on the Efficacy of Epitope Focused Immunotherapy

Gregory E. Holt, Markwin P. Velders, Michael P. Rudolf, Laurie A. Small, Maurizio Provenzano, Sanne Weijzen, Diane M. Da Silva, Marten Visser, Simone A.J. ter Horst, Remco M.P. Brandt and W. Martin Kast

Introduction

Tumor immunotherapy describes the use of the immune system as a tool to eliminate cancer from the stricken patient. The theory contends that immunization against certain proteins either associated with or specific for the tumor will create a potent cytotoxic T lymphocyte (CTL) immunity able to selectively kill the cancer cells. The anti-tumor specificity of the T cells is retained in the ability of the CTLs T cell receptor (TCR) to recognize an eight to twelve amino acid peptide bound to the MHC class I molecules of the tumor cells. Epitope focused immunotherapy (EFIT) is an offshoot of tumor immunotherapy that strives to identify the sequence of protein derived antigenic peptides bound to the MHC molecules and then focus the immunization against these epitopes.

This Chapter deals with the many different epitope based immunization strategies and describes the characteristics of each that contributes or detracts from the overall success of the method. It has been organized into four sections each dealing with an alternative immunization strategy for epitope based vaccines including peptide vaccination, dendritic cell vaccination, DNA vaccination and recombinant virus vaccination. The section on peptide vaccination describes the use of synthetic peptides injected primarily subcutaneously in association with noncellular adjuvants. Dendritic cell (DC) vaccination will detail the ex vivo loading of DCs with antigen using a variety of methods for their eventual reinfusion. The injection of genetic constructs containing "minigenes" encoding each peptide makes up the third part concerning DNA vaccination. Finally, the use of viruses engineered to express the peptides comprises the last section entitled recombinant virus vaccination.

Regardless of the method of vaccination employed, the eventual goal of the research performed in this field is to produce a therapeutic vaccine able to selectively destroy the tumor cells with limited side effects. This vaccine should have a rapid onset of CTL induction to prevent the initiation or exacerbation of tumor related complications and check tumor growth so the kinetics of the tumor cell proliferation does not overwhelm the immune system as per certain viruses.[1] The composition of the vaccine may contain peptides restricted to the patient's MHC haplotypes from every tumor antigen of the targeted cancer arranged consecutively in a

Peptide-Based Cancer Vaccines, edited by W. Martin Kast. ©2000 Eurekah.com.

"beads on a string" construct. The combination of all possible epitopes likely expressed by the tumors should induce a potent and diverse antitumor immunity able to prevent the immune escape mechanism of tumors. Finally, the vaccine should induce a long lasting memory response against the potential recurrence of the tumor.

EFIT predominantly utilizes activated CTLs since their normal effector function is to remove cells containing a pathological antigen which in this case is the tumor antigen. However, the induction of a CTL response is a highly regulated event and requires the concomitant involvement of both DCs and $CD4^+$ T helper cells. The DC is by far the most potent antigen presenting cell and activator of a CTL response[2] and the success or failure of a vaccine will partly depend on its proficiency at delivering sufficient quantities of peptide to them. For the most efficient presentation of the class I restricted peptides to the $CD8^+$ cells, the DCs ultimately must be activated. The in vivo activation of DCs normally occurs through the interaction of $CD4^+$ T helper cells with peptide bound MHC class II molecules and the reciprocal stimulation of the DCs via CD40-CD40L binding. The need for $CD4^+$ T helper responses can be circumvented through the use of other surrogate inducers of DC maturation including lipopolysaccharide(LPS), GM-CSF, or TNFα.[3,4] In either case, the success of EFIT depends on the ability of the vaccine to produce activated DCs loaded with MHC class I restricted peptides.

There are many advantages associated with the epitope focused approach to tumor immunotherapy in comparison to the use of vaccination with the entire antigen. Ishioka et al[5] found that vaccination with a minigene construct consisting of two HBV polymerase peptides induced a more potent CTL response than vaccination with the complete gene. The increased potency of an individual CTL should correlate with a greater anti-tumor efficacy since its cytolytic activity follows the law of mass action.[6] In other words, increases in the affinity of a CTL's TCR for a given peptide-MHC complex decreases the number of peptide-MHC complexes needed on the cell surface for T cell mediated lysis. Since the tumor will probably have a low expression of the peptides targeted by the vaccine, the increased potency will allow for the recognition of these cells by the induced CTLs whereas they would be cryptic to CTLs of lower affinity.

In addition to the increased strength of the immunity, different vaccination strategies involving epitope focusing induce a co-dominant induction of CTLs. The induction of a CTL response using a whole protein or tumor cell immunization predominantly produces an immunity centered around one immunodominant epitope.[7] In contrast, vaccination with a peptide vaccine based on dendritic cells,[8] DNA[5] or recombinant viruses [9,10] has shown to produce strong CTL induction to all included epitopes. This diverse immunity is important in disallowing the tumor an escape mechanism. Since the immunodominant epitope would be contained within one tumor antigen and restricted by one MHC haplotype, an immunity focused solely on that peptide could fail if the tumor mutates that epitope,[11] loses that tumor antigen[12,13] or downregulates that MHC molecule.[14,15] Using a vaccination protocol that elicits a potent CTL response against a broader profile of epitopes from the same or different antigens and different MHC haplotypes found on the tumor cells is unlikely to permit the tumor cells to evade the anti-tumor immunity.

Another advantage to EFIT is that the injected "beads on a string" constructs represent only portions of each tumor antigen that as a whole should not retain their biological activity. This is especially important in DNA based vaccines when the targeted antigen is an activated oncogene or a viral protein necessary for transformation. Even the use of non-tumorigenic proteins can not be considered completely safe since the effect of overexpressing certain tumor antigens is unknown. If an epitope is found to be problematic, it can simply be removed from the vaccine to eliminate its negative influences.

One caveat to the potentially safer use of epitope based vaccines concerns the use of an immunization strategy containing recombinant DNA constructs as a component of the vaccine. Although the expressed minigenes should not contain any ability to transform cells themselves, the injected DNA sequences can still be problematic. To achieve a high level of peptide expression, most DNA vectors employ a constitutively active viral promoter for transcription in all cell types. Should the DNA integrate into the host genome and bring the active promoter adjacent to a proto-oncogene, it could increase its expression and cause tumorigenesis. In a similar way, the disruption of the regulatory regions of certain proto-oncogenes via the integration of the DNA construct within them could have a similar effect.

The role epitope flanking sequences have on the presentation of distinct peptides remains controversial and can be considered both an advantage and a disadvantage. To extract only the minimal epitope from its surrounding sequences could remove it from residues inhibitory or necessary for presentation. Several groups have shown that alterations in the flanking amino acids change the presentation capabilities of the peptide.[16-19] Specifically, the juxtaposition of glycine or proline[19] residues has been found to be detrimental whereas neighboring alanines improve presentation.[17] In contrast to these studies, others have found no difference in the ability of certain peptides to be presented when they have been displaced to other proteins. Although the lack of complete knowledge in this area prevents definitive conclusions, it seems plausible that certain residues do affect the presentation of peptides. The reconciliation of the contrasting reports may be due to the fortuitous placement of the peptides between presentation facillatory residues in the studies where no effect of flanking sequences were seen.[20,21] The omission of the enhancing residues around the minimal epitope would be a disadvantage of EFIT except that many researchers have successfully used it in their vaccines. The discovery of residues with positive and negative influences on peptide presentation can be considered an advantage if this knowledge is used to design enhancing flanking sequences for each epitope in the "beads on a string" construct.

The fact that the peptides used in EFIT vaccines are haplotype restricted and that in the human population there exists a great diversity of MHC alleles indicates that a great amount of research must be performed in order to identify the necessary peptides. The broad spectrum of MHC alleles that a person expresses also complicates matters since each vaccine would need to be tailored to accommodate each patient's haplotype. These two obstacles have limited the application of an EFIT to people with one of the well studied MHC haplotypes which has hampered research in this area.

Peptide Vaccination

Peptide vaccination employs the injection of the minimal T cell epitopes as free peptides admixed with an adjuvant. This method began when Towsend et al[22] discovered that peptides bound to the MHC complex of cells was the recognizable element for CTLs. Simultaneously, Schulz et al[23] and Kast et al[24] independently proved the efficacy of these methods by showing that peptide vaccination could protect mice from challenge with an otherwise lethal dose of virus.

The method relies on the successful utilization of an adjuvant to create a depot after injection for the slow release of peptide in vivo. The peptides that comprise a vaccine include both the desired class I restricted peptides for the induction of the CTLs and class II restricted peptides for the induction of a helper T cell response. For proper induction of a CTL response, Keene et al[25] showed that a helper response was necessary. After the peptides leave the adjuvant depot, they passively move into the interstitial fluid and eventually bind to the MHC molecules found on DCs either in the vicinity or recruited to the area by some non-peptide aspect of the adjuvant itself. Once on the DCs, the class II peptides will induce a $CD4^+$ T helper cell response that will reciprocally activate the DC. Now the activated DC with bound class I peptides will induce the CTL mediated immunity.

In 1993, Feltkamp et al[26] first used free peptide in adjuvant to induce a protective immunity against tumor cells in mice. After an exhaustive search involving the production and testing of every possible nonamer of the E6 and E7 oncoproteins of the human papilloma virus type 16, one peptide corresponding to amino acids 49-57 (RAHYNIVTF) of the E7 protein displayed high affinity for the H-2D^b MHC molecule. This peptide when emulsified in incomplete Freund's adjuvant and injected subcutaneously in mice provided protection against a subsequent challenge with C3, a cell line immortalized with the complete HPV 16 genome and activated EJ-ras. This peptide has proven efficacious in our laboratory in protecting against another cell line of mouse lung fibroblasts transfected with both the E7 and E6 proteins. However, despite the best efforts of many researchers, this peptide when used in a free peptide immunization has failed to show any therapeutic benefit in treating existing tumors.

In hopes of improving the potency of free peptide vaccinations, many alterations of the original immunization protocols have been tested. It could be argued that the passive diffusion of the peptides from the depot to the DCs may be a limiting factor in the vaccination's efficacy. With the recent discovery of class I restricted peptides bound to heat shock proteins[27] and the subsequent induction of tumor specific immunity through the vaccination of in vitro peptide pulsed HSPs, HSPs could serve as a chaperone to directly target the peptides to professional APCs.[28] Since the isolation of HSPs from tumor cells has already been shown to both protect[29] and eradicate existing tumors,[30] further manipulations to optimize the vaccination could have great beneficial effects for EFIT. Therapy has also been shown in a similar situation whereby peptide was complexed to a detoxified cellular invasive Bordatella pertussis adenylate cyclase, CyaA.[31] Theoretically, these methods overcome the therapeutic deficiencies of free peptide vaccination through the direct targeting of peptide to the class I pathway.

In addition to targeting the peptide to the professional APC, it has been found to be beneficial to increase the affinity of the peptides for the MHC molecule without altering its ability to stimulate a CTL response against the normal peptide. Alteration of the anchor residues of weak binding peptides towards the more conserved amino acids for a given haplotype increases the affinity[32-34] and does not affect the immunogenicity due to the deep and hidden nature of their binding.[35-38] Indeed three groups have shown that such a modification did not prevent the peptides from eliciting a CTL response to the normal peptide. Vierboom et al[39] showed that alteration of the anchor residues had no effect on the immunogenicity of the HPV 16 E7 $_{49\text{-}57}$ peptide. Valmori et al[40] showed that the increased HLA affinity resulted in a greater potency of CTL induction in cultures of peripheral blood lymphocytes immunized in vitro. Men et al[41] continued the analysis by showing that some of the peptide analogues induced greater CTL responses in HLA transgenic mice and that the induced CTL were able to cross react with cells pulsed with the normal peptide. In contrast, Clay et al[42] found that although the peptide analogues induced a greater CTL response, they were unable to cross react with the tumor cells. These results indicate that the use of peptide analogues in the future may be beneficial through their increased immunogenicity due to greater MHC affinity but only if the resultant CTLs recognize and lyse the tumor cells.

Another advance in the peptide vaccination protocol involves the use of different tactics to create a better environment for the induction of a CTL response. Due to the necessity for $CD4^+$ T helper cell mediated activation of dendritic cells to achieve optimal CTL priming and the decreased probability that the free class I and class II restricted peptides bind to the same DC, the covalent linkage of the two peptides was attempted. Shirai et al[43] showed in 1994, that the physical linkage of the two peptides produced greater CTL activity than vaccination with both peptides as single entities. Hypothetically, the linkage allows the presentation of both peptides on the same DC allowing its subsequent activation and production of the appropriate Th1 cytokines for efficient CTL induction.

Production of the appropriate cytokine milieu through the inclusion of certain immunostimulatory sequences of DNA to the vaccine's adjuvant has also been shown to augment the induction of the CTLs. Specific DNA sequences found in prokaryotes containing unmethylated cytosine-guanine oligodinucleotides (CpG ODN) seem to act as a so called "danger signal" that alerts the immune system to the presence of an intracellular pathogen. The CpG ODN induce macrophages and DCs to produce IL-12, IL-18 and IFNα[44] of which the IL-12 has been shown to cause IFNγ secretion by natural killer cells.[45] The production of these Th1 type cytokines aids in the induction of a stronger CTL response when added to the adjuvant for peptide immunization.[46]

In addition to the success peptide vaccination has had in the induction of an antitumor immunity, there are many advantages to its usage. Peptides and their subsequent adjuvants tend to be relatively cheap and easy to produce. Since they are mostly injected as a subcutaneous dose, the actual mechanics of the vaccination are also rather simple and minimally invasive. The injected peptide and adjuvant also have been rather well tolerated in the ongoing clinical trials.[47] However, the greater success of the other methods of vaccination presented in this Chapter will probably shift the focus of strategies away from free peptide injections.

One disadvantage to the use of peptide vaccination as a method involves the observations of detrimental immune responses post vaccination. Vaccination of mice with a peptide corresponding to the LCMV is protective if given subcutaneously but tolerizing if given repeatedly i.p in high doses.[48] Although arguments as to the route of vaccination and concentration of cumulative peptide may confound these results, two papers by Toes et al[49,50] concerning tolerance induction by the E1A and E1B peptides of adenoviruses and one by Nieland et al[51] concerning the P1A peptide of P815 tumor cells are undeniable. Vaccination of mice with a peptide s.c. in IFA tolerized the tumor reactive T cells to the point where a normally regressor tumor grew out and killed the mouse whereas control animals survived unaffected after the tumor's characteristic regression. This disturbing observation was proven to be related to the pharmacokinetic behavior of these peptides that allowed them to rapidly spread throughout the body (Weijzen S and Kast WM unpublished observations). The induction of tolerance was also found to be method dependent as vaccination with the same peptides pulsed on DCs[52] or incorporated in a virus like particle (VLP)[51] prevented any outgrowth of tumor. This data reveals an inherent danger to the use of free peptide vaccination that is reversible when using other vaccination methods.

Another disadvantage to the use of peptide vaccination concerns the profile of induced CTL responses when a mixture of peptides are combined in one vaccine. As argued above, the most efficacious immunity should be diverse with strong responses to all epitopes to avoid tumor escape mechanisms through antigen loss, epitope mutation or HLA downregulation. Recent analysis of the immunodominance of five well characterized epitopes showed that injection of free peptides in IFA produced a hierarchy of induced CTL responses. The reversal of this immunodominant behavior between peptides was abrogated when other vaccination methods were employed.[8]

Other disadvantages to the use of peptide vaccination involve quality control issues regarding the actual composition of the vaccine. Gupta et al[53] discuss several difficulties with the use of free peptide emulsified in adjuvant. These problems derive from poor solubility of individual peptides, uncertainty in assuring a homogenous emulsification of peptide and problems of peptide formulation. In addition, to these issues, the relative paucity of approved and efficacious adjuvants for use in humans is also a limiting factor.

The sum extent of this discussion on peptide vaccination is that emulsification of peptides in adjuvant for subcutaneous injection was instrumental in progressing the field of tumor immunotherapy to its current state but may not show great ability to treat preexisting cancers in clinical trials. The passive diffusion of the peptides from the depot to the DCs limits the effectiveness of the method for CTL induction. The increased capacity to induce CTLs with the use of synthetic peptide complexed to HSPs or Bordatella pertussis CyaA probably succeed due to their ability to target the peptides to the cytoplasm of the professional APCs. These two examples display the great importance of efficient loading of the peptides to DCs for therapeutic ability which is ultimately shown in the next section with the direct ex vivo loading of the peptides on isolated DCs.

Dendritic Cell Vaccination

Due to their critical role in the induction of a CTL response, dendritic cells are the logical choice for inclusion in an anti-cancer immunotherapeutic scheme. This approach was confounded by the difficulty in acquiring sufficient numbers of DCs until Inabu et al[54,55] reported on the in vitro production of large DC numbers from the coculturing of either peripheral blood mononuclear cells or bone marrow with high levels of GM-CSF. Improvements on the method that include either the addition of IL-4 to the GM-CSF or the in vivo stimulation of DCs via Flt3 ligand[56] have allowed for great increases in the biology of DC which has been directly applied to their use in EFIT approaches.[57]

Indifferent to the method used for the creation of the DCs, issues concerning their activation and antigen loading predict the eventual success or failure of the vaccine. Both Mackey et al[58] and Labeur et al[59] show that without the proper activation of DCs, they will lose their ability to induce an efficacious anti-tumor effect. Although DCs can be loaded using a variety of methods,[2] Morse et al[57] proved the critical dependence on the sequence of loading and activation depends on the nature of the immunizing antigen. Immunization with genetic material or coincubation of DCs with protein requires loading with subsequent activation whereas the exogenous loading of the DC's MHC molecules with peptides necessitate the opposite order of events. Activation of the DCs with a variety of methods including CD40 ligand, LPS or TNFα results in the upregulation of MHC class II, B7.1 and B7.2 molecules which were found to be necessary for CTL induction.[60] After the appropriate loading and activation of the DCs, they are infused into the patient where the activated DCs home towards secondary lymphoid organs. Herein they interact with the $CD8^+$ cells and cause the subsequent induction of peptide reactive CTLs.

The enormous potential of this method was best shown by Mayordomo et al[61] in 1995. In this landmark paper, synthetic peptide pulsed dendritic cells were able to protect against the injection of lethal doses of tumor cells in three distinct tumor models; MUT1 peptide for the Lewis lung carcinoma, 3LL, $E7_{49\text{-}57}$ for the HPV 16 genome transformed C3 tumor and $OVA_{257\text{-}264}$. In addition, the use of the peptide pulsed DCs were also able to eradicate existing day 7 3LL tumors and day 14 C3 tumors in mice. The sheer strength of the vaccine in the C3 model was shown when vaccination eliminated tumors in 60% and 20% of mice initially vaccinated 21 days and 28 days after tumor challenge. This paper provided the necessary proof to establish the use of peptide pulsed DCs as a forerunner in the quest to determine the ideal method for use as an immunotherapy in humans.

As a result, Murphy et al[62] recently reported the results of a phase II clinical trial where patients suffering from prostate cancer were treated with DCs pulsed with peptides derived from the prostate cancer marker, prostate specific membrane antigen (PSMA). Of the 33 subjects with stage D tumors who all no longer responded to current treatment regimens, 6 showed a partial response and 2 showed a complete response as based on the criteria dictated by the National Prostate Cancer Project. Although the fact that two partial responders and 1 complete responder did not express the haplotype for which the peptides bound, HLA A*0201, confounds these results, the use of the vaccine shows great promise.

Exogenous loading of the MHC molecules with synthetic peptides is not the only loading method with proven efficacy. Using a p53 self peptide, Tuting et al[63] compared the use of exogenous peptide loading and DC transfection with a minigene DNA construct for their abilities to protect against a tumor challenge of a chemically induced p53-positive sarcoma. In their study, they report no difference in the efficacy of either method of loading in protecting the mice from tumor outgrowth.

The effectiveness of peptide loaded DCs vaccines in tumor immunotherapy may be due to one of the advantages of this strategy of immunization. Sandberg et al[8] showed that in contrast to the vaccination of mice with an admixture of five peptides emulsified in IFA, vaccination with DCs pulsed with the same admixture did not show any immunodominance. The exhibited co-dominant induction of CTL activity with peptide pulsed DCs versus peptide in IFA proved that immunodominance was neither a direct result of the biochemical makeup of the peptide nor its location with respect to other epitopes in the whole protein. Thus once the peptide reaches the surface of the DC bound in the MHC molecule, it has an equal chance of inducing a CTL response as every other binding peptide. This production of a diverse but equally potent CTL response may be one of the driving forces behind the great therapeutic ability of peptide pulsed DCs in EFIT.

The ex vivo manipulation is a disadvantage to this method of vaccination. The creation, loading and activation of the cells are time consuming expensive procedures that would require a large laboratory dedicated to the production of these vaccines. This is especially true if the vaccines prove efficacious in a number of cancers since they represent a major cause of illness in the United States. Thus the application of peptide loaded DCs although potentially effective cancer treatments, would involve a great undertaking to commonly apply.

A second disadvantage of DC based vaccination approaches involves the increased invasive nature of the creation and application of the vaccines. Since haplotype mismatched DCs would probably be less efficient vaccines than syngeneic cells, the source of the DCs would derive from the patient and involve either a blood draw or bone marrow tap. After the production and subsequent loading and activation, these DCs would probably be infused through an I.V. Although other vaccination routes may be employed, the relative success of the prostate cancer immunotherapy[62] and the finding that in mice the majority of the injected DCs given s.c. remained at the site of vaccination favor an intravenous application. The invasive nature of these vaccines would not preclude their usage if efficacious but could be less advantageous compared to an equally effective but noninvasive approach.

The use of a peptide loaded DC approach represents the most effective EFIT considering its ability to effectively treat mice with C3 tumors injected three weeks before initiation of therapy. Although the time and expense necessitated for their application are substantial, they represent only logistical impediments that could be overcome if the therapy is ultimately successful. The invasive nature of the method is also undesirable, but it is highly doubtful that a patient would refuse its employment if it meant the cure for their cancer. Thus, DC based immunotherapies currently represent the most effective EFIT to date.

DNA Vaccination

With the discovery by Wolff et al[64] of the potential to inject naked DNA into the muscles of mice for the stable expression of the inserted genes, the field of DNA vaccination was born. The efficacy of such an approach was illustrated by Ulmer et al[65] when they showed that injection of genes from the influenza virus protected mice from subsequent infections. This method was applied to tumor immunology with the demonstration of its protective ability to prevent outgrowth of tumors containing either CEA[66] or MUC1[67] after vaccination with the respective genes. Consequently, DNA vaccination as a method of EFIT is now a heavily studied strategy.

Although the original route of vaccination involved injection of the genetic material i.m., Boyle et al[68] showed that a greater CTL response was induced earlier in intradermal vaccination versus the intramuscular route. Even though the induction of a CTL response via i.m. injection eventually reached the levels of the i.d. approach, it should be obvious that a faster induction of an immunity would be advantageous in terms of cancer treatment. The enhanced speed of induction via i.d. injection is most likely due to the great concentration of DCs in the skin versus the almost absent quantities in the muscle.[69]

The fate of the DNA after injection remains a controversy within the DNA vaccination field. It is an undeniable fact that the antigen produced by the injected DNA eventually reaches a DC for presentation to the T cells.[70,71] It is uncertain whether or not the injected DNA directly transfects DCs or is translated in surrounding tissues for the eventual uptake of the protein by the DCs in the draining lymph nodes.[72,73] In either case, the protein must reach the presentation pathways of the DCs.

The need for $CD4^+$ T helper cell stimulation is another unknown requirement for DNA vaccination. Although some groups report the direct need for $CD4^+$ help[5] others do not.[9] These discrepant results may be reconciled through the discovery of the inherent adjuvanticity of the DNA. It was finally shown that the presence of specific unmethylated cytosine-guanine oligodinucleotide containing sequences provided the immunostimulation necessary for proper CTL induction that is normally provided by $CD4^+$ T helper cells. The mechanism of action entails the production of IL-12 by the reacting DCs[44] that induces IFNγ release from NK cells.[45] Although the inclusion of MHC class II epitopes would not hurt the induction of an immunity via this method, the skewing of the cytokine milieu towards a Th1 pattern by either method allows the induction of peptide reactive CTLs. Once the DC is activated, it will extravasate from the tissues of the vaccination site and home towards the secondary lymphoid organs. Herein, the DCs activate the CTLs in an peptide dependent manner.

Thomson et al[9] was able to apply an epitope focused DNA vaccination towards a tumor immunotherapy. Using either an i.m. or i.d vaccination route, they were able to show that immunization with a DNA construct containing 8 class I restricted minimal epitopes induced co-dominant CTL induction for each peptide. The inclusion of the $_{-264}$ epitope was able to induce a protective response against a lethal challenge of OVA transfected tumor cells. In addition, the inclusion of several viral peptides provided a OVA_{257} protective immunity against later challenge with vaccinia viruses containing the epitopes or the influenza virus.

Other groups also report on the successful use of minimal epitope containing DNA constructs in the induction of CTL responses proving that this is not an isolated phenomenon. Ciernek et al[74] used a minimal epitope containing DNA construct to show the induction of CTLs against a p53 peptide that induced tumor protection against P815 tumors transfected with the p53 peptide. In addition, Ishioka et al[5] recently reported on the successful use of a DNA construct to induce co-dominant CTL responses against HLA-A*0201 and HLA-A*1101 restricted peptides in human haplotype transgenic mice. This last report not only indicates the translation of the response to a quasi-human situation but also describes the use of an ER targeting signal to directly deliver the peptides to the class I MHC molecules.

To augment the induced immune response many different variations of the immunization protocol have been attempted. In another effort to target the peptides to the class I presentation pathway, Wu et al[75] found that the inclusion of an ubiquitin signal enhances the produced immunity theoretically through the increased turnover of the antigenic protein. Since the proper induction of a CTL response requires TH1 type cytokines, many groups have shown that the inclusion of genes for IL-2, IFNγ, IL-15 or GM-CSF have increased the magnitude of the response.[76-78] Finally in an attempt to convert in vivo transfected cells not of the professional APC lineage to inducers of a CTL response, Corr et al[79] attempted to add the necessary costimulatory molecules through the coinjection of the B7.1 and B7.2 molecules with the minigenes. They found that inclusion of the B7.1 molecule gave the greatest potentiation of the CTL response to the peptide found in the minigene.

The potential for DNA vaccination as a method of EFIT is clearly strong. In addition to having measurable efficacy against tumors, the stable nature of the molecules also favors its future usage. With the commonplace use of recombinant DNA technology in today's scientific community, it is possible to easily manipulate the DNA constructs. Also, the defined isolation procedures for DNA make it a cost effective vaccine material since the cost of production is very small compared to other molecules.

In addition to the easy and cost effective attributes of DNA, the produced co-dominant induction of CTL responses to each peptide of a vaccine represents a major advantage. As detailed above in the introduction, the production of a diverse but potent response to all included peptides is key to the production of an efficacious vaccine. The ability of it to induce such a response indicates its potential for future application.

There are several disadvantages to the use of DNA as a vaccine strategy. In laboratories around the world, the integration of DNA into host cell genome is commonly performed through the simple introduction of the DNA into the cell. Since the method of DNA vaccination also relies on the introduction of foreign DNA into the host cells, the potential for integration is real. However, despite the intense effort of many researchers, no instance of integration mediated secondary malignancy induction following DNA vaccination has been reported. Although one strength of EFIT is the use of non-functional constructs to eliminate the introduction of oncogenic genes, the simple disruption of normal gene expression could occur through the interruption of its normal regulatory mechanisms. Although a rare occurrence, the integration into the genome could disregulate the expression of a proto-oncogene and initiate a new malignancy. Plus the use of viral promoters to drive high expression of the included multi-epitope protein could insert close to an oncogene and increase its expression. These plausible situations represent one of the disadvantages to the use of DNA as a method of vaccination.

In some animal models, the injection of subimmunogenic amounts of DNA induced a state of unresponsiveness whereas in other models, the repeated injection of DNA lead to tolerance induction. These reports indicate that the use of DNA as a vaccine needs to be completely tested to insure that the concentration used will allow a normal immunizing response to occur. Other reports of DNA vaccination have shown that the injection of constructs in young animals have elicited a tolerizing effect that was not seen upon injection of the antigenic protein. This further illustrates the need to titrate the dose of DNA with the exact conditions for the immunization.

The antithesis to tolerance induction, autoimmunity, represents one last potential disadvantage to the use of DNA as a vaccine modality. It is possible that the injection and subsequent expression of the gene in muscle cells to elicit an immunity will cause the induced CTLs to destroy the transfected myocytes. Since the amount of muscle or skin transfected is not great, the induction of an autoimmune disease against the small percentage of cells would probably be well tolerated. Skin cells will regenerate from the stem cells found in the basal cell layer while the satellite cells of the muscle will grow and differentiate into new muscle fibers.

A more serious concern is the induction of anti-DNA antibodies via vaccination with DNA. Although mouse models exist where the injection of free DNA does increase the titers of anti-DNA antibodies, normal mice do not produce these antibodies unless the DNA is manipulated in certain artificial ways that would never be used in a DNA vaccination strategy. Thus although studies in animals indicate that the induction of an anti-DNA antibody response and the eventual creation of a Lupus like syndrome is possible, it is not likely to occur with the vaccination of DNA for an immunizing response against the encoded antigen.

The above discussion of DNA as a method of EFIT reveals that it may be possible to elicit anti-cancer therapeutic immunities. Like the use of dendritic cells, the production of a co-dominant response is a major advantage that is partially responsible for its efficacy. The inherent immunostimulatory characteristic of DNA along with the easy addition of costimulatory help in the form of cytokines or secondary signals (i.e. B7.1, B7.2) make it an attractive option for future study. The disadvantages to its use, although potentially serious, may never become clinically relevant if the potential autoimmunity is delineated to small tissue regions around the vaccination site or left for theoretical discussions of anti-DNA antibodies.

Recombinant Virus Vaccination

Theoretically, the strong CTL response to viral proteins produced after infection with a virus should also be induced against proteins artificially introduced into the viruses genome. Indeed, mice were protected from a lethal challenge of LCMV after vaccination with recombinant vaccinia viruses engineered to express a polypeptide containing several LCMV epitopes.[80,81]

The recombinant viruses are engineered through the replacement of the normal genes necessary for viral replication with a genetic construct encoding the peptides of choice. Since these peptide expressing viruses are defective in terms of replication, they must be created in cultured cell lines that produce the viral structural genes replaced in the recombinant virus's genome by the polyepitope construct. These newly created, replication defective viruses are now injected into the host. Since the outer capsid is primarily responsible for the initial act of entering cells, these recombinant viruses still retain the ability to reach the cytoplasm. Once in the cytoplasm, the viral promoters are activated as usual and produce the polypeptide containing all of the peptides in great quantities. If the virus is normally tropic for DCs, the expression of the peptides in the cytoplasm facilitates entry to the class I pathway. For viruses that do not normally infect DCs, the overexpression of antigen can cause cell death and the release of cell debris or apoptotic bodies that are taken up by DCs and presented in a class I restricted manner.[82] In either case the requisite need for added $CD4^+$ T helper cell epitope of previous vaccination strategies may be unnecessary. Viral infections have the innate ability to activate DCs on their own[83-86] due to either strong endogenous helper epitopes or some unknown mechanism of DC activation. Regardless of the mechanism, both instances allow the expressed polyepitope protein to be processed by the dendritic cells for induction of a CTL response.

In an extension of the work by Oldstone[80] and Whitton,[81] Thomson et al[9] created a recombinant vaccinia virus containing epitopes from 5 different viruses, one parasite and the immunodominant peptide from ovalbumin (SIINFEKL). Infection of mice showed an equally potent CTL response to each peptide. Vaccination with this recombinant virus was shown to induce a protective tumor response against EL4 thymoma cells transfected with the ovalbumin protein but not parental EL4 cells. Use of this virus also produced a protective effect against viral challenges of murine cytomegalovirus (MCMV) and Sendai virus due to the inclusion of relevant epitopes from each virus in the epitope construct.

Due to health concerns of vaccinia usage discussed later, Toes et al[10] showed similar results using adenovirus and HPV derived peptides in a safer adenovirus vector. Inclusion of the dominant peptides $E1A_{234\text{-}243}$, $E1B_{192\text{-}200}$ and $E7_{49\text{-}57}$ in a recombinant adenovirus produced equally potent CTL responses to each peptide again displaying co-dominance. Vaccination with the recombinant vaccine was further shown to induce a protective immunity against a subsequent challenge with cell lines transformed with either Ad5E1A and activated EJ-ras or the HPV16 genome and activated EJ-ras. This result also showed that vaccination was almost equally potent if given s.c., i.m., or i.p. which alludes to the great potency afforded this vaccination method.

A new revolutionary strategy that involves the use of the outer virus coat to hypothetically retain the ability to enter the cytoplasm of infected cells like a normal virus employs virus-like particles (VLPs). Using the fact that the L1 and L2 proteins of the HPV virus spontaneously create empty capsids,[87,88] the covalent attachment of peptides to the inside moieties of the capsid proteins have proven efficacious. Peng et al[89] was able to show that bovine papillomavirus VLPs loaded with the $E7^{49-57}$ peptide could prevent the formation of EL4 thymoma tumor cells transfected with the E7 protein. Stronger evidence of the potency of the VLP method was shown by Nieland et al[51] whereby treatment of mice with HPV VLPs containing the P815 tumor peptide, P1A, showed therapy against day 5 and day 10 tumors. Other species viruses have also been used in successful vaccinations such as bovine papillomaviruses, canine oral papillomaviruses, cottontail rabbit papillomaviruses and Ty particles.[90] In addition to the introduction of peptides, Touza et al[91] have been able to load VLPs with DNA and show great ability to deliver the genetic material into cells. With these published reports, the prospect of using VLPs for future cancer immunotherapies is very strong indeed.

In addition to the great potency of the induced CTL response to infection with recombinant viruses expressing a multitude of peptides, there are several other advantages to this vaccination strategy. First, the infection of the host is a relatively simple procedure usually with minimal invasiveness. Although the best vaccination route may be to infect the patient in a similar manner to the normal infection route of the virus, Toes et al[10] showed that at least for adenoviruses, several different routes gave efficient vaccination. In addition, vaccines using adenoviruses have been tested extensively in army recruits[92,93] and found to be generally safe and non toxic.

Like the dendritic cell and DNA approaches, a positive feature of this method of vaccination is the ability to create a co-dominant induction of CTLs against the included epitopes. This diverse and equally potent response should be best suited to disallow the tumor any escape mechanisms. Thus, the use of recombinant viruses seems to be a potential method of future tumor immunotherapies.

There are, however, some disadvantages that need to be addressed before the method is applied. First, several poor outcomes have followed immunization with the vaccinia viruses. Vaccination was shown to produce encephalopathy and post vaccinal encephalitis in some individuals.[94] Its use also had the curious effect of decreasing the immunity of several younger individual's poxvirus immunity.[95] These negative side effects may be potentiated if the replication defective virus reverts to its wild type state. In a therapeutic setting, however, the benefit of eradication of the patient's cancer outweighs these potential risks.

Due to the necessity to produce these vaccines in packaging cells lines that contain the genes necessary for viral production, it is theoretically possible for recombination to create a reverted pathologic virus. Since it is impossible to check every individual virus produced, such an infectious virus may be injected into a human being. Although some of the concerns about reversion to wild type viruses may be alleviated through the use of relatively safer viruses (i.e. vaccinia, adenovirus), all viruses have the potential to cause problems. Even though it may be considered safer to use adenoviral vectors,[92,93] there are rare instances of acute hemorrhagic cystitis associated with adenoviral infections. Thus, perhaps only the use of the virus like particles that contain no genetic material may be considered safe.

A final disadvantage to the use of viral vectors for the induction of a CTL immunity concerns the issue of neutralizing antibody. For the virus to induce the CTL response, it must enter a viable cell to utilize its protein production machinery to make the polyepitope protein. Neutralizing antibody titers either in response to vaccination with the recombinant viruses or subsequent to previous infections with the wild type virus would prevent access to the cytoplasm of the cell. In data from both mouse[96] and human[97] studies, a preexisting immunity to vaccinia reduced the magnitude of the induced immune response after revaccination with recombinant vaccinia virus vectors. This limiting effect of a preexisting immunity was overcome through an alteration in the route of immunization performed with the recombinant viral vaccine.[98] Thus in an oxymoronic manner, the undesirable humoral response to the viral vectors could inhibit the production of the desirable cellular immunity to the included epitopes.

The great immunity induced through the use of the viral vectors make their use in future immunotherapies highly probable. Although their potential negative side effects and the problem of neutralizing antibodies may initially limit their widespread usage, developing technology could make these disadvantages moot points. First, technologic advances may come up with a method of ensuring the exact production of the correct recombinant virus. To overcome the problem of neutralizing antibodies, the use of different strains of the same virus that do not share similar neutralizing antibody epitopes may be beneficial.[99,100] In addition, it has been possible to use similar viruses from different species to deliver the included epitopes without suffering the neutralizing antibody response of the previously used viral vector.[101] Thus, the use of recombinant viral vectors has shown great promise and most likely will eventually show good clinical effect in the immunotherapeutic treatment of cancer.

Conclusions

This Chapter dealt with a comparison of different vaccination strategies used in developing a successful epitope focused immunotherapy for cancer treatment. Since the efficacy of a vaccine will largely depend on the vaccine's ability to load the relevant peptides on the surface MHC class I molecules of activated DCs, each of the four sections attempted to discuss the route it would take to achieve that goal. In every case, the need for both the efficient loading of the peptides and the need for DC activation was discussed.

Free peptide in adjuvant showed success in protecting mice from tumor challenges but no therapy data. Associating free peptide with certain biologically active proteins like HSPs and Bordatella pertussis CyaA did show therapy against preexisting tumors. The latter results are most likely due to the ability of the peptide carriers to deliver the peptide to the DCs, a characteristic lacking in the injection of free synthetic peptide emulsified in adjuvant. It is no surprise then that the strongest data in support of EFIT as a potential clinical tool involved the direct ex vivo loading of synthetic peptides on DC for subsequent reinfusion. DNA vaccination using either the ex vivo transfection of DCs or the i.m or i.d. injection also proved to have great ability in creating potent CTL responses. Finally, the use of the recombinant viral vectors to deliver the "beads on a string" epitope constructs shows great promise most likely since CTL responses are the normal response to these intracellular pathogens.

Which of these methods will eventually become the method of choice for EFIT vaccination remains to be seen. It is interesting to note that the future vaccine may contain aspects of many of the different strategies to exploit the select advantages each method possesses. Schneider et al[102] showed that distinct strategies have unique potentials for satisfying the different requirements for CTL induction. Indeed in their study, they were able to show the increased efficacy of protecting mice against malaria through the initial vaccination with a naked DNA construct followed by a boost with a recombinant vaccinia virus containing the malarial antigens. Reversal of the order of vaccination actually revoked the CTL induction potential. This suggests that DNA vaccination may be better suited for initial priming of a CTL response and recombinant viral immunization for potentiating the T cells. Clear elucidation of the tasks best suited for each vaccination method may aid in the eventual production of a therapeutic vaccine.

It is an exciting time in EFIT with the great success certain investigators have had in treating mice with tumors and the publication of the initial reports from human clinical trials. Reports of absent clinical effect after certain vaccination trials are buoyed by the rationale that many trials are performed on end stage cancer patients whose cancers are refractory to current therapeutic modalities. For example, in contrast to the nonexistent immunity induced in end stage cervical cancer patients immunized with a HPV 16 peptide based vaccine,[103] Muderspach et al[47] report better successes observed in women with CIN lesions using a similar vaccine. In hopes of applying a cancer vaccine earlier in the disease, our lab has shown that the vaccination of mice with a peptide vaccine will still produce an effective immunity given as soon as one day following the last radiation treatment (Small LA and Kast WM unpublished observations 1999). These results indicate that past failures of clinical trials of EFIT may be confounded by the accrual of end stage cancer patients and that future endeavors may benefit from the use of immunizations earlier in the disease progression.

Acknowledgments

This review was partially based on studies supported by grants to W.M. Kast from the NIH (CA74397, CA/AI 78394, CA 74182), the department of defense (PC970131), the Illinois Department of Public Health, The Cancer Research Institute and Wyeth Lederle Vaccines and Pediatrics.

References

1. Moskophidis D, Lechner F, Pircher H et al Virus persistence in acutely infected immunocompetent mice by exhaustion of antiviral cytotoxic effector T cells. Nature 1993; 758-761.
2. Banchereau J, Steinman RM Dendritic cells and the control of immunity Nature 1998; 392:245-252.
3. Ridge JP, Di Rosa F, Matzinger P A conditioned dendritic cell can be a temporal bridge between a CD4⁺ T-helper and a T-killer cell. Nature 1998; 393:474-478.
4. Cella M, Scheidegger D, Palmer-Lehmann K et al Ligation of CD40 on dendritic cells triggers production of high levels of interleukin 12 and enhances T cell stimulatory capacity: T-T help via APC activation J Exp Med 1996; 184:747-752.
5. Ishioka GY, Fikes J, Hermanson G et al. Utilization of MHC class I transgenic mice for development of minigene DNA vaccines encoding multiple HLA-restricted CTL epitopes. J Immunol 1999; 162:3915-3925.
6. Sykulev Y, Cohen RJ, Eisen HN The law of mass action governs antigen-stimulated cytolytic activity of $CD8^+$ cytotoxic T lymphocytes. Proc Natl Acad Sci 1995; 92:11990-11992
7. Sercarz EE, Lehmann PV, Ametani A et al Dominance and crypticity of T cell antigenic determinants. Annu Rev Immunol 1993:729-766.
8. Sandberg JK, Grufman P, Wolpert EZ et al. Superdominance among immunodominant $H\text{-}2K^b$-restricted epitopes and reversal by dendritic cell-mediated antigen delivery. J Immunol 1998; 160:3163-3169.
9. Thomson SA, Sherritt MA, Medveczky J et al. Delivery of multiple CD8 cytotoxic T cell epitopes by DNA vaccination. J Immunol 1998; 160:1717-1723.
10. Toes REM, Hoeben RC, van der voort EIH et al. Protective anti-tumor immunity induced by vaccination with recombinant adenoviruses encoding multiple tumor-associated cytotoxic T lymphocyte epitopes in a string-of-beads fashion. Proc Natl Acad Sci USA 1997; 94:14660-14665.
11. Wang H, Eckels DD. Mutations in immunodominant T cell epitopes derived from the nonstructural 3 protein of hepatits C virus have the potential for generating escape variants that may have important consequences for T cell recognition. J Immunol 1999; 162:4177-4183.
12. Uyttenhove C, Maryanski J, Boon T. Escape of mouse mastocytoma P815 after nearly complete rejection is due to antigen-loss variants rather than immunosuppression J Exp Med 1983; 157:1040-1052.
13. Urban JL, Burton RC, Holland JM et al. Mechanisms of syngeneic tumor rejection. susceptibility of host-selected progressor variants to various immunological effector cells. J Exp Med 1982; 155:557-573.
14. Wallich R, Bulbuc N, Hammerling GJ et al. Abrogation of metastatic properties of tumour cells by de novo expression of H-2K antigens following H-2 gene transfection. Nature 1985; 315:301-305.
15. Hui K, Grosveld F, Festenstein H Rejection of transplantable AKR leukaemia cells following MHC DNA-mediated cell transformation. Nature 1984; 311:750-752.

16. Yellenshaw AJ, Eisenlohr LC Regulation of class I restricted epitope processing by local or distal flanking sequence. J Immunol 1997; 158:1727-1733.
17. Del-Val M, Schlicht HJ, Ruppert T et al. Efficient processing of an antigenic sequence for presentation by MHC class I molecules depends on its neighboring residues in the protein. Cell 1991; 66:1145-1153.
18. Eggers M, Boesfabian B, Ruppert T et al, The cleavage preference of the proteasome governs the yield of antigenic peptide. J Exp Med 1995; 182: 1865-1870.
19. Niedermann G, Butz S, Ihlenfeldt HG et al. Contribution of proteasome-mediated proteolysis to the hierarchy of epitopes presented by major histocompatibility complex class I molecules. Immunity 1995; 2:289-295.
20. Lippolis JD, Mylin LM, Simmons DT et al. Functional analysis of amino acid residues encompassing and surrounding two neighboring H-2D^b-restricted cytotoxic T-lymphocyte epitopes in simian virus 40 tumor antigen. J Virol 1995; 69:3134-3146.
21. Weidt G, Deppert W, Buchhop S et al. Antiviral protective immunity induced by major histocompatibility complex class I molecule-restricted viral T-lymphocyte epitopes inserted in various positions in immunologically self and nonself proteins. J Virol 1995; 69:2654-2658.
22. Townsend ARM, Rothbard J, Gotch FM et al. The epitopes of influenza nucleoprotein recognized by cytotoxic T lymphocytes can be defined with short synthetic peptides. Cell 1986; 44:959-968.
23. Schulz M, Zinkernagel RM, Hengartner H. Peptide induced antiviral protection by cytotoxic T cells. Proc Natl Acad Sci USA 1991; 88:991-993.
24. Kast WM, Roux L, Joseph C et al. Protection against lethal sendai virus infection in vivo priming of virus specific cytotoxic T lymphocytes with a free synthetic peptide. Proc Natl Acad Sci USA 1991; 88:2283-2287.
25. Keene JA, Forman J Helper activity is required for the in vivo generation of cytotoxic T lymphocytes. J Exp Med 1982; 155:768-782.
26. Feltkamp MC, Smits HL, Vierboom MP et al. Vaccination with cytotoxic T lymphocyte epitope containing peptide protects against a tumor induced by human papillomavirus type 16-transformed cells. Eur J Immunol 1993; 23:2242-2249.
27. Ishii T, Udono H, Yamano T et al. Isolation of MHC class I-restricted tumor antigen peptide and its precursors associate with heat shock proteins hsp70, hsp90 and gp96. J Immunol 1999; 162:1303-1309.
28. Suto R, Srivastava PK A mechanism for the specific immunogenicity of heat shock protein-chaperoned peptides. Science 1995; 269:1585-1588.
29. Udono H, Srivastava PK Heat shock protein 70-associated peptides elicits specific cancer immunity. J Exp Med 1993; 178:1391-1396.
30. Tamura Y, Peng P, Liu K et al. Immunotherapy of metastatic lung cancer by heat shock protein preparations. Science 1997; 278:117-120.
31. Fayolle C, Ladant D, Karimova G et al. Therapy of murine tumors with recombinant bordetella pertussis adenylate cyclase carrying a cytotoxic T cell epitope. J Immunol 1999; 162:4157-4162.
32. Falk K, Rotschke O, Rammensee HG Allele-specific motifs revealed by sequencing of self peptides eluted from MHC molecules. Nature 1991; 351:290-296.
33. Hunt DF, Henderson RA, Shabanowitz J et al. Characterization of peptide bound to the class I MHC molecule HLA-A2.1 by mass spectrometry. Science 1992; 255:1261-1263.
34. Jardetzky TS, Lane WS Robinson RA et al. Identification of self peptides bound to purified HLA-B27. Nature 1991; 353:326-329.
35. Fremont DH, Matsumara M, Stura EA et al. Crystal structure of two viral peptides in complex with murine class I H-2K^b Science 1992; 257:919-927.
36. Madden DR, Gorga JC, Stromlinger JL et al. The three dimensional structure of HLA-B27 at 2.1 A resolution suggests a general mechanism for tight peptide binding to MHC. Cell 1992; 70:1035-1048.
37. Matsumara M, Fremont DH, Peterson PA et al. Emerging principle for the recognition of peptide antigens by MHC class I molecules. Science 1992; 257:927-934.
38. Zhang W, Young ACM, Imarai M et al. Crystal structure of the MHC class I H-2K^b molecule containing a single viral peptide: implications for peptide binding and T cell receptor recognition Proc Natl Acad Sci USA 1992; 89:8403-8407.

39. Vierboom MPM, Feltkamp MCW, Neisig A et al. Peptide vaccination with an anchor-replaced CTL epitope protects against human papillomavirus type 16-induced tumors expressing the wild type epitope. J Immunother 1998; 21(6):399-408.
40. Valmori P, Fonteneau JF, Lizana CM et al. Enhanced generation of specific tumor reactive CTL in vitro by selected melan-A/MART-1 immunodominant peptide analogues. J Immunol 1998; 160:1750-1758.
41. Men Y, Miconnet I, Valmori D et al. Assessment of immunogenicity of human melan-A peptide analogues in HLA-A*0201/K^b transgenic mice. J Immunol 1999; 162:3566-3573.
42. Clay TM, Custer MC, McKee MD et al Changes in the fine specificity of gp100(209-217)-reactive T cells in patients following vaccination with a peptide modified at an HLA-A2.1 anchor residue. J Immunol 1999; 162:1749-1755.
43. Shirai M, Pendelton CD, Ahlers J et al. Helper-cytotoxic T lymphocyte (CTL) determinant linkage required for priming of anti-HIV $CD8^+$ CTL in vivo with peptide vaccine constructs. J Immunol 1994; 152:549-555.
44. Roman M, Martin-Orozco E, Goodman JS et al. Immunostimulatory DNA sequences function as T-helper-1 promoting adjuvants. Nature Med 1997; 3:849-854.
45. Trinchieri G, Interleukin-12: a proinflammatory cytokine with immunoregulatory functions that bridge innate resistance and antigen specific adaptive immunity. Ann Rev Immunol 1995; 13:251-276.
46. Davis HL, Weeranta R, Waldschmidt TJ et al. CpG DNA is a potent enhancer of specific immunity in mice immunized with recombinant hepatitis B surface antigen. J Immunol 1998; 160:870-876.
47. Muderspach LI, Roman LD, Facio G. Phase I trial of a HPV-16 E7 peptide vaccine in women with high-grade cervical and vulvar dysplasia. Abstract at 30th annual meeting of the Society of Gynecologic Oncologists 1999; 447.
48. Aichele P, Brduscha-Riem K, Zinkernagel RM et al. T cell priming versus T cell tolerance induced by synthetic peptides. J Exp Med 1995; 182: 261-266.
49. Toes REM, Blom RJJ, Offringa R et al. Enhanced tumor outgrowth after peptide vaccination - functional deletion of tumor specific CTL induced by peptide vaccination can lead to the inability to reject tumors. J Immunol 1996; 156:3911-3918.
50. Toes REM, Offringa R, Blom RJJ et al. Peptide vaccination can lead to enhanced tumor growth through the specific T-cell tolerance induction. Proc Natl Acad Sci USA 1996; 93:7855-7860.
51. Nieland JD, Da Silva DM, Velders MP et al. Chimeric papillomavirus-like particles induce a murine self-antigen-specific protective and therapeutic antitumor immune response J Cell Biochem 1999; 73:145-152.
52. Toes REM, van der Voort EIH, Schoenberger SP et al. Enhancement of tumor outgrowth through CTL tolerization after peptide vaccination is avoided by peptide presentation on dendritic cells. J Immunol 1998; 160:4449-4456.
53. Gupta RK, Siber GR Adjuvants for human vaccines-current status, problems and future prospects. Vaccine 1995; 13:1263-1276.
54. Inaba K, Steinman RM, Pack MW et al. Identification of proliferating dendritic cell precursors in mouse blood. J Exp Med 1992; 175:1157-1167.
55. Inaba K, Inaba M, Romani N et al. Generation of large numbers of dendritic cells from mouse bone marrow cultures supplemented with granulocyte/macrophage colony stimulating factor. J Exp Med 1992; 176:1693-1702.
56. Drakes ML, Lu L, Subbotin VM et al. In vivo administration of flt3 ligand markedly stimulates generation of dendritic cell progenitors from mouse liver. J Immunol 1997; 159:4268-4278.
57. Morse MA, Lyerly HK, Gilboa E et al. Optimization of the sequence of antigen loading and CD40-ligand-induced maturation of dendritic cells. Cancer Res 1998; 38:2965-2968.
58. Mackey MF, Gunn JR, Maliszewski C et al. Cutting Edge: dendritic cells require maturation via CD40 to generate protective antitumor immunity.
59. Labeur MS, Roters B, Pers B et al. Generation of tumor immunity by bone marrow-derived dendriticcells correlates with dendritic cell maturation stage. J Immunol 1999; 162:168-175.
60. Caux C, Massacrier C, Vanbervliet B et al. Activation of human dendritic cells through CD40 crosslinking. J Exp Med 1994; 180:1263-1272.

61. Mayordomo JI, Zorina T, Storkus WJ et al. Bone marrow-derived dendritic cells pulsed with synthetic tumour peptides elicit protective and therapeutic antitumour immunity. Nature Med 1995; 12(1):1297-1302.
62. Murphy GP, Tjoa BA, Simmons SJ et al. Infusion of dendritic cells pulsed with HLA-A2 specific prostate-specific membrane antigen peptides: a phase II prostate cancer vaccine trial involving patients with hormone-refractory metastatic disease. Prostate 1999; 38:73-78.
63. Tuting T, DeLeo AB, Lotze MT et al. Genetically modified bone marrow-derived dendritic cells expressing tumor-associated viral or "self" antigens induce antitumor immunity in vivo. Eur J Immunol 1997; 27:2702-2707.
64. Wolff JA, Malone RW, Williams P et al. Direct gene transfer into mouse muscle in vivo. Science 1990; 247:1465-1468.
65. Ulmer JB, Deck RR, DeWitt CM et al. Protective immunity by intramuscular injection of low doses of influenza virus DNA vaccines. Vaccine 1994; 12:1541-1544.
66. Conry RM, LoBuglio AF, Loechel F et al. A carcinoembryonic antigen polynucleotide vaccine has in vivo antitumor activity. Gene Ther 1995; 2:59-65.
67. Graham RA, Burchell JM, Beverley P et al. Intramuscular immunization with MUC1 cDNA can protect C57 mice challenged with MUC1-expressing syngeneic mouse tumor cells. Int J Cancer 1996; 65:664-670.
68. Boyle JS, Silva A, Brady JL et al. DNA immunization: induction of higher avidity antibody and effect of route on T cell cytotoxicity. Proc Natl Acad Sci USA 1997; 94:14626-14631.
69. Maurer D, Stingl G. Dendritic cells in the context of skin immunity. In: Lotze MT, Thomson AW, eds. Dendritic Cells 1st ed. New York: Academic Press, 1999:111-122.
70. Akbari O, Panjwani N, Garcia S et al. DNA vaccination: transfection and activation of dendritic cells as key events for immunity. J Exp Med 1999; 189:169-177.
71. Doe B, Selby M, Barnett S et al. Induction of cytotoxic T lymphocytes by intramuscular immunization with plasmid DNA is facilitated by bone marrow-derived cells. Proc Natl Acad Sci 1996; 93:8578-8583.
72. Kovacsovics-Bankowski M Rock KL A phagosome-to-cytosol pathway for exogenous antigens presented on MHC class I molecules. Science 1995; 243-246.
73. Reis e Sousa C, Germain RN Major histocompatibility complex class I presentation of peptides derived from soluble exogenous antigen by a subset of cells engaged in phagocytosis J Exp Med 1995 182:841-851.
74. Ciernik IF, Berzofsky JA, Carbone DP Induction of cytotoxic T lymphocytes and anti-tumor immunity with DNA vaccines expressing single T cell epitopes. J Immunol 1996; 156:2369-2375.
75. Wu Y, Kipps TJ Deoxyribonucleic acid vaccines encoding antigens with rapid proteosome-dependent degradation are highly efficient inducers of cytotoxic T lymphocytes. J Immunol 1997; 159:6037-6043.
76. Barouch DH, Santra S, Steenbeke TD et al. Augmentation and suppression of immune responses to an HIV-1 DNA vaccine by plasmid cytokine/Ig administration. J Immunol 1998; 161:1875-1882.
77. Kim JJ, Trivedi NN, Nottingham LK et al. Modulation of amplitude and direction of in vivo immune responses by co-administration of cytokine gene expression cassettes with DNA immunogens. Eur J Immunol 1998; 28:1089-1103.
78. Chow Y-H, Chiang B-L, Lee Y-L et al. Development of Th1 and Th2 populations and the nature of immune responses to hepatitis B virus DBA vaccines can be modulated by co-delivery of various cytokine genes. J Immunol 1998; 160:1320-1329.
79. Corr M, Tighe H, Lee D et al. Costimulation provided by DNA immunization enhances antitumor immunity. J Immunol 1997; 159:4999-5004.
80. Oldstone MB, Tishon A, Eddleston M et al. Vaccination to prevent persistent viral infection. J Virol 1993; 67:4372-4378.
81. Whitton JL, Sheng N, Oldstone MB et al. A 'string of beads' vaccine comprising linked minigenes confers protection from lethal-dose virus challenge. J Virol 1993; 67:348-353.
82. Albert ML, Sauter B, Bhardwaj N Dendritic cells acquire antigen from apoptotic cells and induce class I-restricted CTLs. Nature 1998; 392:86-89.
83. Ahmed R, Butler LD, Bhatti L. $T4^+$ T helper cell function in vivo: differential requirement for induction of antiviral cytotoxic T cell and antibody responses. J Virol 1988; 62:2102-2106.

84. Nash AA, Jayasuriya A, Phelan J Different roles for L3T4+ and Lyt2+ T cell subsets in the control of an acute herpes simplex virus infection of the skin and nervous system. J Gen Virol 1987; 68:825-833.
85. Buller RM, Holmes KL, Hugin A. Induction of cytotoxic T-cell responses in vivo in the absence of CD4 helper cells. Nature 1987; 328:77-79.
86. Liu Y, Mullbacher A The generation and activation of memory class I MHC restricted cytotoxic T cell responses to influenza A virus in vivo do not require CD4+ T cells. Immunol Cell Biol 1987; 67:413-420.
87. Kirnbauer R, Booy F, Cheng N et al. Papillomavirus L1 major capsid protein self-assembles into virus-like particles that are highly immunogenic. Proc Natl Acad Sci USA 1992; 89:12180-12184.
88. Kirnbauer R, Taub J, Greenstone HL et al. Efficient self-assembly of human papillomavirus type 16 L1 and L1-L2 into virus like particles J Virol 1993; 67:6929-6936.
89. Peng S, Frazier IH, Fernando GJ et al. Papillomavirus virus-like particles can deliver defined CTL epitopes to the MHC class I pathway. Virology 1998; 240:147-157.
90. Gilbert SC, Plebanski M, Harris SJ et al. A protein particle vaccine containing multiple malaria epitopes. Nature Biotechnol 1997; 15:1280-1284.
91. Touza A, Coursaget P In vitro gene transfer using human papillomavirus-like particles. Nucleic Acids Res 1998; 26:1317-1323.
92. Top FH, Buescher EL, Bancroft WH et al. Immunization with live type 7 and 4 adenovirus vaccines. II. antibody response and protective effect against acute respiratory disease due to adenovirus type 7. J Infect Dis 1971; 124:155-160.
93. Imler J-L Adenovirus vectors as recombinant viral vaccines. Vaccine 1995; 13:1143-1151.
94. Behbehani AM The smallpox story: life and death of an old disease. Microbiol Rev 1983; 47:455-599.
95. Gurvich EB The age dependent risk of postvaccination complications in vaccines with smallpox vaccine. Vaccine 1992; 10:96-97.
96. Rooney JF, Wohlenberg C, Cremer K et al. Immunization with a vaccinia virus recombinant expressing herpes simplex virus type 1 glycoprotein D: long term protection and effect of revaccination. J Virol 1988; 62:1530-1534.
97. Cooney EL, Collier AC, Greenberg PD et al. Safety of and immunological response to a recombinant vaccinia virus vaccine. Lancet 1991; 337:567-672.
98. Belyakov IM, Moss B, Strober W et al. Mucosal vaccination overcomes the barrier to recombinant vaccinia immunization caused by preexisting poxvirus immunity. Proc Natl Acad Sci USA 1999; 96:4512-4517.
99. Roden RB, Greenstone HL, Kirnbauer R et al. In vitro generation and type-specific neutralization of a human papillomavirus type 16 virion pseudotype. J Virol 1996; 70:5875-5883.
100. Madrigal M, Janicel MF, Sevin BU et al. In vitro antigene therapy targeting HPV-16 E6 and E7 in cervical carcinoma. Gyenecol Oncol 1997; 64:18-25.
101. Breitburd F, Kirnbauer R, Hubbert NL et al. Immunization with virus-like particles from cottontail rabbit papillomavirus (CRPV) can protect against experimental CRPV infection. J Virol 1995; 69:3959-3963.
102. Schneider J, Gilbert SC, Blanchard TJ et al. Enhanced immunogenicity for CD8+ T cell induction and complete protective efficacy of malaria DNA vaccination by boosting with modified vaccinia virus Ankara. Nature Med 1998; 4(4):397-402.
103. Ressing ME, van Driel WJ, Brandt RMP et al. Detection of T helper responses, but not of HPV-specific CTL responses, following peptide vaccination of cervical carcinoma patients. J Immunother; 2000; 23:255-266.

Chapter 5

Cancer Peptide Vaccines in Clinical Trials

Jeffrey S. Weber

Introduction

The revelation that protein antigens were processed into peptides by a pathway of intracellular degradation and presented on the surface of antigen presenting cells for recognition by T-cells in association with class I and II MHC molecules created a new paradigm for the generation and detection of antigen-specific immune responses in humans.[1-3] The subsequent discovery, cloning and identification of several classes of tumor-associated and tumor-specific antigens from human melanomas and other cancers has facilitated the performance of a number of clinical trials of peptide vaccines with and without adjuvants in patients with metastatic and resected melanoma and several other malignancies and pre-cancerous conditions. In this review I will summarize the different classes of melanoma and human papillomavirus (HPV) antigens that have been defined and describe the available data on recent attempts to boost immunity directed against defined melanoma and HPV antigens using peptides and detail their clinical significance. I will conclude with a proposal for an "optimal" vaccine schema and a call for an expansion of current peptide vaccine efforts in melanoma and other histologies.

Melanoma Antigens

The development of cancer vaccines has been most advanced in melanoma, which is unique among human tumors in the existence of compelling evidence for its immunogenicity. Spontaneous regression of primary melanomas is quite common, and the prognosis of cutaneous melanomas varies directly with the lymphocytic infiltrate.[4-5] Vitiligo is an autoimmune destruction of melanocytes that commonly occurs in melanoma patients, especially those who have been treated with interferon alpha or interleukin-2, and it is known to correlate with a favorable response to immunotherapy.[6-8] Tumor-reactive lymphocytes from patient peripheral blood or those which infiltrate metastatic melanoma lesions can be propagated in vitro as long term T cell lines or clones.[9-11] The ease with which such melanoma specific T-cells can be grown in vitro allowed Boon and colleagues in 1991 to describe the cloning of an antigen derived from a mutagenized melanoma cell line which was recognized by T cells.[12,13] This antigen was called MAGE, and was shown to define a family of antigens that had not previously been identified. MAGE-1 and several members of its multigene family were shown to be present on a significant proportion of melanoma cell lines and fresh tumors but were also found on a variety of tumors of epithelial and neuroectodermal origin as well as normal testis and placental tissue, but no other normal tissue.[14] MAGE, GAGE, BAGE and RAGE defined X-chromosome linked families of genes that were found respectively on melanoma, gastrointestinal, breast and renal cell tumors, many of which encoded antigens that were recognized by T-cells

Peptide-Based Cancer Vaccines, edited by W. Martin Kast. ©2000 Eurekah.com.

and defined the "cancer/testis" class of tumor antigens. Many melanoma lines and fresh tumors were shown to express cancer/testis antigens.

After the cloning of the first cancer/testis antigens, several groups defined members of a second group of antigens that were melanosome-related and are considered "neo-antigens" derived from gene products produced in normal cells. The first melanosome-related antigen cloned was MART-1/Melan A, defined through its recognition by CTL clones from melanoma patient peripheral blood and by lymphocytes infiltrating tumors (TIL). The TIL that recognized MART-1 as well as TIL from a number of other melanoma patients reacted with virtually all melanoma cell lines that expressed HLA-A2, and transfection of the A2 gene into other non-A2 expressing melanoma lines increased their sensitivity to TIL lysis.[15] This suggested that MART-1 was a common A2 restricted melanoma antigen recognized by CTL. MART-1 was expressed by virtually all metastatic melanoma lesions, a majority of cell lines derived from metastatic melanomas, and also by melanocytes, but not by any other normal tissue. The MART-1 gene encoded a putative protein of 26,000 MW including a number of sequences that matched the known HLA-A2 binding motifs. The nonamer sequence AAGIGILTV, representing residues 27-35 of the MART-1 protein bound most strongly to HLA-A2.[16] This peptide stimulated the growth of specific CTL from the peripheral blood mononuclear cells (PBMC) of melanoma patients and of normals.[17] Multiple re- stimulations of PBMC with MART-1:27-35 peptide resulted in cultures of MART-1 specific CTL derived from 11 of 12 melanoma patients.[18,19] These CTL lysed fresh uncultured melanoma cells and were 100 fold more lytically active against melanoma cells than TIL grown in high dose IL-2. The majority of TIL grown from patients with melanoma are capable of recognizing the MART-1:27-35 peptide, and some of those TIL cultures induced regression of metastatic melanoma after adoptive transfer with IL-2.[20] The repertoire of Vß T cell receptor molecules from TIL and peripheral blood derived CTL lines that are MART-1 specific are quite skewed.[21-22]

Peptides derived from MART-1 were eluted from melanoma cells, suggesting that MART-1 is a naturally occurring antigen on fresh tumors.[23] A protein database analysis demonstrated that sequences conforming to the MART-1 A2 binding motif and possessing features important for CTL recognition occurred frequently in proteins,[24,25] and that a peptide derived from glycoprotein C of herpes simplex virus could sensitize targets cells to lysis by MART-1 specific CTL.[26] These data suggest that epitope mimicry by normal or other commonly occurring proteins may account for the frequency of CTL detected against melanoma antigens like MART-1.

Greater MART-1 reactive CTL activity has been demonstrated in the peripheral blood of melanoma patients compared to normals, suggesting that a tumor related "priming" effect has occurred.[27] In a clinical study of patients with metastatic melanoma who had lymphnodal disease that was resected, a novel technology using peptide-bound class I tetramers was employed to detect MART-1 and tyrosinase specific CTL in tumor-infiltrating lymph nodal tissue. From 0.1 to 3% of $CD8^+$ lymphocytes infiltrating tumor-involved lymph nodes were MART-1 specific, and in contrast the proportion of MART-1 specific cells in non-tumor containing lymph nodes was not different from background.[28] The infiltrating tumor antigen specific CTL were functional, since they were capable of secreting cytokines after contact with or could lyse peptide-pulsed targets or antigen-expressing tumor cell lines.

The overlapping MART-1 26-35 peptide has been shown to be more immunogenic than the 27-35 epitope, and a single amino acid modification to the 26-35 peptide rendered it a stronger binder to A2.1 and even more immunogenic.[28] This peptide is a candidate for future clinical vaccine trials. The MART-1 26-35 and 27-35 peptides have been shown to bind to multiple other A2 subtypes with a significant degree of heterogeneity,[29,30] as well as allele A45,[31] but no other MART-1 specific peptides have been shown to elicit specific immune responses in patients bearing other HLA class I alleles.[32]

pMel17/gp100 is another antigen defined through its recognition by CTL clones from melanoma patient peripheral blood and by tumor infiltrating lymphocytes (TIL). The gp100 antigen is a transmembrane glycoprotein of approximately 100 kilodaltons that is recognized by the HMB-45 and NKI-beteb monoclonal antibodies on melanocytes and melanoma cells.[33,34] CTL clones derived from melanoma patients and TIL grown from a melanoma patient recognized melanoma cells that expressed gp100 in association with HLA-A2.1, but not gp100 negative lines.[35] Multiple peptides derived from the gp100 sequence that fit the consensus motif for binding to HLA-A2 can be recognized by TIL from melanoma patients, including gp100 209-217 (ILDQVPSFV), gp100 154-162 (TKTWGQYWQV) as well as gp100 457-466 (LLDGTAATLRL).[36] A naturally processed gp100 peptide 280-288 (YLEPGPVTA) was defined by elution of peptides from melanoma cells and purification by tandem mass spectroscopy. This peptide redirected HLA-A2 restricted lysis by CTL lines from a melanoma patient.[37] TIL from a melanoma patient also recognized gp100 280-288. All three gp100 peptides are recognized by self reactive T cells, and bind with moderate or low affinity to HLA A2, suggesting that the gp100 specific TIL or CTL may have escaped from tolerance induction because of their low affinity for TcR, whereas the autoreactive CTL that bear high affinity TcR may be deleted in the thymus. It is unclear how melanoma patients possess apparently autoreactive T cells, like the TIL line that recognized gp100 154-162, without the induction of autoimmune disease. It is of some interest that in melanoma patients gp100 specific TIL have been reported to induce regression of metastatic melanoma in patients receiving TIL and IL-2 therapy. Four of a group of 14 TIL that were analyzed were reactive with one or more epitopes of gp100; all four induced tumor regression.[35] Peptides derived from gp100 that are recognized by CTL in association with members of the HLA-A3 superfamily have also been defined, suggesting that the repertoire of HLA haplotypes eligible for vaccination trials with gp100 may be expanded.[38,39] These data suggest that gp100 may be a promising target for immunotherapy using peptide vaccination strategies and/or adoptive therapy of antigen-specific cytolytic T cells.

Tyrosinase is a membrane bound protein involved in the melanin synthesis pathway.[40] Tyrosinase is expressed by virtually all primary cutaneous melanomas, and by up to 90% of metastatic lesions. [40]It encodes several epitope peptides that are presented by HLA-A2 to CTL reactive to human melanomas.[41] A peptide derived from tyrosinase, aa 368-376, YMNGTMSQV, was shown to be post-translationally modified by deamidation of asparagine to aspartic acid resulting in a sequence recognized by human CTL, YMDGTMSQV, known as tyrosinase 368-376 (370D).[42] The tyrosinase 368-376 (370D) peptide is felt to encode a biologically important epitope and can induce CTL in vitro.

TRP-1(gp75) and TRP-2 are tyrosinase related proteins that are integral parts of the melanosome apparatus, and they have been shown to be recognized by CTL from patients with melanoma, although these neo-antigens have not yet reached the stage of clinical trials.[43-45]

Recently, a "cancer-testis" antigen distinct from the MAGE, GAGE, BAGE and RAGE families called NY-ESO-1 was discovered that has the property of eliciting both a strong humoral response yet encodes epitope peptides recognized by CTL clones and TIL clones from patients with melanoma and breast cancer.[46,47] The antigen was found to be expressed by a variety of squamous and adenocarcinomas as well as melanoma, suggesting a wide potential utility.

A final group of melanoma antigens are mutated normal genes uniquely present on individual tumors; p16, beta-catenin and mum-1 are examples.[48-50] Because they are unique to individual tumors, their clinical utility for vaccines is limited and will not be pursued in this Chapter.

Clinical Trials with Peptide Vaccines

The initial trials of peptide vaccines for melanoma were performed in patients with metastatic disease to test the toxicity of and immune responses to the vaccine. The peptide vaccines showed few toxicities other than local pain and granuloma formation, and escalating doses of peptides in aqueous solution or combined as in prior mouse experiments with incomplete Freund's adjuvant (IFA or Montanide ISA 51, an oil-in-water emulsion) at doses up to 10 mg showed no serious side effects. Initial experiments with the HLA-A1 restricted MAGE-3 peptide EVDPIGHLY given in aqueous solution at low (100-300 ug/injection) doses to 12 patients with metastatic melanoma indicated that six were able to complete three injections at monthly intervals, and three of the six had objective clinical responses.[51] No evidence of boosted immunity was detected in the peripheral blood. When the same epitope peptide was combined with IFA and administered to patients with resected melanoma, immune responses were detected by CTL assays in five out of 16 patients tested at our institution, and DTH skin responses to peptide were also observed.[52] No clear correlation was observed in this small trial between immune response by specific cytokine or chromium release after exposure to peptide-pulsed T2 cells and time to relapse.

The MART-1 27-35 and 26-35, tyrosinase 1-9 and 368-376 as well as several gp100 peptides (280-288 and 457-467) have been tested alone and in combination in patients with metastatic or resected melanoma. Jaeger, Knuth and colleagues immunized 6 metastatic melanoma patients intradermally with multiple peptides from MART-1, gp100 and tyrosinase in aqueous solution at 100 ug each weekly for four immunizations, and observed augmented cytolysis of antigen-restimulated PBMC directed against one but not necessarily all three antigens. [53] 3/6 patients showed augmented cytolysis to MART-1, 2/6 to a tyrosinase peptide but 0/6 demonstrated augmented reactivity to gp100 peptides. When GM-CSF was injected at 75 micrograms per dose subcutaneously as an adjuvant for 3 days prior to and 2 days after immunization with multiple antigen peptides, significant boosting of immune reactivity was seen compared to vaccination with peptides alone.[54] Three out of three patients showed increased immune reactivity to tyrosinase peptide 1-9, and 1/3 had increased reactivity to MART-1 26-35. 3/3 objective clinical responses were seen, including two partial regressions in involved lymph nodal, cutaneous and liver lesions, and 1 complete regression in patient with subcutaneous disease. An infiltrate of $CD8^+$ T cells was observed at vaccination sites after GM-CSF injections.

A number of small pilot studies have been conducted at the National Cancer Institute in which patients with metastatic melanoma received multiple subcutaneous injections of a single peptide emulsified with IFA at three week intervals. MART-1 27-35, gp100 209-217, 154-162 and 280-288 have been used in these trials.[55] In one study, escalating doses of the gp100 209 (ITDQVPSFY), 280 (YLEPGPVTA) or 154 (KTWGQYWQV) peptides at doses from 1 to 10 mg were administered subcutaneously every three weeks with IFA.[56] Immune assays were performed using the above "native" peptides for antigenic stimulation and substituted 209-2M (IMDQVPFSY) and 280-9V (YLEPGPVTV) peptides. 90-100% of patients had strong evidence of boosted immune reactivity post-vaccination as shown by an assay in which release of gamma interferon from PBMC restimulated one to three times in the presence of IL-2 and peptide antigen was measured by ELISA. Seven out of seven patients had boosted gp100 reactivity post-vaccine after only one-restimulation with the 209-2M peptide, and five out of six were boosted with one restimulation with the 280-9V peptide. Higher release of cytokine was seen after four immunizations than two in most patients, and a greater level of reactivity was observed when substituted peptides were used in assays contrast with "native" peptides. Boosted cytokine release was shown to correlate with cytolytic responses. When tumor cell lines expressing the correct MHC restriction element and antigen was used as a stimulator in cytokine release assays, lower levels of cytokine were observed compared with T2 cells pulsed with the relevant peptide, suggesting that peptide density on the target was important for recognition by effector cells in PBMC. Objective partial and complete remissions were uncommon, with one out of 20 patients having a complete regression. No clear correlation was observed between the level of immune response and the rare clinical responses, so no statement about clinical benefit could be made. In a second study of escalating doses of the MART-1 27-35 peptide administered subcutaneously every three weeks with IFA, 15 out of 16 patients had evidence of boosting of immunity directed against the "native" MART-1 27-35 peptide, with increased reactivity after four compared with fewer vaccinations.[57] No objective clinical responses were seen. At USC/Norris, we have treated 25 melanoma patients with high risk resected stages III/IV disease with increasing doses of the MART-1 27-35 peptide emulsified with IFA every three weeks subcutaneously (Wang et al, submitted). Toxicity consisted of local inflammation and granuloma formation, and low grade fevers, fatigue, headache and myalgias were common. Ten out of 22 patients had evidence of boosted immunity by cytokine release assays, and a correlation was observed between the absolute level of gamma interferon released after multiple re-stimulations of peptide-pulsed patient PBMC post-vaccination and time to relapse, with a p value of 0.003. The correlation of immune reactivity with time to relapse suggests that a positive response to peptide vaccination may incur clinical benefit. Figure 5.1 shows the level of gamma interferon release post-vaccination in the "positive" patient 10180 compared to a "no" response patient 10127. In this experiment, cultures were split and half were stimulated with the HLA-A2 restricted FLU matrix peptide M1 58-66, and half with the MART-1 27-35 peptide. Cultures of T cells after three stimulations in vitro were reacted with MART-1 peptide-pulsed targets or FLU peptide-pulsed targets to test the specificity of the CTL that develop. Gamma interferon is measured on the ordinate in pg/mL released per 10e5 cells in 24 hours. The data suggest that both patients mount an immune response to FLU before and after vaccination and that the development of MART-1 specific CTL is antigen specific, since MART-1 stimulated CTL do not react with FLU targets, and vice versa. All 9 of the relapses and all 3 deaths have occurred in the minimal or no response group thus far, supporting the idea that development of an immune response directed against MART-1 is clinically beneficial.

The anchor amino acids that form hydrogen bonds between epitope peptides and class I MHC molecules can be modified to strengthen their binding, resulting in greater immunogenicity in vitro and in vivo. When such heteroclitic peptides derived from gp100 were used to restimulate PBMC from patients with metastatic melanoma immunized with a wild type gp100 peptide, immune reactivity was detected with greater frequency and sensitivity.[58] Patients immunized with the heteroclitic gp100 peptide demonstrated a higher level of immune reactivity compared with the wild type peptide. The MART-1 27-35 peptide has been shown to be "naturally" processed and is immunodominant. However, the MART-1 26-35 peptide has been shown to be a better MHC binder and more immunogenic vitro,[59] and was more effective at the detection and quantitation of MART-1 specific CTL in a flow cytometry assay when used for the generation of MHC/peptide tetramers.[28] Heteroclitic gp100 peptides combined with IFA have been used to immunize patients with metastatic melanoma, and more than 90% of the patients in one vaccine study had evidence of boosted immunity detected after one restimulation in vitro, suggesting that the substituted peptide was more immunogenic than "natural" peptide in vivo.[60] When the same substituted gp100 peptide was injected followed within several days by high dose intravenous IL-2, surprisingly no detectable augmented immune response was observed, but 13 out of 31 patients had an antitumor response, including 12 PRs and 1 CR for an overall response rate of 42%, significantly greater than rates observed with IL-2 alone (15-20%).[61,62] The utility of a peptide vaccination added to high dose IL-2 as an "adjuvant" is currently being explored in two large multi-center randomized trials. At our institution, the heteroclitic gp100 209-217 (210M) and tyrosinase 368-376 (370D) peptides emulsified in IFA with or without IL-12 as an adjuvant are being examined in a randomized phase II trial in patients with high risk resected stage III/IV melanoma where immune response

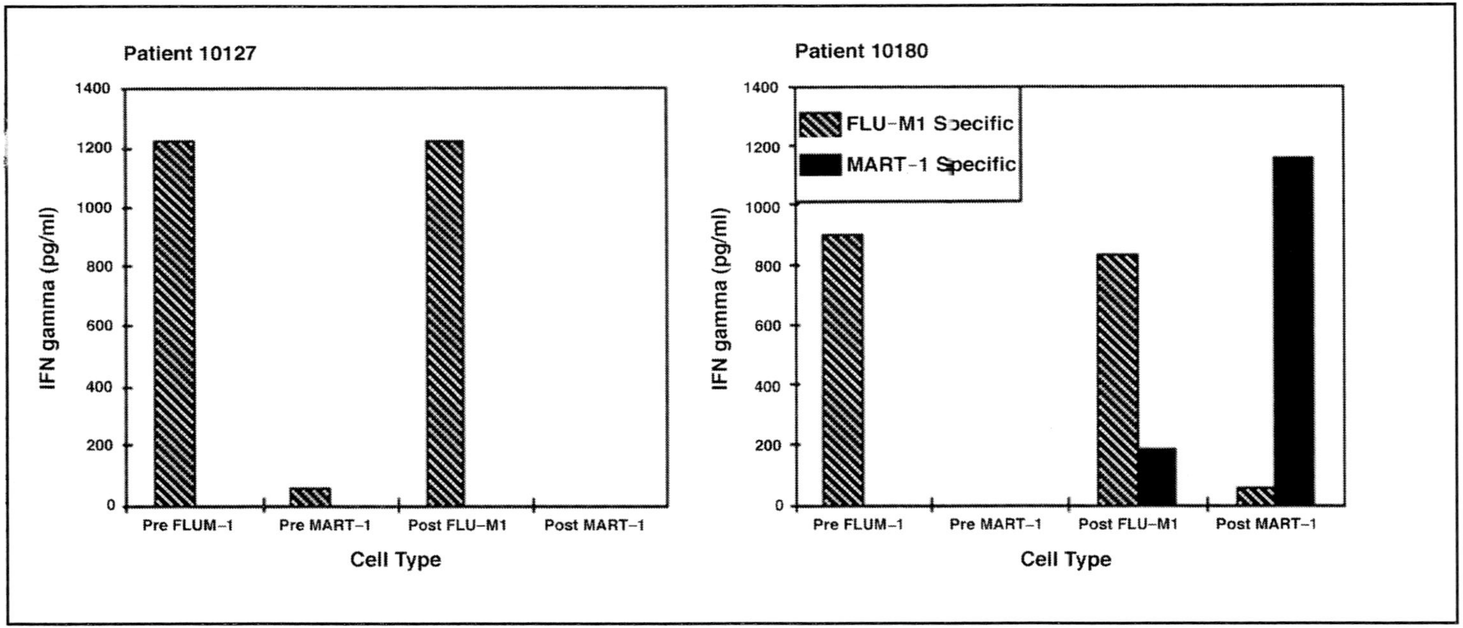

Fig. 5.1. Immune response by ELISA to MART-1 vaccinations. Effector cells were prepared by incubation of PBMC with 10 ug/ml peptide then addition of 50 IU/mL IL-2 48 hours later followed by weekly restimulation with peptide pulsed irradiated PBMC. 100,000 resulting effector cells after the fourth restimulation in vitro were plated with 100,000 T2, FLU or MART-1 peptide-pulsed T2, or 624-mel target cells in a 24 well plate in complete media for 18 hours in a volume of 1 mL. The supernatant was harvested then spun in a microfuge at 14,000 g for 30 seconds to pellet cells and debris. Supernatants were removed and used to measure gamma interferon release using a commercial ELISA kit as described in Materials and Methods. Figures shown are the means of duplicate values. Gamma interferon released per 10e5 cells per 24 hours is shown on the ordinate, and condition of incubation on the abcissa. FLU specific cytokine release is shown in cross hatched bars, and MART-1 specific cytokine release in solid bars. Similar results were obtained in repeated experiments for each patient.

Table 5.1. Immune response to gp100/tyrosinase peptides with IFA and IL-12

Patient PRE/POST	**Cytokine release/ FLU stim.**			**Cyto. release/ gp100 stim.**			**Cyto. release/ tyr stim.**		
	T2 FLU	T2 gp100	T2 tyr	T2 FLU	T2 gp100	T2tyr	T2 FLU	T2 gp100	T2tyr
10236 PRE	**860**	0	222	0	0	40	12	0	**370**
10313 POST	**346**	96	14	0	**830**	0	0	190	**1181**

is one endpoint and time to relapse is the clinical endpoint. Strong DTH skin test responses to gp100 but not tyrosinase have been seen in most patients, and augmented immune responses to both gp100 and tyrosinase have been observed (Weber et al, unpublished). Table 5.1 indicates gp100 and tyrosinase-specific responses by release of gamma interferon in a patient's restimulated PBMC after vaccination with peptide emulsified in IFA and intradermally delivered IL-12; antigen specific responses are indicated in bold face type.

In one study, a HLA-A2.1 restricted tyrosinase peptide combined with adjuvant QS-21 was used to vaccinate patients with metastatic melanoma, and no tumor responses were seen, although in two of nine patients there was evidence of boosted tyrosinase-specific immunity.[63] Toxicity was modest, with fevers and local injection site pain as the only side effects.

Peripheral blood mononuclear antigen-presenting cells prepared by purifying adherent cells and exposing them to 1000 U/mL GM-CSF for seven days were pulsed with the MAGE-1 A1 restricted peptide EVDPIGHLY and injected into HLA-A1 patients with metastatic melanoma.[64] In a small pilot trial, T lymphocytes were grown from the vaccination site of two patients and found to be MAGE-1 reactive. No significant toxicity was observed in this trial. In one patient, T lymphocytes grown from a peripheral tumor were found to be MAGE-1 specific and could lyse autologous tumor cells. When a DTH assay was performed with peptide-pulsed APCs injected intradermally, a CD8 and $CD4^+$ infiltrate was seen, but no patients had an anti-tumor response. It was suggested that growth of vaccine site infiltrating lymphocytes (VIL) and the demonstration that they were tumor specific constituted a surrogate assay for immune response.

Rosenberg and colleagues used adenoviral vectors encoding the cDNAs for gp100 and MART-1 to immunize patients with metastatic melanoma.[65] They performed a phase I trial of escalating doses of recombinant virus from 10e7 to 10e11 particles per injection, given either intradermally or subcutaneously, either alone or with the addition of high dose interleukin-2. 36 patients received the MART-1 vector, and 18 the gp100 construct. Immunizations were given four times at monthly intervals. Mild and transient erythema at injection sites were the only side effects observed. 1/16 patients who was given the MART-1 vector had a complete response, but zero out of six patients who received the gp100 adenovirus alone responded. Of the 20 patients who were given IL-2 plus the MART-1 vector, there were two PRs and two CRS. No consistent augmentation of antigen specific T-cell reactivity was observed in any subgroup of patients, in contrast to the significant reactivity seen with the use of peptides with IFA by the same investigators. Only five out of 23 patients had augmented MART-1 immunity detected by release of cytokines by restimulated PBMC, and zero out of six had gp100 boosting, but 34 of 45 sera tested for adenoviral antibodies showed significant increases in anti-viral IgG. These data suggested to the investigators that adenoviral serologic reactivity might limit

Table 5.2. Immune response to gp100/tyrosinase peptide-pulsed dendritic cell therapy

Patient #/ PRE/POST	Cytokine release/ FLU stim.			Cyto. release/ gp100 stim.			Cyto. release/ tyr stim.		
	T2 FLU	T2 gp100	T2 tyr	T2 FLU	T2 gp100	T2tyr	T2 FLU	T2 gp100	T2tyr
10236 PRE	**670**	0	0	0	**2188**	0	0	0	0
10313 POST	**1148**	0	0	0	**1822**	0	0	0	**919**

the utility of engineered adenoviral vectors as vaccines, and they concluded that current recombinant adenoviral vaccines were not superior to peptide based immunizations.

Dendritic cells (DC) are "nature's adjuvant," can prime naive T cells and are potent stimulators of cytolytic and helper T cell immunity.[66-70] Dendritic cells pulsed with melanoma antigen peptides can stimulate the development of antigen specific CTL in vitro more potently than peptide-pulsed PBMC or macrophages.[71] DC have been prepared from the peripheral blood monocytes of patients by growth in IL-4 and GM-CSF[72,73] and pulsed with multiple melanoma peptides. In vitro, DC pulsed with melanoma peptides can stimulate the development of potent antigen-specific CTL responses.[74] These DC have been injected subcutaneously and intra-lymphnodally in patients with metastatic melanoma, and have been shown to induce antigen-specific immune reactivity and clinical responses. In one published study, 1 complete remission and three partial remissions were observed in patients treated with multiple peptide-pulsed monocyte-derived DC without evidence of significant toxicity.[75] At our institution we have treated 12 melanoma patients with intravenous injections of monocyte-derived DC pulsed with multiple heteroclitic melanoma antigen peptides from gp100 and tyrosinase, and have seen objective clinical responses and boosting of immunity in responding patients, as shown in Table 5.2. The data indicate that tyrosinase but not gp100 specific CTL cytokine release was observed from a responding patient's PBMC that were restimulated after immunization with peptide-pulsed DC. In contrast to patients who receive peptides with IFA, no peptide-related DTH skin test reactivity has been observed in patients treated with peptide-pulsed DC.

The MART-1, gp100 and tyrosinase antigens are all expressed by normal melanocytes[68-70], and the use of neo-antigen peptide vaccines have the potential to induce autoimmune reactions. It is not known whether normal melanocytes effectively present the MART-1, gp100 or tyrosinase epitope peptides to T cells in vivo, and previous clinical experience with the adoptive transfer of large numbers of CTL that were highly MART-1 or gp100 reactive and mediated regression of tumor did not indicate the onset of any autoimmune damage to skin, brain, inner ear and retina, where melanocyte lineage cells are located. Patients with metastatic or resected melanoma that received MART-1, gp100 or tyrosinase peptide vaccines at our institution (58, Wang et al submitted, Weber et al, unpublished) or described in the published literature did not demonstrate any evidence of ocular or other toxicity. No ocular problems nor any evidence of autoimmune pathology have occurred in any patients on our MART-1, gp100 or tyrosinase peptide trials with a median follow-up of 16 months.

One potential pitfall of any antigen-specific immunization strategy is immunoselection *in* vivo after peptide vaccination.[76-84] Western blotting as well as immunohistochemical staining using MART-1 antibodies have established that MART-1 is a transmembrane protein component of the melanosome complex.[76,77] RT-PCR analysis has shown that MART-1 mRNA is present in virtually 100% of metastatic melanoma lesions, yet immunohistochemical staining has shown that there is considerable heterogeneity in MART-1 expression on primary and metastatic lesions, with 60-90% of all lesions staining positively.[78-84] In one study, deletion of MART-1 expression as well as TAP transporter expression rendered cells transparent to CTL recognition, suggesting that loss of MART-1 may be a mechanism for immune evasion.[83] In one pilot clinical trial of multiple peptides from MART-1, gp100 and tyrosinase, two patients that were repeatedly immunized with peptides in aqueous solution intradermally had tumors biopsied prior to and after vaccination. One patient developed MART-1 reactive T cells detected in the peripheral blood but had a progressing lesion biopsied showing loss of MART-1 expression. Another patient had an initial biopsy showing expression of all three antigens but after immunization and development of tyrosinsase specific reactivity, a progressing tumor expressed gp100/MART-1 but not tyrosinase.[84] A number of analyses of antigen expression have been performed on biopsied metastatic melanoma lesions, suggesting that a significant level of antigenic heterogeneity occurs within lesions as well as complete loss of antigen expression on individual lesions.[68-82] The results of those studies suggest that a successful vaccine would consist of multiple antigenic peptides to minimize the loss of any individual antigen.

An ideal candidate for an antigen-specific immunization strategy with peptides is HPV disease, which constitutes a spectrum from anogenital warts caused by HPV 6 and 11, to low grade cervical intraepithelial neoplasia caused by HPV 16, 18, 31, 33 and 45, to high grade dysplasia, carcinoma in situ and frank invasive cervical and vulvar carcinoma in which high risk HPV 16 and 18 have been implicated.[85] HPV 16 and 18 encode two early antigens called E6 and E7 which are important for the development and maintenance of the malignant state.[86] E6 binds and inactivates the Rb tumor suppressor, and E7 causes the degradation of p53, promoting cell transformation.[87] Both E6 and E7 encode epitopes that bind to HLA-A2 and a number of peptides stimulate the development of antigen specific cytolytic T cells in vitro, including E7 11-20, 12-20, 86-93 and 83-92.[88] The 86-93 E7 epitope constructed as a lipopeptide linked to a pan-DR synthetic helper sequence with a lipid tail has been used to immunize patients with advanced cervical cancer.[89] Twelve HLA-A2+ patients were treated with escalating doses of the lipopeptide subcutaneously every three weeks.[90] Toxicity was modest, consisting of local pain and inflammation. 11/12 patients had evidence of anti-FLU 58-66 reactivity as a baseline to measure immune competence. One patient was HLA-A0205, and did not react to FLU 58-66 or HPV E7 86-93. Only three out of 10 patients tested had evidence of augmented anti-HPV 86-93 CTL reactivity as shown by at least a doubling of cytokine release after incubation of re-stimulated PBMC with peptide-pulsed T2 targets. One additional patient had anti-E7 86-93 CTL activity both pre- and post-vaccine. No anti-tumor responses were seen.

At USC/Norris, we have chosen to immunize women with high grade cervical/vulvar dysplasia that are HPV 16 positive and HLA-A2(+) with peptides derived from the E7 protein. Our hypothesis is that women with a pre-malignant condition without systemic spread of disease would have a greater likelihood of mounting a T cell immune response against a viral

Table 5.3. Cytokine release assays in women receiving a HPV peptide vaccine

Patient #/ Dose level	Cytokine release after FLU stimulation			Cytokine release after E7 12-20 stimulation		
	T2 no pep	T2 FLU	T2 E7	T2 no pep	T2 FLU	T2 E7
300 ug						
10165 PRE	0	**1290**	18	5	0	**5**
10206 POST	0	**425**	0	25	0	**1140**
10193 PRE	118	**1298**	205	205	0	845
10216 POST	82	**412**	65	0	32	765
10198 PRE	365	**1432**	365	15	0	**492**
10218 POST	168	**1418**	225	0	0	**1012**
10224 PRE	213	**1307**	393	173	140	**187**
10246 POST	207	**1573**	140	573	427	**3684**

associated antigen that is moderately immunogenic. The E7 12-20 peptide was administered with IFA, and the 86-93 lipopeptide used in the aforementioned cervical cancer trial was employed in combination. 16 women have been treated with escalating doses of peptides thus far, with no grade III/IV toxicity seen, and the vaccine was well tolerated. Women with biopsy-proven high grade dysplasia had removal of their lesions delayed so that they would receive four vaccinations of the peptide vaccine, each three weeks apart. Pheresis samples were removed prior to and after the series of vaccinations for analysis of PBMC samples to test whether E7-specific immune responses had been boosted. Table 5.3 below shows immune response assays for patients receiving 300 ug of the E7 12-20 peptide/IFA, indicating that three of the four patients in that cohort displayed boosted immunity with good antigen specificity.

All 4 patients had good FLU reactivity before and after vaccinations, indicating at least a minimal degree of immunocompetence. Other endpoints of this trial included detection of virus in pap smears by DNA PCR and measurement of CIN/VIN lesions pathologically and clinically. One of the first 12 patients had complete regression of a CIN lesion after vaccination, and eight out of 12 patients had disappearance of HPV 16 by virapap PCR assay by the time of LEEP removal of their dysplastic lesion. In contrast to the PCR findings on the pap smears of the cervix/vulva, all 12 resected lesions still had detectable HPV 16 by DNA PCR indicating that the uninvolved areas might have cleared the virus, but that intact virus or integrated E7 DNA may remain in the dysplastic tissue. These data suggest that HPV specific immune responses may be seen after peptide vaccination of patient with high grade cervical/vulvar dysplasia, and that disappearance of virus from non-dysplastic tissue may correlate with boosted immunity.

In summary, melanoma and HPV peptides alone, combined with cytokines or other adjuvants, or pulsed onto dendritic cells have been shown to have clinical activity in carefully selected patients with metastatic melanoma, cervical cancer and cervical dysplasia. Evidence of boosted immunity by ELISPOT, cytokine release or chromium release of re-stimulated PBMC has been observed in several trials. In one small study of patients with resected melanoma, there was a correlation between augmented antigen-specific T-cell reactivity and relapse-free survival. In another trial of peptides with high-dose IL-2 in patients with metastatic melanoma, high response rates were seen compared with that expected for IL-2 alone, although antigen-specific responses were not observed. These data and encouraging preliminary data from ongoing trials support the notion that evidence of increased immunity and clinical benefit will result from vaccination of cancer patients with peptides.

Conclusions and Proposals for Future Trials

One of the biggest problems in the cancer vaccine field today is the abundance of good ideas that are chasing a limited supply of patients. A second critical issue is the lack of a good surrogate laboratory assay for a successful immune response. Peptides with adjuvants pulsed onto dendritic cells, DNA plasmids encoding peptide mini-genes or whole antigens and viral vectors encoding peptides/whole antigens are all promising candidates that must be tested. Dose, schedule, route and type of adjuvant are all variables that must be worked out in patients in small phase II pilot trials that should be devised based on the best information that we have from mouse models and using the best available immune assay. Although achieving a clinical response to a particular vaccine regimen in patients with measurable metastatic melanoma is one gold standard, the significant tumor burden and well documented immune suppression present in many patients with bulk metastatic melanoma suggest that they are not the ideal population in which to test new vaccines. However, it should be accepted that any phase II cancer vaccine trial must incorporate an endpoint of clinical benefit that should be correlated with assays of immune response. I propose that future vaccine trials be performed in patients with resected high risk melanoma who have a greater than 50% chance of relapse and death from metastatic disease, but a microscopic disease burden. The clinical endpoint would be time to relapse, which should be correlated with any immune response assay. Although immune assays using re-stimulated PBMC have been used, new assays using intracellular cytokine release and class I-peptide tetramer binding combined with spectratype TcR analysis followed by multi-color flow cytometry may give a more sensitive and accurate assessment of antigen specific reactivity with fresh peripheral blood cells.

An "optimal" peptide vaccine trial which I would like to propose would include HLA-A2.1 patients who have resected stage IV melanoma and are rendered free of disease. Patients with stage III disease and two or more palpable lymph nodes, extracapsular extension, or in transit metastases would also be included due to their poor outcome with surgery alone. Multiple heteroclitic peptides from tyrosinase, gp100, MART-1 and gp75 at a dose of 300 to 1000 ug for each peptide would be included with a pan-DR tetanus peptide and IFA as the adjuvant. Treatment with Flt-3 ligand for 7 days in the limb to be immunized would precede each injection, and a single dose of IL-12 would be given intradermally after vaccination. Several days later, low doses of IL-2 at "immune reconstituting" doses of 1-2 mU/M2 would be given subcutaneously for 5 days after each vaccination. A total of 8-10 injections would be given over a period of 6-12 months at increasing intervals of time. The clinical endpoints are time to relapse and overall survival. The immune endpoints are tetramer and cytokine release assays on PBMC isolated from fresh blood and propagated for 24-48 hours ex vivo in the presence of cytokines but without antigen. A similar trial has been proposed to begin at our institution in late 1999.

Abbreviations Used

PBMC:	peripheral blood mononuclear cells
MHC:	major histocompatibility locus
HLA:	human leucocyte antigen
HPV:	human papillomavirus
kD:	kilodalton
HIV:	human immunodeficiency virus
TIL:	tumor infiltrating lymphocyte
APC:	antigen presenting cell
DC:	dendritic cell
GM-CSF:	granulocyte-macrophage colony stimulating factor
IU:	international unit
R:	Rad
DTH:	delayed type hypersensitivity
PHA:	phytohemagglutinin
IFA:	incomplete Freund's adjuvant
CTL:	cytolytic T cell
Cr:	Chromium
E:T:	effector: target
cpm:	counts per minute
TcR:	T-cell receptor
PCR:	polymerase chain reaction
CR:	complete regression
PR:	partial regression
CIN:	cervical intraepithelial neoplasia
LEEP:	loop electrocautery excision procedure

References

1. Townsend ARM, Gotch RM, Davey J. Cytotoxic cells recognize fragments of the influenza nucleoprotein Cell 1985;42:457-467.
2. Townsend ARM, Rothbard J, Gotch FM et al. The epitopes of influenza nucleoprotein recognized by cytotoxic T lymphocytes can be defined with short synthetic peptides. Cell 1986;44:959-968.
3. Maryanski JL, Paola P, Coradin G et al. H-2 restricted cytotoxic T cells specific for HLA can recognize a synthetic HLA peptide. Nature 1986;324:578-579.
4. Clark WH, Elder DE, Guerry DE et al. A model predicting survival in stage I melanoma based on tumor progression J Natl Cancer Inst 1989;81:1893-1898.
5. Paladugu RR, Yonemoto RH. Bilogic behavior of thin melanomas with regressive changes Arch Surgery 1983;118:41-45.
6. Bystryn JC, Rigel D, Friedman RJ, Kopf A. Prognostic signficance of hypopigmentation in malignant melanoma. Arch Dermatology 1987;123:1053-1055.
7. Richards JM, Mehta M, Schroeder L et al. Sequential chemoimmunotherapy for metastatic melanoma J Clin Oncol 1992;9:1152-1160.
8. Nordlund JJ, Kirkwood JM, Forget BM et al. Vitiligo in patients with metastatic melanoma: A good prognostic sign. J Am Acad Derm 1983;9:689-697.
9. Anichini A, Mazzocchi A, Fossatti G, Parmiani G. Cytotoxic T lymphocyte clones from peripheral blood and from tumor sites detect intra-tumoral heterogeneity of melanoma cells: Analysis of specificity and mechanisms of interaction. J Immunol 1989;142:3692-3701.
10. Wolfel T, Klehmann E, Muller C et al Lysis of human melanoma cells by autologous cytolyitc T cell clones. Identification of human histocompatibility leucocyte antigen A2 as a restriction element for three different antigens. J Exp Med 1989; 170:797-805.
11. Topalian SL, Solomon D, Rosenberg SA. Tumor specific cytolysis by lymphocytes infiltrating human tumors. J Immunol 1989;142: 3714-3725.
12. Van Der Bruggen P, Traversari C, Chomez P et al A gene encoding an antigen recognized by cytolytic T lymphocytes on a human melanoma Science 1991;254:1643-1647.
13. Traversari C, van der Bruggen P, Luescher, I et al. A nonapeptide encoded by human gene MAGE-1 is recognized on HLA-A1 by cytolytic T lymphocytes directed against tumor antigen MZ2-E. J Exp Med 1992;176:1453-1457.
14. Gauler B, van den Eynde B, van der Bruggen P et al Human gene MAGE-3 codes for an antigen recognized on melanoma cells by autologous lymphocytes. J Exp Med 1994;179: 921-929.
15. Kawakami Y, Eliyahu S, Delgado C et al Cloning of the gene coding for a shared melanoma antigen recognized by autologous T cells infiltrating into tumor Proc Natl Acad Sci USA. 1994;96:3515-3519.
16. Coulie PG, Brichard V, Van Pel A et al A. New Gene Coding for a Differentition Antigen Recognized by Autologous Cytolytic T Lymphocytes on HLA-A2 Melanomas. J Exp Med 1994;180:35-42.
17. Kawakami Y, Eliyahu S, Sakaguchi K et al. Identification of the Immunodominant Peptides of the MART-1 Human Melanoma Antigen Recognized by the Majority of HLA-A2-restricted Tumor infiltrating lymphocytes. J Exp Med 1994;180:347-352.
18. Stevens E, Jacknin L, Robbins PF et al. The generation of tumor specific cytotoxic lymphocytes from melanoma patients using peripheral blood stimulated with allogeneic melanoma tumor cell lines: fine specificity and MART-1 melanoma antigen recognition. J Immunol 1995;154:762-767.
19. Romero P, Gervois N, Schneider J et al. Cytolytic T lymphocyte recognition of the immunodominant HLA-A*0201-restricted Melan-A/MART-1 antigenic peptide in melanoma. J Immunol 1997;159:2366-2374.
20. Rivoltini L, Kawakami Y, Sakaguchi K et al. Induction of Tumor Reactive CTL from Peripheral Blood and Tumor-Infiltrating Lymphocytes of Melanoma Patients by In vitro Stimulation with an Immunodominant Peptide of the Human Melanoma Antigen MART-1 J Immunol 1995;154:2257-2265.
21. Cole DJ, Weil DP, Shilyansky J et al. Characterization of the Functional Specificity of a Cloned T-Cell Receptor Heterodimer Recognizing the MART-1 Melanoma Tumor Antigen Cancer Res 1995;55:748-752.

22. Sensi M, Salvi S, Castelli C et al. T cell receptor structure of autologous melanoma reactive cytotoxic T lymphocyte (CTL) clones: Tumor infilrating lymphocytes overexpress in vivo the TCR chain sequence used by an HLA-A2 restricted and melanocyte lineage-specific CTL clone. J Exp Med 1993;178:1231-1248.
23. Hunt DF, Henderson RA, Shabanowitz J et al. Chracterization of peptides bound to the class I molecule HLA-A2.1 by mass spectrometry. Science. 1992;255:1261-1264.
24. Storkus WJ, Zeh HJ, Maeurer MJ et al. 1993 Identification of human melanoma peptides recognized by class I restricted tumor infiltrating lymphocytes. J Immunol 1993; 151:3719-3726.
25. Cox AL, Skipper J, Chen Y et al. Identification of a peptide recognized by five melanoma- specific human cytotoxic T cell lines. Science. 1994; 264:716-719.
26. Loftus DJ, Castelli C, Clay TM et al. Identification of epitope mimics recognized by CTL reactive to the melanoma/melanocyte-derived peptide MART-1(27-35). J Exp Med 1996; 184:647-657.
27. Marincola FM, Rivoltini L, Salgaller ML et al. Differential anti-MART-1/MelanA CTL activity in peripheral blood of HLA-A2 melanoma patients in comparison to healthy donors: Evidence of in vivo priming by tumor cells. J Immuno 1996; 19:266-77.
28 Romero P, Dunbar PR, Valmori D et al. Ex vivo staining of metastatic lymph nodes by class I major histocompatibility complex tetramers reveals high numbers of antigen-experienced tumor specific cytolytic T lymphocytes. J Exp Med 1998; 188 1641-1650.
29. Fleischhauer K, Tanzarella S, Wallny HJ et al. Multiple HLA-A alleles can present an immunodominant peptide of the human melanoma antigen Melan-A/MART-1 to a peptide-specific HLA-A*0201+ cytotoxic T cell line. J Immunol 1996; 157:787-97.
30. Rivoltini L, Loftus DJ, Barracchini K et al. Binding and presentation of peptides derived from melanoma antigens MART-1 and glycoprotein-100 by HLA-A2 subtypes. Implications for peptide-based immunotherapy. J Immunol 11996; 56:3882-3891.
31. Schneider J, Brichard V, Boon T et al. Overlapping peptides of melanocyte differentiation antigen Melan-A/MART-1 recognized by autologous cytolytic T lymphocytes in association with HLA-B45.1 and HLA-A2.1. Int J Cancer 1998; 75:451-8.
32. Marincola FM. Stringent allele/epitope requirements for MART-1/Melan A immunodominance: implications for peptide-based immunotherapy. J Immunol 1998; 161:877-889.
33. Jimenez M, Maloy WL, Hearing VJ. Specific identification of an authentic clone for mammalian tyrosinase. J Biol Chem 1989; 264:3397-3403.
34. Adema GJ, DeBoer AJ, Vogel AM et al. Molecular characterization of the melanocyte lineage specific antigen gp100. J Biol Chem 1994; 269:20126-20133.
35. Bakker ABH, Schreurs WJ, de Boer AJ et al. Melanocyte lineage specific antigen gp100 is recognized by melanoma-derived tumor-infiltrating lymphocytes. J Exp Med 1994; 179:1005-1011.
36. Kawakami Y, Eliyahu S, Delgado C et al. Identification of a human melanoma antigen recognized by tumor-infiltrating lymphocytes associated with in vivo tumor rejection. Proc Natl Acad Sci USA 1994; 91:6458-6462.
37. Bakker ABH, Schreurs MWJ, Tafazzul G et al. Identification of a novel peptide derived from the melanocyte specific gp100 antigen as the dominant epitope recognized by an HLA-A2.1 restricted anti-melanoma CTL line. Int J Cancer 1995; 62:97-102.
38. Skipper JC, Kittlesen DJ, Hendrickson RC et al. Shared epitopes for HLA-A3 restricted melanoma-reactive human cytolytic T lymphocytes include a naturally processed epitope from pMel17/gp100. J Immunol 1996; 157:5027-5033.
39. Kawashima I, Tsai V, Southwood S, Takesako K et al. Idenitification of gp100-derived melanoma-specific cytolytic T lymphocyte epitopes restricted by HLA-A3 supertype molecules by primary in vitro immunization with peptide-pulsed dendritic cells. Int J Cancer 1998; 78:518-524.
40. Brichard V, Van Pel A, Wolfel T et al. The tyrosinase gene encodes for an antigen recognized by autologous cytolytic T lymphocytes on HLA-A2 melanomas. J Exp Med 1993; 178:489-495.
41. Wolfel T, Van Pel A, Brichard V et al Two tyrosinase nonapeptides recognized on HLA-A2 melanomas by autologous cytolytic T lymphocytes. Eur J Immunol 1994; 24:759-764.
42. Skipper JCA, Hendrickson RC, Gulden PH et al. An HLA-A2 restricted tyrosinase antigen on melanoma cells results from post-translational modification and suggests a novel processing pathway for membrane proteins. J Exp Med 1996; 183:527-534.

43. Wang RF, Robbins PF, Kawakami Y et al. Identification of a gene encoding a melanoma tumor antigen recognized by HLA-A31 restricted tumor-infiltrating lymphocytes. J Exp Med 1995; 181:799-806.
44. Wang R-F, Parkhurst MR, Kawakami Y et al. Utilization of an alternative open reading frame of a normal gene in generating a human cancer antigen. J Exp Med 1996; 183:1131-1138.
45. Wang R-F, Appella E, Kawakami Y et al. Identification of TRP-2 as a human tumor antigen recognized by cytotoxic T lymphocytes. J Exp Med 1996; 184:2207-2214.
46. Jaeger E, Chen Y-T, Drijfhout JW, Karbach J et al. Simulataneous humoral and cellular immune response against cancer testis antigen NY-ESO-1: Definition of human histocompatibility leucocyte antiegn (HLA)-A2 binding peptide epitopes. J Exp Med 1998; 187:265-274.
47. Wang R-F, Johnston SL, Zeng G et al. A breast and melanoma-shared tumor antigen: T cell responses to antigenic peptides translated from different open reading frames. J Immunol 1998; 161:3596-3606.
48. Wolfel T, Hauer M, Schneider J et al. A p16INK4a-insensitive CDK4 mutant targetted by cytotoxic T lymphocytes in a human melanoma. Science 1995; 269:1281-1285.
49. Robbins PF, El-Gamil M, Li YF et al. A mutated beta-catenin gene encodes a melanoma specific antigen recognized by tumor-infiltrating lymphocytes. J Exp Med 1996; 183:1185-1192.
50. Coulie PG, Lehmann F, Lethe B et al. A mutated intron sequence codes for an antigenic peptide recognized by cytolytic T lymphocytes on a human melanoma. Proc Natl Acad Sci USA 1995; 92:7976-7980.
51. Marchand M et al Tumor regression responses in melanoma patients treated with a peptide encoded by MAGE-3. Int J Cancer 1995; 63:883-885.
52. Weber JS, Hua FL, Spears L et al. A Phase I Trial of a HLA-A1 Restricted MAGE-3 Epitope Peptide with Incomplete Freund's Adjuvant in Patients with Resected High-Risk Melanoma. J Immunother 1999; in press.
53. Jaeger E, Bernhard H, Romero P et al. Generation of cytotoxic T cell responses with synthetic melanoma-associated peptide in vivo: Implications for tumor vaccines with melanoma-associated antigens. Int J Cancer 1996; 66:162-170.
54. Jaeger E, Ringhoffer M, Dienes H-P et al. Granulocyte macrophage colony stimulating factor enhances immune responses to melanoma associated peptides in vivo. Int J Cancer 1997; 67:54-62.
55. Salgaller MM, Marincola FM, Cormier JN et al. Immunization against epitopes in the human melanoma antigen gp100 following patient immunization with synthetic peptides. Cancer Res 1996; 56:4749-4757.
56. Salgaller ML, Afshar A, Marincola FM et al. Recognition of multiple epitopes of the human melanoma antigen gp100 by peripheral blood lymphocytes stimulated in vitro with synthetic peptides. Cancer Res 1995; 55:4972-4977.
57. Cormier JN, Salgaller ML, Prevette T et al. Enhancement of cellular immunity in melanoma patients immunized with a peptide from MART-1/Melan A. Canc J Scient Amer 1997; 3:37-44.
58. Parkhurst MR, Salgaller ML, Southwood S et al. Improved induction of melanoma reactive CTL with peptides from melanoma antigen gp100 modified at HLA-A0201 binding residues. J Immunology 1996; 157:2536-2548.
59. Valmori D, Fonteneau JF, Lizana CM et al. Enhanced generation of specific tumor-reactive CTL in vitro by selected Melan-A/MART-1 immunodominant peptide analogues. J Immunol 1998; 160:1750-8.
60. Rosenberg SA, Yang JC, Schwartzentruber DJ et al. Immunologic and therapeutic evaluation of a synthetic peptide vaccine for the treatment of patients with metastatic melanoma. Nat Med 1998; 4:321-327.
61. Rosenberg SA, Lotze MT, Yang JC et al. Experience with the use of high-dose interleukin-2 in the treatment of 652 cancer patients. Ann Surgery 1990; 210:474-480.
62. Parkinson DR, Abrams JS, Wiernik PH et al. Interleukin-2 therapy in patients with metastatic melanoma: A phase II study. J Clin Oncol 1990; 8:650-1659.
63. Lewis JJ, Janetski S, Wang S et al. Phase I trial of vaccination with tyrosinase peptide plus QS-21 in melanoma. Proc Amer Soc Clin Oncol 1998;17:1650 428a.
64. Mukherji B, Chakraborty NG, Yamasaki S, Okino T et al. Induction oif antigen-specific cytolytic T cells in situ in human melanoma by immunization 1995; 92:8079-8082.

65. Rosenberg SA, Zhou Y, Yang JC et al. Immunizing patients with metastatic melanoma using a recombinant adenovirus encoding MART-1 or gp100 melanoma antigens. J Natl Canc Inst 1998; 90:1894-1900.
66. Steinman RM. The dendritic cell system and its role in immunogenicity. Annu Rev Immunol 1991; 9:271-290.
67. Van Voorhis WC, Hair SL, Steinman RM et al. Human dendritic cells: Enrichment and characterization from human blood. J Exp Med 1982; 155:1172-1183.
68. Steinman RM, Witmer MD. Lymphoid dendritic cells are potent stimulators of the primary mixed leukocyte response in mice. Proc Natl Acad Sci USA 1978; 75: 5132-5136.
69. Inaba K., Metlay JP, Crowley MT et al. Dendritic cells pulsed with protein antigens in vitro can prime antigen-specific, MHC-restricted T cells in situ. J Exp Med 1990; 172: 631-640.
70. Inaba K, Young JW, Steinman RM. Direct activation of CD8+ cytolytic T lymphocytes by dendritic cells. J Exp Med 1987; 166:182-194.
71. Macatonia SE, Taylor PM, Knight SC et al. Primary stimulation by dendritic cells induces antiviral proliferative and cytotoxic T responses in vitro. J Exp Med 1988; 169:1255-1264.
72. Sallustro F, Lanzavecchia A. Efficient presentation of soluble antigen by cultured human dendritic cells is maintained by granulocyte/macrophage colony-stimulating factor plus interleukin 4 and downregulated by tumor necrosis factor. J Exp Med 1994; 179:1109-1122.
73. Romani N, Gruner S, Brang D et al. Proliferating dendritic cell progenitors in human blood. J Exp Med 1994; 180: 83-93
74. Van Elsas A, Van der Burg SH, Van Der Minne CE et al. Peptide-pulsed Dendritic cells induce tumoricidal cytolytic T cells from healthy donors against stable HLA-A0201 binding peptides from Melan A/MART-1 self antigen. Eur J Immunol 1996; 26:1683-1689.
75. Nestle FO, Alijagic S, Gilliet M et al. Vaccination of melanoma patients with peptide- or tumor lysate-pulsed dendritic cells. Nature Medicine 1998; 4:378-332.
76. Chen YT, Stockert E, Jungbluth A et al. Serological analysis of Melan-A (MART-1), a melanocyte-specific protein homogeneously expressed in human melanomas. Proc Natl Acad Sci USA. 1996; 93:5915-5919.
77. Marincola FM, Hijazi YM, Fetsch P et al. Analysis of expression of the melanoma-associated antigens MART-1 and gp100 in metastatic melanoma cell lines and in in situ lesions. J Immunotherapy 1996; 19:192-205.
78. Kageshita T, Kawakami Y, Hirai S et al. Differential expression of MART-1 in primary and metastatic melanoma lesions. J Immuno 1997; 20:460-465.
79. Fetsch PA, Cormier J, Hijazi YM. Immunocytochemical detection of MART-1 in fresh and paraffin-embedded malignant melanomas. J Immuno 1997; 20:60-64.
80. de Vries TJ, Fourkour A, Wobbes T et al. Heterogeneous expression of immunotherapy candidate proteins gp100, MART-1, and tyrosinase in human melanoma cell lines and in human melanocytic lesions. Cancer Res 1997; 57:3223-3229.
81. Dalerba P, Ricci A, Russo V et al. High homogeneity of MAGE, BAGE, GAGE, tyrosinase and Melan-A/MART-1 gene expression in clusters of multiple simultaneous metastases of human melanoma: Implications for protocol design of therapeutic antigen-specific vaccination strategies. Int J Cancer. 1998; 77:200-4.
82. Cormier JN, Abati A, Fetsch P et al. Comparative analysis of the *in* vivo expression of tyrosinase, MART-1/Melan-A, and gp100 in metastatic melanoma lesions: implications for immunotherapy. J Immun 1998; 21:27-31.
83. Maeurer MJ, Gollin SM, Martin D et al. Tumor escape from immune recognition: lethal recurrent melanoma in a patient associated with downregulation of the peptide transporter protein TAP-1 and loss of expression of the immunodominant MART-1/Melan-A antigen. J Clin Invest 1996; 98:1633-1641.
84. Jaeger E, Ringhoofer M, Karbac J et al Inverse relationship of melanoma differentiation antigen expression in melanoma tissues and CD8+ cytotoxic T cell responses: Evidence for immunoselection of antigen loss variants in vivo. Int J Cancer 1996; 66: 470-476.
85. Fu YS, Huang I, Beaudenon S et al. Correlative study of HPV, DNA, histopathology and morphometry in cervical condyloma and intraepithelial neoplasia. Int J Gynecol Pathol 1988; 7: 297-303.

86. Crook T, Morgenstern JP, Crawford L et al. Continuous expression of the HPV E7 protein is required for the maintenance of the transformed phenotype of cells co-transformed by HPV 16 plus EJ-ras. EMBO J 1989; 8:513-519.
87. Munger K, Scheffner M, Huibregste JM et al. Interactions of HPV E6 and E7 proteins with tumor suppressor gene products. Cancer Surv 1992; 12:197-217.
88. Ressing ME, Sette A, Brandt R et al. Human CTL epitopes encoded by human papillomavirus type 16 E6 and E7 identified through in vivo and in vitro immunogenicity studies of HLA-A0201 binding peptides. J Immunol 1995; 154:5934-5943.
89. Alexander M, Salgaller ML, Celis E et al. Generation of tumor-specific CTL from the peripheral blood of cervical cancer patients by in vitro stimulation with a synthetic HPV-16 E7 epitope. Am J Ob Gyn 1996; 175:1586-1593.
90. Steller MA, Gursk KJ, Murakami M et al. Cell mediated immunologic responses in cervical and vaginal cancer patients immunized with a lipidated epitope of human papillomavirus type 16 E7. Clin Can Res 1998; 4:2103-2109.

Chapter 6

Carcinoembryonic Antigen (CEA) Peptides and Vaccines for Carcinoma

Jeffrey Schlom

Introduction

This Chapter addresses the current status of the development of recombinant vaccines employing carcinoembryonic antigen (CEA) as the target antigen. Included is an over view of preclinical studies and the pros and cons of CEA as a vaccine target. Several CEA peptides that are recognized by human T cells have now been identified. This Chapter will elucidate the identification and utilization of these CEA peptides in (a) vaccine design and development, (b) the assay of human immune responses to an array of CEA vaccines, and (c) the characterization of human T-cell lines directed against CEA. Future directions in vaccine design and development, including the use of agonist CEA epitopes, also are discussed.

The CEA Gene Family

The carcinoembryonic antigen (CEA) gene family consists of 29 genes which are located within on the long arm of chromosome 19. CEA was first described in 1965.[1,2] The isolation of cDNAs or genes for CEA family members led to their identification in 1986 as members of the immunoglobulin (Ig) supergene family.[3-9] All CEA family members are glycoproteins composed of an N-terminal Ig variable region-like domain, followed by 0,2,3,4, or 6 Ig constant region-like domains of subtype A or B, and terminated by a processed hydrophobic C-terminal domain. The CEA family may be subdivided into two groups based on sequence comparisons. The first is CEA and the CEA cross-reacting molecules, including nonspecific cross-reacting antigen (NCA), biliary glycoprotein (BGP) and CGM-6; the second consists of the pregnancy-specific glycoproteins. The CEA subgroup is subdivided further by structural characteristics into those members attached to the outer cell membrane by glycophosphatidyl inositol (CEA, NCA, CGM-6) and those that have both transmembrane and cytoplasmic domains (BGP splice variants). Studies have shown that CEA mediates $Ca2^{+}$- and temperature- independent cell aggregation.[10] Some studies have demonstrated that CEA is an intercellular adhesion molecule. However, other studies have shown that CEA may act as a signal protein that inhibits intercellular contact in the malignant process.[11] In terms of its tissue distribution, biochemistry and molecular structure, CEA is one of the most well-characterized tumor associated antigens (TAA). CEA is extensively expressed in the vast majority of human colorectal, gastric and pancreatic carcinomas, in approximately 50% of breast cancers and in 70% of non-small-cell lung cancers. To a lesser extent, CEA is also expressed on normal colon epithelium and in some fetal tissue. The relative tissue specificity of this antigen thus makes CEA a potential target antigen for active specific immunotherapy. In addition, since CEA is considered an adhesion molecule, it might play an important role in the metastatic process by mediating attachment of tumor cells to normal cells. Thus, active immunotherapy targeted to CEA might be particularly beneficial in preventing metastasis.

At the amino acid level, CEA shares approximately 70% homology with NCA, which is found on normal granulocytes.[12-14] This immediately raises some provocative questions. Will this situation render patients tolerant to CEA and thus incapable of inducing CEA immune responses? If immune responses are generated, will this lead to autoimmune responses against normal tissues, or will the quantitative and/or qualitative differences in expression of CEA in tumor vs. normal tissues make this inconsequential? Can immunodominant CEA peptides that share little or no homology to CEA-related sequences found on normal tissues be identified and recognized by human T cells? This review will attempt to address some of these issues.

Quantitative Analyses of CEA Expression

In a recent study, the degree of CEA expression in colorectal carcinomas, adjacent "histologically normal" tissue, benign colon lesions, and colonic mucosa from healthy individuals was evaluated by quantitative RIA.[15] Tissues and sera from 110 patients diagnosed with primary colorectal carcinoma, 20 patients with benign colorectal disease, and 31 healthy donors were subjected to quantitative CEA analysis. Multiple samples from tumor lesions and autologous, histologically normal mucosa (10 cm from the tumor) were obtained at the time of surgery (cancer patients) or endoscopy (benign patients and healthy volunteers). CEA content was measured in protein extracts obtained from these tissues using a quantitative RIA method. An arbitrary limit of "normality" for CEA content was established as 300 ng/mg of protein. Using this cut-off, 94.5% of carcinomas had elevated CEA levels (Fig. 6.1). A statistically significant difference between CEA content in tumor lesions and in histologically normal mucosa from cancer patients was observed (p= -0.001). Moreover, CEA content was statistically higher in the normal mucosa from cancer patients than in that from healthy donors (p= 0.005). No statistical correlation between CEA content in carcinoma tissues and serum CEA levels (r= 0.195, p= 0.13) was found.[15] Therefore, in considering diagnosis or therapy with anti-CEA monoclonal antibodies (MAb) for colorectal carcinoma patients or potential therapies with anti-CEA recombinant vaccines, serum CEA levels should not be taken as the only indication of CEA expression in tumor lesions.

Although fewer numbers of benign lesions were available for analysis, a trend in quantitative CEA expression appears to be evident.[15] All six hyperplastic lesions analyzed showed CEA levels below the 300 ng/mg protein limit. However, values were between 300 and 600 ng/mg protein for all eight tubulovillous adenomas with low or moderate dysplasia, and values exceeded 600 ng/mg protein for all six tubulovillous adenomas with severe dysplasia (Fig. 6.1). The fact that CEA is overexpressed in potentially premalignant lesions leaves open the future possibility that CEA vaccines might be given to patients with "benign" disease to eliminate these lesions and, by extension, the risk of carcinoma development.

The quantitative differences in CEA expression in carcinoma vs. normal tissues has also been demonstrated by the clinical use of anti-CEA MAb. Numerous studies have now demonstrated the selective targeting of radiolabeled MAb to carcinoma, while no targeting to normal tissues was observed.[16-18]

Animal Models

Much thought needs to be given to the use of appropriate model systems to determine the immunogenicity of a given antigen or epitope for humans. At first glance, rodent models may appear attractive. However, since human class I and class II MHC alleles are distinct from

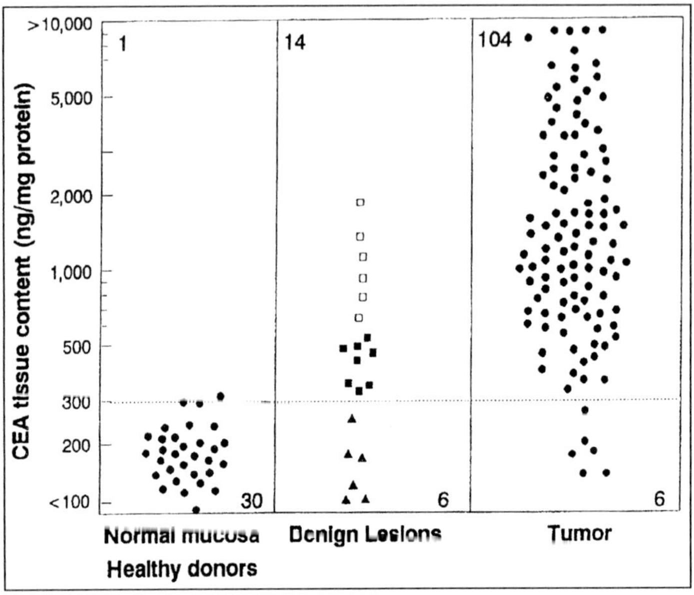

Fig.6.1.CEA content in colorectal tissues from healthy donors (n=31), in patients diagnosed with benign lesions (n=20) [i.e., hyperplastic polyps (▲); tubulovillous adenomas with low or moderate dysplasia (■); and tubulovillous adenomas with severe dysplasia (□)], and in patients diagnosed with colorectal cancer (n=110) (15).

those of any other species, it is inappropriate to try to define whether immunogenicity of a human TAA in a murine model will predict immunogenicity in a human. This is also true in the use of transgenic (Tg) mice containing a human tumor-antigen transgene. Such Tg mice still have murine MHC class I and II alleles as well as murine TCR and, thus, cannot predict immunogenicity in humans. Certain Tg models are very useful to help define the principles of the immunogenicity of a "self-antigen," such as CEA.[19,20] However, it is unclear how the specific Tg murine model reflects the human in terms of degree of expression of the gene in question during embryonic development and actual levels of expression in normal tissues and tumor tissues.[21] This is supported by recent experiments and theories further examining the concept of "self" versus "non-self".[22-26]

A Tg mouse has recently been developed bearing the human MHC class I HLA-A2 transgene. This has been shown to be an extremely useful and important model. However, studies have shown that a correlation does not always exist between an immune response in the HLA-A2 Tg mouse and human in vitro immune responses for some epitopes.[27] Moreover, the HLA-A2 Tg mouse still contains a murine TCR repertoire.[27] Thus, one is faced with the dilemma of finding an appropriate in vivo model to determine the immunogenicity of a given human antigen and/or epitope. One point of view is that a small Phase I clinical trial to define immunogenicity of a given antigen or epitope is without equal. This, however, should absolutely not diminish the extensive use of animal models to investigate basic concepts such as advantages or disadvantages of vaccine delivery methodologies, mechanisms of T-cell costimulation, use of cytokines to enhance immune responses, the advantage of diversified prime and boost protocols, and principles for the optimal utilization of adoptive immunotherapy.

Initial preclinical studies compared the relative degree of potency and type of immune response achieved in mice that received different CEA vaccines. These vaccines included native CEA protein obtained from human metastases, recombinant CEA protein expressed in baculovirus, recombinant vaccinia virus expressing the human CEA gene (rV-CEA), an anti-idiotype (Id) MAb directed against an anti-CEA MAb, a CEA polynucleotide vaccine, and a replication-defective avipox virus recombinant expressing CEA.[13,28-36] The advantage of employing cytokines such as IL-2 and granulocyte-macrophage colony-stimulating factor (GM-CSF) in CEA vaccines in preclinical models has also been demonstrated.[37] In addition, the advantage of the use of T-cell costimulatory molecules in anti-CEA vaccines has been determined by admixing rV-CEA with recombinant vaccinia viruses expressing the B7.1, B7.2 or CD70 costimulatory genes or by placing both the CEA and B7.1 gene on the same vaccinia vector.[38-40] Finally, the preclinical models have shown the advantage of using diversified prime and boost strategies for CEA vaccines. For example, priming with rV-CEA and boosting with either avipox-CEA or CEA protein was shown to be more efficacious than using one of these vaccines alone.[30, 41]

Recently, Tg mice that express CEA (CEA.Tg) as a self-antigen with a tissue distribution similar to that of humans have been developed.[42,43] A recent study has compared the immune responsiveness of the CEA.Tg mice to CEA protein administered in adjuvant with the presentation of CEA via a recombinant vaccinia virus (rV-CEA).[44] The nonvaccinated CEA.Tg mice were unresponsive to CEA, as defined by the lack of detectable CEA-specific serum antibodies and the inability to prime an in vitro splenic T-cell response to CEA. Furthermore, vaccination with whole CEA protein in adjuvant failed to elicit either anti-CEA IgG titers or CEA-specific T-cell responses. Only weak anti-CEA IgM antibody titers were found. In contrast, CEA.Tg mice immunized with rV-CEA generated relatively strong anti-CEA IgG antibody titers and demonstrated evidence of Ig class switching. Those same mice also developed T_H1-type CEA-specific $CD4^+$ responses and CEA peptide-specific cytotoxicity. The ability to generate CEA-specific host immunity correlated with protection against challenge with CEA-expressing tumor cells. This protection against tumor growth was accomplished with no apparent immune response directed at CEA-positive normal tissues as defined by detailed histochemical and immunohistochemical analyses. The results demonstrate the ability to generate an effective antitumor immune response to a self tumor-antigen by immunizing with a recombinant vaccinia virus.[44] CEA.Tg mice should be an excellent experimental model in which to study the effects of more aggressive immunization schemes.

Identification of Human CEA-Specific T-Cell Epitopes

Since the entire amino acid sequence of human CEA is known and human HLA class I-A2 consensus motifs have been described, studies were undertaken to identify a series of peptides that would potentially bind class I-A2 molecules.[45,46] A2 was chosen since it is the most common HLA class I molecule, being represented in approximately 50% of North American Caucasians and 34% of African-Americans.[47] The peptide sequence of CEA was thus examined for matches to the consensus motifs for HLA-A2 binding peptides. Moreover, peptides were selected only if their sequence diverged sufficiently from the CEA-related NCA and BGP sequences. The amino acid sequence of human CEA (GeneBank Accession #M17303) was scanned using a predictive algorithm that combines a search for anchor residues with numerical assignments to all residues at all positions.[48] Six peptides ranging in length from 9 to 11 amino acids contained the HLA-A2 binding motif of leucine or isoleucine at position 2, and valine or leucine at the C-terminal (Table 6.1). They were given the designation Carcinoembryonic Antigen Peptide (CAP). Another peptide, designated CAP-7, also possessed the motif for binding to HLA-A3.[49] All peptides were selected to have minimal homology to the parallel regions of NCA and BGP after optimal alignment of the latter sequences with CEA.

The T2 cell-binding assay has been used to predict human HLA-A2 consensus motifs.[50] In this assay, the binding of an appropriate peptide results in the upregulation of surface HLA-A2 on the T2 cells, which can be qualified via FACScan using an anti-HLA-A2 antibody. As seen in Table 6.1, the CEA peptides (CAP1 through CAP6) scored positive for T2 binding. The order of T2 cell-peptide binding did not always correspond to the predictive algorithm.[48] Since peptide 571-579 (designated CAP1) demonstrated the highest level of T2 binding, the peptide reflecting the NCA analog (the corresponding NCA peptide obtained after optimal alignment of NCA and CEA) was also synthesized and tested; this peptide, designated NCA-1, showed only background binding to T2 cells (Table 6.1). The low level of binding was consistent with the fact that an amino acid substitution in NCA had abolished one of the A2 anchor residues (Arg for Leu at position 2).

Phase I studies are primarily designed as toxicity studies and are traditionally conducted in patients with advanced disease. Of course, these patients comprise the least desirable population in which to examine the efficacy of a vaccine to initiate T-cell responses, since defects in TCR zeta chains have been observed.[51-53] Initial immunization with rV-CEA resulted in a clinical "take" as measured by the local erythematous reaction in 23 of 26 patients. Four of seven patients in Cohort 1 (lowest dose level) and all 19 patients in Cohorts 2 and 3 demonstrated erythema. Thus, despite the fact that all patients had previously received the smallpox vaccine, most patients displayed the classical acute manifestations of the vaccinia immunization, confirming the capacity to be reimmunized. Lesion size correlated with the dose of rV-CEA given but was reduced with subsequent vaccinations at the same dose level. At all dose levels, no toxicity was apparent other than that normally observed with the administration of a smallpox vaccine.

Establishment of T-Cell Lines to CEA Peptides

In an attempt to establish T-cell lines from patients who had received the rV-CEA construct, PBMC were obtained from several patients with the HLA-A2 allele pre- and postvaccination. PBMC were alternately pulsed with CEA peptides and interleukin (IL)-2 in the presence of autologous PMBC as antigen presenting cells (APC).[14] T-cell lines could be established from five of five HLA- A2 patients from postvaccination PBMC; these T cells were cytotoxic for T2 cells when pulsed with the CAP1 peptide. The cell lines from these patients were primarily

Table 6.1. Binding of CEA peptides to the HLA class I-A2 molecule*

Peptide	Amino Acid Position in CEA	Sequence	T2 Binding
CAP-1	571-579	YLSGANLNL	561
CAP-2	555-579	VLYGPDTPII	515
CAP-3	87-89	TLHVIKSDLV	480
CAP-4	1-11	KLTIESTPFNV	441
CAP-5	345-354	TLLSVTRNDV	405
CAP-6	19-28	LLVHNLPQHL	381
NCA-1	571-579	YRPGENLNL	252
Positive control	-	ALAAAAAAV	632
No peptide	-	-	280

$CD8^+$ or $CD8^+/CD4^+$ double positive. T-cell lines could not be established from these same patients when prevaccination PBMC were used.

At the time, no studies had demonstrated that human APC could endogenously process the CEA "self" molecule in a manner to bind class I molecules for presentation at the cell surface. Epstein-Barr virus-transformed B (EBV-B) cells of patients were transduced with the human CEA gene using a retroviral vector. Unlike the untransfected control cells, only EBV-B cells could serve as targets for lysis by the T cells generated by CAP1 from the same patients. The question also remained as to whether human carcinoma cells could act in the same manner as APC and could serve as potential targets for T-cell lysis. Studies revealed that carcinoma cells that were both CEA- and HLA-A2-positive could indeed be lysed by such T cells, while tumor cells that were either HLA-A2- or CEA-negative could not be lysed.[14] These studies thus demonstrated for the first time that CEA could be processed by professional APC and, just as important, by human carcinoma cells for lysis by epitope-specific T cells. To further show the MHC-restricted nature of this lysis, non-HLA-A2 human carcinoma cells were infected with either wild-type vaccinia virus or a recombinant vaccinia virus containing the human HLA-A2 gene. Only the cells infected with the HLA-A2 transgene and expressing CEA were susceptible to lysis.[14]

Another CEA peptide, designated CAP2 (amino acid positions 555-563) has been employed to successfully generate CEA-specific CTL lines from a patient immunized with rV-CEA.[54]

Several groups have now identified CEA T-cell epitopes for a variety of HLA alleles. One study identified a set of 34 CEA-specific peptides that fit with a specified HLA-A*0301-binding motif and a set of six peptides with high binding affinity to this allele.[55] These peptides can be regarded as potential CTL epitopes.

Some 73 CEA-derived peptides that fulfill the HLA-A*0201 motif have been described.[56] Peptides with a high binding affinity and a low peptide-MHC dissociation rate were subsequently tested for their immunogenicity in HLA-A*0201K^b Tg mice. One CEA-derived peptide was shown to induce peptide-specific CTL in these mice.

Several CEA-specific CTL epitopes and CEA epitope analogs have now been identified.[57] Of a total of 18 motif-containing peptides (9- and 10-mer) tested for HLA-A2.1 binding, nine bound to HLA-A2.1 with an IC_{50} of 500 nM or less. Interestingly, the second highest binder, peptide CEA[9_{605}] was previously reported as a CTL epitope in cancer patients after vaccination with a recombinant CEA construct.[14]

The HLA-A2.1-binding CEA peptide CAP1 was studied for its capacity to elicit CTL using dendritic cells (DC) as APC. It was demonstrated that CTL could be generated in vitro employing the CAP1 peptide and DC obtained from PBMC of either healthy individuals or cancer patients.[58] These CTL were capable of killing both CAP1-pulsed targets and CEA-expressing tumor cells. CTL have also been generated using CEA mRNA-transfected DC from PBMC of healthy individuals and cancer patients as APC.[59] In other studies, three of five peptides tested were found to stimulate CTL responses that recognized peptide-sensitized target cells.[57] Moreover, after restimulation with antigen and APC, killing of CEA-expressing tumor cells was demonstrated for CEA [9_{605}]- and CEA[9_{691}]- specific CTL. The CTL line reactive with peptide CEA[9_{691}] was cloned and studied further in terms of specificity and recognition of various CEA-expressing tumor cells. Cloned CTL specific for CEA[9_{691}] had high, specific lytic activity toward tumor target cells (colon and gastric) that express HLA-A2 and CEA. Two analog peptides from CEA[9_{24}] were prepared and tested for binding to purified HLA-A2.1 molecules. These two analogs differed by the residue incorporated in anchor position 2. The analog peptides bound to purified HLA-A2.1 molecules with an approximately tenfold and fortyfold increased affinity as compared to the natural sequence. It was notable that although peptide CEA[9_{24}] was not capable (according to the experimental protocol) of triggering

CTL responses in vitro, both peptide analogs were immunogenic in terms of CTL induction. CTL lines induced by both analogs specifically recognized and killed SW403 colon cancer cells expressing CEA and HLA-A2.

The HLA-A24 allele occurs in 60% of the Japanese population and frequently in the Caucasian population. CEA-encoded HLA-A24 binding peptides were tested for their capacity to elicit anti-tumor CTL in vitro.[60] CD8$^+$ T lymphocytes from PBMC of a healthy donor and autologous peptide-pulsed DC were used as APC. This approach resulted in the identification of two peptides, QYSWFVNGTF and TYACFVSNL, which were capable of eliciting CTL lines that lysed tumor cells expressing HLA-A24 and CEA. The cytotoxicity to tumor cells by the CTL lines was antigen-specific since it was inhibited by peptide-pulsed cold target cells as well as by anti-class I MHC and anti-CD3 MAb. Induction of CEA and HLA-A24-specific CTL has been demonstrated by culturing human PBMC on formalin-fixed autologous adhesive PBMC loaded with CEA-bound latex beads.[61] The CTL killed CEA-producing tumor cells. Nine other A24 peptides were active to a lesser degree. The lysis observed was shown to be MHC-restricted.[61] The identification of these novel CEA epitopes for CTL also offers the opportunity to design and develop epitope-based immunotherapeutic approaches for treating both HLA-A24$^+$ and HLA-A2 patients who have tumors that express CEA.

CEA epitopes have also been identified employing an anti-Id MAb to a CEA MAb as immunogen.[33,62] The cDNA encoding the variable heavy and light chains of 3H1 were cloned and sequenced to study the cellular immunity invoked by the 3H1 anti-Id MAb, and the amino acid sequence of the heavy and light chains was deduced.[62] Several regions of homology in 3H1 heavy and light chain variable regions and framework, with that of CEA, were found.[62] A number of peptides were synthesized and used to stimulate PBMC from patients immunized with the 3H1 anti-Id MAb. Two peptides (designated LCD-2 and CEA-B) were identified by strong stimulation responses in 10 of 21 patients. No correlation with class I MHC was observed, and responding T cells were predominantly CD4$^+$ and of the Th_1-type.[62]

CTL lines have now been generated against defined peptides of a range of human TAA. One of the potential uses of these epitope-specific CTL is in adoptive transfer immunotherapy. This is a modality, however, that will require long-term in vitro culture of CTL. To date, little has been reported concerning the phenotypic stability of human epitope-specific CTL to self-antigens such as CEA as a consequence of long-term in vitro propagation via peptide stimulation. The serial phenotypic characterization of a CTL line directed against the immunodominant CEA epitope CAP1 has been reported.[63] This CTL line was derived from PBMC of a patient with metastatic carcinoma who had been vaccinated with rV-CEA. The CTL line was analyzed through 20 in vitro cycles of stimulation with CAP1 peptide and IL-2 in the presence of autologous APC. The CTL line was shown to be phenotypically stable in terms of high levels of cytokine (interferon [IFN], tumor necrosis factor, and GM-CSF) production, expression of homing-adhesion molecules, ability to lyse peptide-pulsed targets, and ability to lyse human carcinoma cells endogenously expressing CEA in a MHC-restricted manner. V T-cell receptor (TCR) gene usage was also analyzed.[63] These studies thus present a rationale for the use of long-term cultured epitope-specific human CTL that are directed against a human self-TAA for potential adoptive transfer immunotherapy protocols.

Identification of a CEA Enhancer Agonist CTL Peptide

One strategy to enhance the immunogenicity of a self-antigen such as CEA, or indeed any antigen, would be to slightly modify a known CTL epitope such as CAP1. The key, of course, is to ensure that the resultant CTL derived from such an analog would retain specificity for the native antigen as presented in the context of MHC on tumors. Recent studies have shown some enhanced immunogenicity in vitro after modification of anchor sequences to MHC.[64-67]

These studies were intended to increase peptide binding to the MHC since anchor residues of those peptides were not optimal. In the case of CAP1, however, anchor residues conformed to optimal motifs.[68] In recently described studies, a different approach has been taken to improve the immunogenicity of a CTL peptide.[68] It was proposed that by altering non-anchor amino acid residues expected to contact the TCR, one could generate a TCR agonist (i.e., an analog with substitutions at non-MHC anchor positions that stimulates CTL more efficiently than the native peptide). The rationale for this approach was derived from previous findings involving the identification of peptide antagonists.[69-76] In these studies, inhibition of the T-cell response by modified peptides was shown to be TCR-mediated and could not be explained by MHC competitive binding. By analogy, the strictest definition of a peptide TCR agonist would be an analog that increased effector function without accompanying increases in MHC binding.

Several factors were considered in deciding which positions of CAP1 to examine for effects on TCR interactions. Sequencing and mapping experiments have defined a binding motif in which position 2 and the C-terminal (position 9 or 10) are critical for peptide presentation by HLA-A2.[77] In addition, Tyr at position 1 has been identified as an effective secondary anchor. Therefore, CAP1 residues at these positions were not altered. X-ray crystallographic studies of several peptides bound to soluble HLA-A2 suggest that all binding peptides assume a common conformation in the peptide binding groove.[78] When five model peptides were examined, residues 5 through 8 bulge away from the binding groove and are potentially available for binding to a TCR. Studies therefore focused on modifying these residues in an attempt to define CAP1 analogs that would more efficiently stimulate human CEA-specific cytotoxic T cells. A panel of 80 CAP1 analog peptides, in which the residues at positions 5 through 8 were synthesized with each of the 20 natural amino acids, were produced by pin technology. The effects of these amino acid substitutions on TCR recognition were studied using the CAP1-specific, HLA-A2-restricted human CTL line designated V8T. V8T was generated by CAP1 and IL-2 stimulation of PBMC from a patient vaccinated with rV-CEA.[14] For initial screening, V8T was used in a cytotoxicity assay employing C1R-A2 target cells incubated with each member of the peptide panel (at three peptide concentrations).[63] Of the 80 single amino acid substitutions, all but six failed to activate cytotoxicity of V8T.

Subsequent studies revealed that one peptide, designated CAP1-6D (substitution of Asn by Asp at position 6), was the best candidate agonist. CAP1-6D (YLSGADLNL) was compared to native CAP1 in a CTL assay over a more extended range of peptide concentrations, using two different cell lines as targets. Analog CAP1-6D was over 100 times more effective in mediating lysis by V8T than native CAP1 (Fig. 6.2).[68] CAP1 and the CAP1-6D analog were tested for binding to HLA-A2 by measuring cell surface HLA-A2 in the transport-defective human cell line T2; there were no differences in binding (Fig. 6.3).[68] Thus, the improved effectiveness of CAP1-6D in the CTL assays suggests a better engagement to the TCR. The CAP1-6D agonist potentially could be useful in both experimental and clinical applications if it could stimulate growth of CEA-specific CTL from patients with established carcinomas more efficiently than CAP1. Postimmunization PBMC from a cancer patient (designated Vac8) were stimulated in vitro with CAP1-6D and were assayed for CTL activity against targets coated with CAP1 or CAP1-6D. This new line demonstrated peptide-dependent cytotoxic activity against target cells coated with either CAP1-6D or the native CAP1 peptide.

PBMC from patients immunized with rV-CEA were shown to produce CTL activity when stimulated with CAP1, while preimmunization PBMC were negative.[14,63] New attempts to stimulate CTL activity from healthy, nonimmunized donors using CAP1, using this protocol, were also unsuccessful. However, CTL could be generated from healthy, nonimmunized donors by in vitro stimulation with the agonist peptide CAP1-6D. Several peptide-specific CTL lines were obtained when generated with CAP1-6D, but not with CAP1. The CAP1-6D-derived CTL line was tested against a panel of human tumor cells. The CTL line was shown to

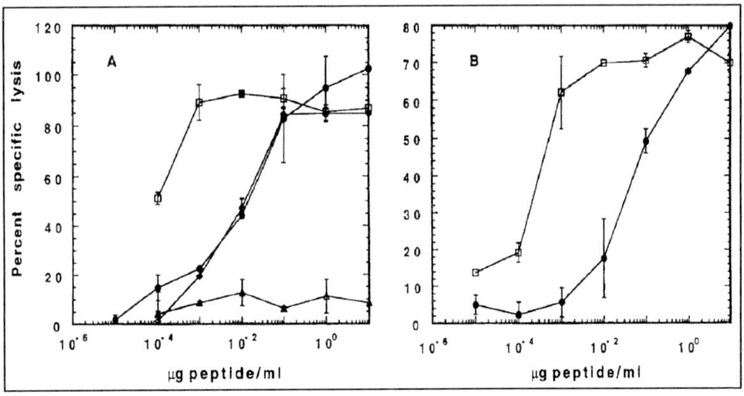

Fig. 6.2. CAP1 and analogs show different sensitivity to CEA CTL T-Vac8 cytotoxicity. T2 (A) and C1R-A2 (B) target cells were labeled with Cr_{51} and incubated in round- bottom, 96-well plates (10,000/well) with CAP1 (l) or substituted peptide CAP1-6D (£) at the indicated concentrations. After 1 h, T-Vac8 CTL were added at E:T= 2.5, and isotope release was determined after 4 h. All assays were done in triplicate. NCA571 gene NCA (68).

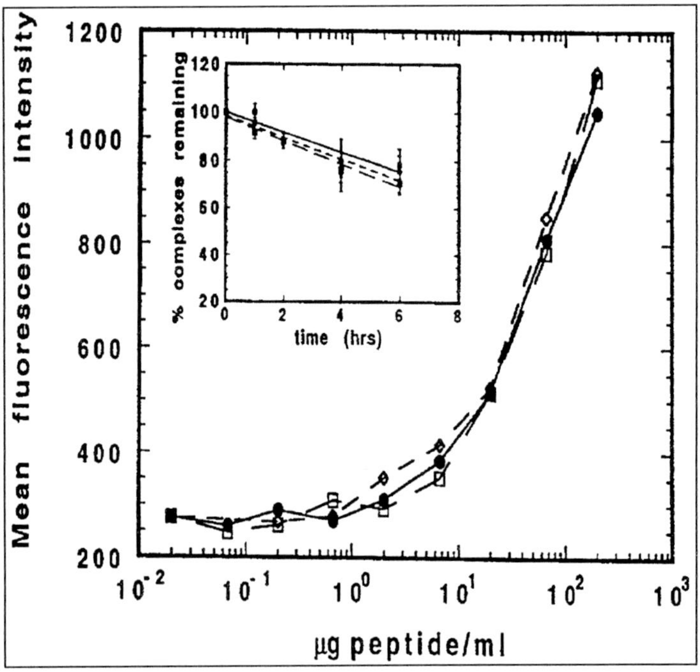

Fig. 6.3. Effect of single amino acid substitutions in CAP1 peptide on binding to and stability of HLA-A2 complexes. T2 cells were collected in serum-free medium and then incubated overnight (10^6 cells/well) with peptides CAP1 (l) or CAP1-6D (£) at the indicated concentrations. Cells were collected and assayed for cell surface expression of functional HLA-A2 molecules by staining with conformation-sensitive MAb BB7.2, HLA-specific antibody W6/32 (data not shown) and isotype-control Ab MOPC-195 (data not shown). MFI was determined on a live, gated cell population. Insert: Cells were incubated with peptide at 100 &g/ml overnight, then washed free of unbound peptide and incubated at 37°C. At the indicated times, cells were stained for the presence of cell-surface peptide-HLA-A2 complexes. The error bars indicate SEM for two experiments (68).

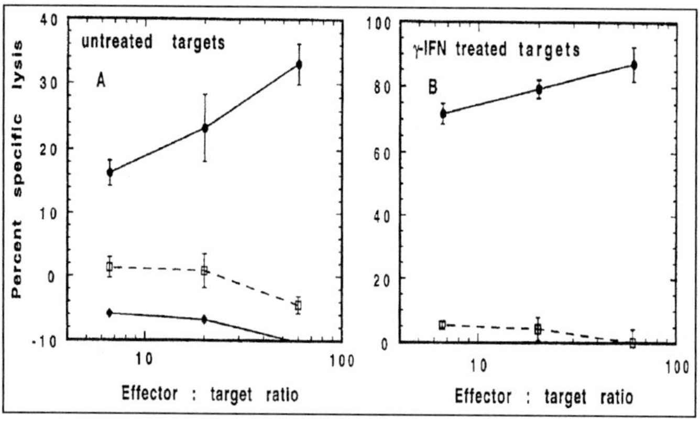

Fig. 6.4. CTL generated with CAP1-6D lyse CEA-positive, HLA-A2-positive tumors: effect of IFN-. The T-N1 CTL generated with CAP1-6D were assayed against various tumor cell lines: SW480 (CEA$^+$ and HLA-A2$^+$; l), SW1116 (CEA$^+$ but HLA-A2$^-$; £), and CaOV3 (CEA$^-$ but HLA-A2$^+$; $^-$). Tumor cells were cultured for 72 h in the absence (A) or presence (B) of IFN-, trypsinized, labeled with Cr_{51} and then incubated (5,000 cells/well) with T-N1 CTL at increasing E:T ratios. Cultures were incubated for 4 h, and the amount of isotope release was determined in a gamma counter. Values were determined from triplicate cultures (68).

kill tumor targets expressing both endogenous CEA and class I HLA-A2 (Fig. 6.4).[68] All human tumor lines negative for either HLA-A2 or for CEA were not lysed.

Immunogenicity of CEA in Humans

The ability of a CEA vaccine to induce a CEA-specific immune response in humans is no longer in question. As seen in Table 6.2, a wide range of immunogens have been employed to induce CEA-specific CTL responses, T-cell lymphoproliferative responses, and antibody responses in cancer patients. The vast majority of CEA-specific CTL responses were identified employing the CAP1 peptide to stimulate T cells and/or to pulse target cells. The following immunogens have now been shown to induce CEA-specific CTL in Phase I trials: (a) rV-CEA; (b) CAP1-pulsed or CEA RNA-transfected DC; (c) avipox-CEA; (d) rV-CEA followed by avipox-CEA boosts and, most recently, (e) avipox-CEA-B7.[14,59,79-81] The assays used to define these T-cell responses were CTL precursor frequency, ELISPOT, FastImmune and, most recently, the tetramer assay. A number of groups have also shown that human CEA-specific T cells can be generated in vitro from PBMC of apparently healthy individuals or from cancer patients (Table 6.2). These studies included the use of peptide-pulsed DC or RNA-transfected DC as APC, and the use of the CAP1-6D enhancer agonist peptide. It should be emphasized that in the vast majority of the studies outlined above, the CEA-specific T cells derived from either immunized patients or from stimulation in vitro were shown to be capable of lysing (a) CEA-expressing allogeneic tumors and/or autologous tumors; (b) CEA peptide-pulsed T2 cells, C1R-A2 cells, or autologous B cells; and/or (c) autologous B cells transduced with the CEA gene. All instances of lysis mentioned above were shown to be MHC-restricted. Therefore, these studies demonstrated not only the ability of CEA vaccines to induce CTL responses, but also the ability of human carcinoma cells to process endogenous CEA, transport CEA peptides to the cell surface in the form of MHC-peptide complexes, and to be rendered lytic by the appropriate CEA-specific CTL.

Table 6.2. Immunogenicity of CEA in humans

Clinical Trials Immunogen	Type of Immune Response	Reference
rV-CEA	CTL*	14, 63
Anti-Id	Ab, LP	33
CEA protein + GM-CSF	Ab, LP	82
CAP1-pulsed DCs	CTL	59
CEA-RNA transfected DCs	CTL	59
Avipox-CEA	CTL	79
rV-CEA + avipox CEA	CTL	79
Avipox CEA-B7	CTL	80, 81
In Vitro Studies		
CAPI-6D agonist	CTL	68
CEA peptides (A3)	CTL	55
CAP-2	CTL	54
CAP1-pulsed	CTL	58
CEA peptides (A2)	CTL	56, 57
CEA peptides (A24)	CTL	60

Antibody and $CD4^+$ T-cell responses have also been demonstrated in patients that have received CEA vaccines (Table 6.2). As mentioned previously, it was demonstrated that colorectal cancer patients receiving 5FU regimens (5FU and leucovorin or levamisole) simultaneously with CEA anti-Id vaccine in adjuvant generated high-titer polyclonal anti-CEA responses that mediate ADCC.[33] Moreover, many patients generated CEA-specific T-cell responses.[33] These studies are extremely important since they demonstrate that CEA vaccines can be administered in the adjuvant setting.

A recent study has also demonstrated the potential importance of cytokines in the clinical applications of CEA vaccines. The use of recombinant CEA in colorectal patients with advanced cancer, with and without the use of GM-CSF at the injection site has been reported.[82] Rather weak responses were noted with CEA protein with adjuvant; patients receiving the vaccine in the presence of GM-CSF, however, developed strong IgG anti-CEA responses and CEA-specific proliferative T-cell responses. Both class I- and class II-restricted responses were noted. As with all of the other CEA vaccines described above, no signs of autoimmunity were noted.

Future Directions in Vaccine Design and Development

The immunogenicity of CEA in humans is now established. A number of vaccines have now demonstrated the induction of $CD8^+$ CTL, $CD4^+$ T-cell, and antibody responses specific for CEA. Several epitopes within the CEA molecule have now been identified as capable of inducing CTL responses. Moreover, the ability of human APC and human tumors to process CEA peptides, and the ability of CEA-specific CTL to lyse both allogeneic and autologous tumors in a MHC-restricted fashion, has also been established by several investigators. Thus far, these findings have been achieved principally in Phase I trials without evidence of any acute or chronic toxicity. However, the quantity (and quality, perhaps) of T-cell responses observed in these Phase I studies is moderate at best. At this point, it is not known whether this is due to the inability of patients with advanced cancer to mount aggressive T-cell responses, to the potency of the vaccines, or to both. Phase II trials in patients with less advanced disease and the use of more potent vaccines and vaccine strategies should provide answers to these questions.

The identification of agonist epitope peptides, such as CAP1-6D, may well enhance future vaccine efficacy. Since CEA is a "self" gene product, one would hypothesize the induction of only weak or moderate T-cell responses. The demonstration that an agonist epitope can enhance these responses may be crucial. Agonist epitopes can be exploited as peptide vaccines with adjuvant, with pulsed DC, or via the modification of the CEA gene to express the agonist epitope in vector-based vaccines such as vaccinia, avipox, DNA, etc. Moreover, preclinical studies and a recent clinical trial have demonstrated that diversified immunization protocols using two different forms of the immunogen are more potent than the repeated use of the same immunogen. The use of CEA vaccines in combination with cytokines may also be crucial. Preclinical and clinical studies have now demonstrated that the use of local GM-CSF clearly enhances CEA-specific responses. Preclinical studies also demonstrate that the use of low-dose IL-2 increases T-cell and antitumor responses. Finally, the use of T-cell costimulation may be crucial in enhancing CEA-specific T-cell responses. This can be achieved by the use of DC to present CEA or CEA peptides, or by the use of vectors engineered to express both CEA, or CEA epitopes, and one or more costimulatory molecules. A recent study employing a vector expressing three different T-cell costimulatory molecules has shown a great enhancement of T-cell responses, especially under conditions of low levels of signal 1.[83] Thus, the use of this type of vector in concert with a self-antigen such as CEA, with or without epitope modifications, may bring host T-cell responses to a level of therapeutic efficacy.

Acknowledgments

The author thanks Dr. Alfred Tsang for helpful discussions and Nicole Ryder for editorial assistance.

References

1. Gold P, Freedman SO. Demonstration of tumor-specific antigens in human colonic carcinomata by immunological tolerance and absorption techniques. J Exp Med 1965; 121:439.
2. Gold P, Freedman SO. Specific carcinoembryonic antigens of the human digestive system. J Exp Med 1965; 122:467.
3. Gold P, Goldenberg NA. The Carcinoembryonic Antigen (CEA): Past, Present and Future. McGill J Med 1997; 3:46-66.
4. Pignatelli M, Durbin H, Bodmer WF. Carcinoembryonic antigen functions as an accessory adhesion molecule mediating colon epithelial cell-collagen interactions. Proc Natl Acad Sci USA. 1990; 87:1541-1545.
5. Thompson JA. Molecular cloning and expression of carcinoembryonic antigen gene family members. Tumor Biol 1995; 16:10-16.
6. Schrewe H, Thompson J, Bona M et al. Cloning of the complete gene for carcinoembryonic antigen: Analysis of its promoter indicates a region conveying cell type-specific expression. Mol Cell Biol 1990; 10:2738-2748.
7. Thompson J, Zimmerman W. The carcinoembryonic gene family: structure, expression and evolution. Tumor Biol 1988; 9:63-83.
8. Thompson JA, Pande H, Paxton RJ et al. Molecular cloning of a gene belonging to the carcinoembryonic antigen gene family and discussion of a domain model. Proc Natl Acad Sci USA 1987; 84:2965-2969.

9. Thompson JA, Grunert F, and Zimmerman W. Carcinoembryonic antigen gene family: Molecular biology and clinical perspectives. J Clin Lab Anal 1991; 5:344-366.
10. Benchimol S, Fuks A, Jothy S et al. Carcinoembryonic antigen, a human tumor marker, functions as an intercellular adhesion molecule. Cell 1989; 57:327-334.
11. von Kleist S, Migule I, Halla B. Possible function of CEA as cell-contact inhibitory molecule. Anticancer Res 1995; 15:1889-1894.
12. Conry RM, LoBuglio AF, Kantor J et al. Immune response to a carcinoembryonic antigen polynucleotide vaccine. Cancer Res 1994; 54:1164-1168.
13. Kantor J, Irvine K, Abrams S, et al. Anti-tumor activity and immune responses induced by a recombinant vaccinia-carcinoembryonic antigen (CEA) vaccine. J Natl Cancer Inst 1992; 84:1084-1091.
14. Tsang KY, Zaremba S, Nieroda CA et al. Generation of human cytotoxic T cells specific for human carcinoembryonic antigen epitopes from patients immunized with recombinant vaccinia-CEA vaccine. J Natl Cancer Inst 1995; 87:982-990.
15. Guadagni F, Roselli M, Cosimelli M et al. Quantitative analysis of CEA expression in colorectal adenocarcinoma and serum: Lack of correlation. Int J Cancer 1997; 72:949-954.
16. Sharkey RM, Goldenberg DM, Goldenberg H et al. Murine monoclonal antibodies against carcinoembryonic antigen: immunologica, pharmacokinetic and targeting properties in humans. Cancer Res 1990; 50:2823-2831.
17. Siccardi AG, Buraggi GL, and Callegaro L. Immunoscintigraphy of adenocarcinomas by means of radiolabeled F(ab')2 fragments of an anti-carcinoembryonic antigen monoclonal antibody: A multicenter study. Cancer Res 1989; 49:3095-3103.
18. Mach JP, Forni M, Ritschard J et al. Use and limitations of radiolabeled anti-CEA antibodies and their fragments for photoscanning detection of human colorectal carcinomas. Oncodevelopmental Biol Med 1980; 1:49-69.
19. Frelinger J, Wei C, Willis R et al. Targeted CTL-mediated immunity for prostate cancer: Development of human PSA-expressing transgenic mice. Proc Am Assoc Cancer Res 1996; 37:3027.
20. Hasegawa T, Isobe K, Nakashima I et al. Quantitative analysis of antigen for the induction of tolerance in carcinoembryonic antigen transgenic mice. Immunol 1992; 77: 577-581.
21. Sinclair NRS. The trouble with transgenic mice. Immunol Cell Biol 1995; 73:169-173.
22. Matzinger P. Tolerance, danger, and the extended family. Annu Rev Immunol 1994; 12:991-1045.
23. Nanda NK, Sercarz EE. Induction of anti-self-immunity to cure cancer. Cell 1995; 82:13-17.
24. Ridge JP, Fuchs EJ, Matzinger P. Neonatal tolerance revisited: Turning on newborn T cells with dendritic cells. Science 1996; 271:1723-1726.
25. Fenton RG, Longo DL. Danger versus tolerance: Paradigms for future studies of tumor- specific cytotoxic T lymphocytes. J Natl Cancer Inst 1997; 89:272-275.
26. Tjoa BA, Kranz DM. Generation of cytotoxic T-lymphocytes to a self-peptide/class I complex: A model for peptide-mediated tumor rejection. Cancer Res 1994; 54:204-208.
27. Theobald M, Biggs J, Dittmer D et al. Targeting p53 as a general tumor antigen. Proc Natl Acad Sci USA 1995; 92:11993-11997.
28. Irvine K, Kantor J, Schlom J. Comparison of a CEA-recombinant vaccinia virus, purified CEA, and an anti-idiotypic antibody bearing the image of a CEA epitope in the treatment and prevention of CEA- expressing tumors. Vaccine Res 1993; 2:79-94.
29. Bei R, Kantor J, Kashmiri SVS et al. Serological and biochemical characterization of recombinant baculovirus carcinoembryonic antigen. Mol Immunol 1994; 31:771-780.
30. Bei R, Kantor J, Kashmiri SVS et al. Enhanced immune responses and anti-tumor activity by baculovirus recombinant CEA in mice primed with the recombinant vaccinia CEA. J Immunother 1994; 16:275-282.
31. Salgaller ML, Bei R, Schlom J et al. Baculovirus recombinant expressing the human carcinoembryonic antigen gene. Cancer Res 1993; 53:2154-2161.
32. Kaufman H, Schlom J, Kantor J. A recombinant vaccinia virus expressing human carcinoembryonic antigen (CEA). Int J Cancer 1991; 48:900-907.
33. Foon KA, John WJ, Chakraborty M et al. Clinical and immune responses in advanced colorectal cancer patients treated with anti-idiotype monoclonal antibody vaccine that mimics the carcinoembryonic antigen. Clin Cancer Res 1997; 3:1267-1276.

34. Lou D, Kohler H. Enhanced molecular mimicry of CEA using photoaffinity crosslinked C3d peptide. Nature Biotech 1998; 16:1458-1462.
35. Conry RM, LoBuglio AF, Loechel F et al. A carcinoembryonic antigen polynucleotide vaccine for human clinical use. Cancer Gene Ther 1995; 2:33-38.
36. Hodge JW, McLaughlin JP, Kantor JA et al. Diversified prime and boost protocols using recombinant vaccinia virus and recombinant nonreplicating avian pox virus to enhance T-cell immunity and antitumor responses. Vaccine 1997; 16:759-768.
37. McLaughlin JP, Schlom J, Kantor JA et al. Improved immunotherapy of a recombinant CEA vaccinia vaccine when given in combination with Interleukin-2. Cancer Res 1996; 56:2361-2367.
38. Lorenz MGO, Kantor JA, Schlom J et al. Antitumor immunity elicited by a recombinant vaccinia virus expressing CD70 (CD27L). Human Gene Ther 1999; in press.
39. Hodge JW, McLaughlin JP, Abrams S et al. The admixture of a recombinant vaccinia virus containing the gene for the costimulatory molecule B7 and a recombinant vaccinia virus containing a tumor associated antigen gene results in enhanced specific T-cell responses and antitumor immunity. Cancer Res 1995; 55:3598-3603.
40. Kalus RM, Kantor JA, Gritz L et al. The use of combination vaccinia vaccines to enhance antigen-specific T-cell immunity via T-cell costimulation. Vaccine 1999; 17:893-903.
41. Hodge JW. Carcinoembryonic antigen as target for cancer vaccines. Cancer Immunol Immunother 1996; 43:127-134.
42. Eades-Perner AM, van der Putten, H, Hirth A et al. Mice transgenic for the human carcinoembryonic antigen gene maintain its spatiotemporal expression pattern. Cancer Res 1994; 54:4169-4176.
43. Thompson JA, Eades-Perner AM, Ditter M et al. Expression of transgenic carcinoembryonic antigen (CEA) in tumor-prone mice: an animal model for CEA-directed tumor immunotherapy. Int J Cancer 1997; 72:197-202.
44. Kass E, Schlom J, Thompson J, et al. Induction of protective host immunity to carcinoembryonic antigen (CEA), a self-antigen in CEA transgenic mice, by immunizing with a recombinant vaccinia-CEA virus. Cancer Res 1999; 59:676-683.
45. Falk K, Rotzschke O, Stevanovic S et al. Allele-specific motifs revealed by sequencing of self-peptides eluted from MHC molecules. Nature 1991; 351:290-296.
46. Hunt DF, Henderson RA, Shabanowitz J et al. Characterization of peptides bound to the class I MHC molecule HLA-A2.1 by mass spectometry. Science 1992; 255:1261-1263.
47. Lee J. The HLA System: A New Approach. New York: Springer-Verlag, 1990, p. 154.
48. Parker KC, Bednarek MA, Coligan JE. Scheme for ranking potential HLA-A2 binding peptides based on independent binding of individual peptide side-chains. J Immunol 1994; 152:163-175.
49. DiBrino M, Parker KC, Shiloach J et al. Endogenous peptides bound to HLA-A3 possess a specific combination of anchor residues that permit identification of potential antigenic peptides. Proc Natl Acad Sci USA 1993; 90:1508-1512.
50. Nijman HW, Houbiers JG, Vierboom MP et al. Identification of peptide sequences that potentially trigger HLA-A2.1-restricted cytotoxic T lymphocytes. Eur J Immunol 1993; 23:1215-1219.
51. Mizoguchi H, O'Shea JJ, Longo DL et al. Alterations in signal transduction molecules in T lymphocytes from tumor-bearing mice. Science 1992; 258:1795-1798.
52. Finke JH, Zea AH, Stanley J et al. Loss of T-cell receptor z chain and $p56^{lck}$ in T-cells infiltrating human renal cell carcinoma. Cancer Res 1993; 53:5613-5616.
53. Nakagomi H, Petersson M, Magnusson I et al. Decreased expression of the signal- transducing z chains in tumor-infiltrating T-cells and NK cells of patients with colorectal carcinoma. Cancer Res 1993; 53:5610-5612.
54. Zhu M, Zaremba S, Correale P et al. Generation of specific anti-human carcinoembryonic antigen (CEA) cytotoxic T lymphocytes from a colon carcinoma patient immunized with recombinant vaccinia-CEA (rV-CEA) vaccine by stimulation with a CEA synthetic peptide (CAP2) in vitro. J Immunother 1996; 19:459.
55. Bremers AJA, van der Burg SH, Kuppen PJK et al. The use of Epstein-Barr virus- transformed B lymphocyte cell lines in a peptide-reconstitution assay: identification of CEA-related HLA-A*0301-restricted potential cytotoxic T lymphocyte epitopes. J Immunother 1995; 18:77-85.

56. Ras E, van der Burg SH, Zegveld ST et al. Identification of potential HLA-A *0201- restricted CTL epitopes derived from the epithelial cell adhesion molecule (Ep-CAM) and the carcinoembryonic antigen (CEA). Human Immunol 1997; 53:81-89.
57. Kawashima I, Hudson SJ, Tsai V et al. The multi-epitope approach for immunotherapy for cancer: identification of several CTL epitopes from various tumor-associated antigens expressed on solid epithelial tumors. Human Immunol 1998; 59:1-14.
58. Alters SE, Gadea JR, Sorich M et al. Dendritic cells pulsed with CEA peptide induce CEA- specific CTL with restricted TCR repertoire. J Immunother 1998; 21:17-26.
59. Nair SK, Boczkowski D, Morse M et al. Induction of primary carcinoembryonic antigen (CEA)-specific cytotoxic T lymphocytes in vitro using human dendritic cells transfected with RNA. Nature Biotech 1998; 16:364-369.
60. Nukaya I, Yasumoto M, Iwasaki T et al. Identification of HLA-A24 epitope peptides of carcinoembryonic antigen which induce tumor-reactive cytotoxic T lymphocyte. Int J Cancer 1999; 80:92-97.
61. Kim C, Matsumura M, Saijo K et al. In vitro induction of HLA-A2402-restricted and carcinoembryonic antigen-specific cytotoxic T lymphocytes on fixed autologous peripheral blood cells. Cancer Immunol Immunother 1998; 47:90-96.
62. Chatterjee SK, Tripathi PK, Chakraborty M et al. Molecular mimicry of carcinoembryonic antigen by peptides derived from the structure of an anti-idiotype antibody. Cancer Res 1998; 58:1217-1224.
63. Tsang KY, Zhu MZ, Nieroda CA et al. Phenotypic stability of a cytotoxic T cell line directed against an immunodominant epitope of human carcinoembryonic antigen. Clinical Cancer Res 1997; 3:2439-2449.
64. Parkhurst MR, Salgaller ML, Southwood S et al. Improved induction of melanoma- reactive CTL with peptides from the melanoma antigen gp100 modified at HLA-A*0201-binding residues. J Immunol 1996; 157:2539-2548.
65. Bakker ABH, Vanderburg SH, Huubens RJF et al. Analogs of CTL epitopes with improved MHC class-I binding capacity elicit anti-melanoma CTL recognizing the wild type epitope. Int J Cancer 1997; 70:302-309.
66. Pogue RR, Eron J, Frelinger JA et al. Amino-terminal alteration of the HLA-A*0201- restricted human immunodeficiency virus pol peptide increases complex stability and in vitro immunogenicity. Proc Natl Acad Sci USA 1995; 92:8166-8170.
67. Lipford G, Bauer S, Wagner H et al. Peptide engineering allows cytotoxic T cell vaccination against human papilloma virus tumor antigen E6. Immunity 1995; 84:298-303.
68. Zaremba S, Barzaga E, Zhu MZ et al. Identification of an enhancer agonist CTL peptide from human carcinoembryonic antigen. Cancer Res 1997; 57:4570-4577.
69. DeMagistris MT, Alexander J, Coggeshall M et al. Antigen analog-major histocompatability complexes act as antagonists of the T cell receptor. Cell 1992; 68: 625-634.
70. Bertoletti A, Sette A, Chissari FV et al. Natural variants of cytotoxic epitopes are T cell receptor antagonists for antiviral cytotoxic T cells. Nature 1994; 369:407-410.
71. Klenerman P, Rowland-Jones S, McAdam S et al. Cytotoxic T cell activity antagonized by naturally occurring HIV-1 gag variants. Nature 1994; 369:403-407.
72. Kuchroo VK, Greer JM, Kaul D et al. A single TCR antagonist peptide inhibits experimental allergic encephalomyelitis mediated by a diverse T cell repertoire. J Immunol 1994; 153:3326-3336.
73. Jameson SC and Bevan MJ. T cell receptor antagonists and partial agonists. Immunity 1995; 2:1-11.
74. Meier U-C, Kleerman P, Griffin P et al. Cytotoxic T lymphocyte lysis inhibited by viable HIV mutants. Science 1995; 270:1360-1362.
75. Chen A, Ede NJ, Jackson DC et al. CTL recognition of an altered peptide associated with asparagine bond rearrangement: Implications for immunity and vaccine design. J Immunol 1996; 157:1000-1005.
76. Rammensee H-G, Friede T, Stevanovic S. MHC ligands and peptide motifs: First listing. Immunogenetics 1995; 41:178-228.
77. Madrenas J, Germain RN. Variant TCR ligands: New insights into the molecular basis of antigen-dependent signal transduction and T-cell activation. Seminars Immunol 1996; 8:83-101.
78. Madden DR, Garboczi DN, Wiley DC. The antigenic identity of peptide-MHC complexes: A comparison of the conformations of five viral peptides presented by HLA-A2. Cell 1993; 75:693-708.

79. Marshall JL, Hawkins MJ, Tsang KY et al. A phase I study in cancer patients of a replication defective avipox (ALVAC) recombinant vaccine that expresses human carcinoembryonic antigen (CEA). J Clin Oncol 1999; 17:332-337.
80. von Mehren M, Davies M, Rivera V et al. Phase I trial with ALVAC-CEA B7.1 immunization in advanced CEA-expressing adenocarcinomas. Proc Amer Soc Clin Oncol 1999; (submitted).
81. Lee DS, Conkright W, Horig HE et al. Preliminary results of ALVAC-CEA-B7.1 phase I vaccine trial in patients with metastatic CEA-expressing tumors. Proc Amer Soc Clin Oncol 1999; (submitted).
82. Samanci A, Yi Q, Fagerberg J et al. Phamacological administration of granulocyte/macrophage-colony-stimulating factor is of significant importance for the induction of a strong humoral and cellular response in patients immunized with recombinant carcinoembryonic antigen. Cancer Immunol Immunother 1998;47:131-142.
83. Hodge JW, Sabzevari H, Lorenz MGO et al. A triad of costimulatory molecules synergize to amplify T-cell activation. Proc Natl Acad Sci USA 1999; in press.

Chapter 7

Studies of MUC1 Peptides

Vasso Apostolopoulos, Geoffrey A. Pietersz and Ian FC McKenzie

Introduction

There have been more studies of Mucin 1 (MUC1) peptides in breast cancer than of any other peptides in this disease, and it is appropriate that the use of MUC1 peptides and vaccines be reviewed here. In contrast to melanoma peptides (discussed elsewhere), where the peptide epitopes were defined by CTLs at the beginning of the studies, mucin peptides were examined for different reasons. Firstly, there is a great increase in mucins- up to 100 fold in cancers, compared to normal tissues, and therefore could provide an appropriate target for the differential activity of CTLs.[1] Thus, if the increase in total amount of mucins is reflected with an increased amount of peptide presented by Class I molecules, then CTLs would be 100 times more effective against cancer than against normal tissues; whether this is so remains to be seen. Secondly, there is much information available on MUC1 and its peptides; MUC1 was the first mucin to be cloned and the protein structure deduced, it is a large cell surface molecule which, by definition, is highly O-glycosylated and therefore contains many serine, threonine and proline residues. Of interest, not only for MUC1, but for all the mucins (MUC1-8 described thus far) is a 20 amino acid repeat in the extracellular region (the VNTR), and for MUC1 it is repeated up to 40 times in different alleles.[1] Thirdly, MUC1 is an attractive molecule for immunological studies, it is highly immunogenic in mice, shown initially by immunizing mice with human cancers to make monoclonal antibodies, when almost all antibodies were found to react with the highly immunogenic VNTR, indeed with the APDTR amino acids within this region.[2,3] This immunogenicity in mice would not be of great interest or relevance other than for Olja Finn's findings that the lymph nodes of patients with breast cancer contains cells that could be stimulated to give non-MHC restricted CTLs.[4] While the in vivo relevance of these unusual CTLs is not known in protecting patients against cancer, these studies were of great importance as they galvanized breast cancer "vaccinologists" into using MUC1 as a potential vaccine and preceded other studies using MUC1 peptides in breast cancer. However, such non- restricted cells cannot be lightly dismissed, clearly they are specifically selected (while MHC restricted cells are selected against) in their mode of identification, by re- stimulating with different MUC1+ cell lines with different HLA phenotypes. However, it is of interest that a patient with breast cancer later became pregnant (MUC1 is greatly increased in the lactating breast) and had a severe case of mastitis accompanied with heavy lymphocyte infiltration, a high frequency of CTLs to MUC1 and there has been no recurrence of the cancer; it is possible that the non-MHC restricted cells were protective (see below).[5]

Thus, there are many reasons why MUC1 should be examined as a potential target for immunotherapy, however, the initial statements must be re-emphasized. At this time, while we can report on the induction of CTLs in mice by immunizing with human MUC1 and more recently, that MUC1 transgenic mice also make MUC1 CTLs which have anti- tumor effects in mice. There have been no examples, as yet, where immunization with MUC1 peptides has led to a significant anti-tumor response in humans. The ultimate goal of all vaccine or immunotherapeutic trials. This Chapter will review the studies performed with MUC1 peptides and it will be seen that almost all the studies, with the exception of one, (which examined non-VNTR peptides) have concentrated on the immunogenicity of the VNTR peptides. While this volume concentrates on peptides, it should be noted here that a substantial amount of work has been done with MUC1 carbohydrates, particularly TF and STn, and it has been suggested that immunizing with these carbohydrates, associated with MUC1 leads to improved survival in patients with breast cancer, however, this is not the focus of this Chapter, nor this volume.[6] We will concentrate on MUC1 peptides and discuss a number of novel delivery systems used for the satisfactory induction of CTLs, and draw attention to the unusual way MUC1 peptides are presented by Class I molecules.

MUC1 in Cancer

The immunotherapy of solid cancers requires a suitable target antigen and the production of a cytotoxic T cell response. Potential target antigens for different tumors includes the glycoproteins and glycolipids, developmental antigens such as CEA and over expressed antigens (gp75, MAGE, tyrosinase, Mart1/melan-A, gp100, MUC1) and mutated oncogenes (i.e., p53, HER-2/neu, ras) (see other chapters). For reasons described above, we have been focusing on MUC1 as a target for tumor immunotherapy. MUC1 (also called Episialin, PEM, EMA, CA15-3, DF3 antigen) is a membrane associated glycoprotein of high molecular weight (>200Kd), which protrudes above the cell surface (>200nm). The extracellular domain consists of 40-90 homologous 20 amino acid repeats (VNTR) (one repeat sequence: PDTRPAPGSTAPPAHGVTSA). In cancer MUC1 is over expressed, and is often present over the entire cell surface, whereas it is restricted to the apical surface of secretory cells in normal tissues. Further in cancer, there can be an alteration in the structure of MUC1 so that new carbohydrate epitopes, (Tn, STn, TF), and peptide epitopes (APDTRPA in the VNTR) are exposed and are not detected in normal tissues or secretions. Most of the immunogenicity (with regard to antibody production), resides in the VNTR region.[1,3,4] As mentioned above, MHC non-restricted CTLs have been described, MHC restricted CTLs have also been detected in breast cancers and in healthy multiparous women.[7,8] In addition to cellular immunity, antibodies have been demonstrated to selected epitopes of MUC1 within the VNTR, from patients with ovarian, breast, pancreatic or colon cancers[9-11]; furthermore, circulating immune complexes of MUC1/antibody have been documented in ovarian and breast carcinoma patients.[12] In mice, MUC1 VNTR peptides are immunogenic, forming the basis for the use of MUC1 as a target for tumor immunotherapy in humans (see below). We will now discuss studies of the immunogenicity of MUC1 in mice, particularly for the induction of cellular immunity, studies which are conventionally a prelude to clinical studies.

MUC1 Tumor Growth in Mice (Table 7.1)

A number of studies of human MUC1 have been performed in mice and in this setting, MUC1 is clearly immunogenic. In addition, human MUC1 has been used in MUC1 transgenic mice and murine, and monkey Mucin 1 have also been examined. MUC1+ 3T3 was used in BALB/c mice to demonstrate that it is highly immunogenic, leading to tumor resistance after challenge with $5x10^7$ MUC1+3T3 cells.[13] The immune response induced by these live tumor cells was predominantly cellular in type with little antibody produced, and tumor rejection was mediated by $CD3^+/CD8^+$ cells, with $CD4^+$ cells playing little role, although IFN was secreted and there was a high CTLp frequency (1/14,600) present, i.e., the response to MUC1 or tumor cells was of the T1 type.[13] Thus, efforts to immunize with recombinant or synthetic

MUC1 to inhibit tumors, should also produce $CD8^+$ CTLs cells and little antibody. In other studies (Table 7.1), immunizing mice with the mouse mammary epithelial tumor cell 410.[4] transfected with human MUC1, led to a reduction in tumor incidence after challenging with 10^3 MUC1+ 410.4 cells.[14] Further MUC1 RMA cells (C57BL/6 T cell thymoma) expressing human MUC1 were used to immunize syngeneic and transgenic mice for human MUC1, and MUC1 specific CTL generated in both.[15] Such approaches could be used in humans using autologous cells expressing MUC1 for vaccination. This strategy has been used in the past without great success, however it would be preferable to use synthetic agents rather than autologous tumors.

MUC1 Peptides in Mice (Table 7.1)

There have been a number of studies examining synthetic MUC1 to induce immune responses similar to those seen with tumor immunizations of mice. It was known that synthetic peptides linked to a carrier were able to immunize mice for the production of MUC1 antibodies, but could they be used to induce cellular responses? A MUC1 synthetic peptide containing 20 amino acids of MUC1 VNTR - linked to either KLH or diphtheria toxoid or a fusion protein containing 5 repeats of the MUC1 VNTR, were used to immunized mice. Strong antibodies (IgG1) were produced, $CD4^+$ DTH reactions, but there were no CTLs and only weak anti-tumor effects with protection occurring only with a low dose of tumors for challenge (10^6 cells),[13,16] the protection disappeared when 5 times the dose of tumor cells was used. Similar results were noted when mice were immunized with whole native MUC1 molecule (HMFG).[13] Thus, an effective CTL response could not be produced by MUC1 peptides conjugated either to KLH or diphtheria toxoid, although there was a measurable frequency of CTL precursors.[13] In other studies, a synthetic peptide derived from the 20 amino acid sequence of human MUC1 VNTR coupled to KLH and emulsified in Ribi adjuvant, gave strong DTH reactions and antibody responses in mice.[17] Others have demonstrated that a 30 amino acid peptide spanning the VNTR coupled to KLH and mixed with QS-21 or Bacillus Calmette-Guerin induced strong antibody responses but again there were no T cell responses such as T cell proliferation, delayed-type hypersensitivity, or CTLs.[18] Recent studies suggest that glycosylated peptides should be used in preference to simple synthetic peptides as these are how the native peptides are presented (see below). Furthermore, a synthetic peptide which has been O-glycosylated at the T in the APDTRP sequence of the VNTR has been made and immune responses are currently being examined,[19] there is currently no evidence which demonstrates that peptides presented by Class I molecules are glycosylated.

It is apparent from these studies that human MUC1 peptides induce strong antibody responses, but that CTL responses have been difficult to induce in mice using MUC1 peptides. Other MUC1 constructs have been designed with the aim of inducing strong cellular responses.

MUC1 Constructs Used in Mice To Induce Cellular Immunity (Table 7.1)

Vaccinia Virus MUC1

Recombinant viruses have been shown to induce immune responses to antigens in a number of experimental models. For example in a proportion of DBA/2 mice, vaccinia virus-MUC1 protected mice from growth of MUC1+P815 tumors;[20] and CTL and antibody responses could also be induced in DBA/2 MUC1 transgenic mice (Acres et al , manuscript submitted). In another model, vaccinia virus- MUC1 in rats protected 60-80% against a challenge with MUC1+ tumor cells.[21] A recombinant vaccinia virus containing a modified MUC1 gene with 10 VNTR repeats (to minimize vaccinia mediated rearrangement) was made and immunized

Table 7.1. Experimental studies using MUC1 based immunogens

MUC1 Formulation		Cellular	Humoral Response	Tumor Response	Reference Protection
Inbred Mice					
Tumors	MUC1+3T3	Yes	Weak	Yes	13
	MUC1+P815	Yes	Weak	Yes	13
	MUC1+410.4	ND	ND	Yes	14
	MUC1+RMA	Yes	ND	Yes	15
Whole native	MUC1 (HMFG)	No	Yes	No	13
Peptides	20mer-KLH	No	Yes	No	13
	FP	No	Yes	No	13
	20mer-KLH+Ribi	Weak	Yes	ND	17
	30mer-KLH+QS21	No	Yes	ND	18
	30mer-KLH+BCG	No	Yes	ND	18
	9mer-mannan	Yes	No	Yes	35
	MUC1 mimic	Yes	Yes	Yes	50
Other constructs	Vaccinia virus	ND	ND	Yes	20
	Vaccinia virus+B7	Yes	ND	Yes	22
	Liposome	Yes	Yes	Yes	26
	cDNA	Yes	Yes	Yes	15
	DC-MC38 fusion	Yes	ND	Yes	27
	Oxidized mannan	Yes	Weak	Yes	31, 32
	Reduced mannan	No	Yes	No	31, 32
	Mannan-HMFG	Yes	Weak	ND	(Submitted)
MUC1 Transgenic mice					
	MUC1+RM	Yes	ND	Yes	15
	DC-MC38 fusion	Yes	Yes	Yes	27
	Oxidized mannan	Yes	Weak	Weak	(Submitted)
	Vaccinia virus	Yes	Weak	ND	(Submitted)
Rats					
	Vaccinia virus	ND	ND	Yes	21
Primates					
Chimpanzees	Dendritic cells	No	Yes	ND	51
Chimpanzees	EBV-B cells	Yes	No	ND	52
Rhesus macaques	Oxidized mannan	Yes	Yes	ND	(Submitted)

DC-MC38 fusion, dendritic cells fused with MUC1+MC38 cells; HMFG, human milk fat globule which contains native MUC1; ND, not determined; FP, MUC1 fusion protein consisting of 5 VNTR repeats.

with vaccinia virus- MUC1 and vaccinia virus containing the T cell costimulatory molecule B7-1.[22] Such mice were protected from pulmonary metastasis and developed MUC1 CTLs which were enhanced by the vaccinia virus-B7-1.[22]

Retroviral Vectors

A retroviral vector (MFG-MUC1) containing 22 VNTR repeats led to MUC1 expression by murine 3T3 cells, and dendritic cells which could elicit MUC1 responses in mice.[23] Human $CD34^+$ dendritic cells have also been transduced with MUC1 and it will be of interest to determine what type of immune responses will be generated to these cells.[24]

Liposomes

A 24 amino acid synthetic peptide spanning the MUC1 VNTR was encapsulated with monophosphoryl lipid A adjuvant in multilamellar liposomes and low dose injections in C57BL/6 mice provided protection against a tumor challenge, which was accompanied by T-cell proliferation, IFN-γ production and IgG2a antibodies.[25] The mode of association of the peptide with liposomes was an important determining factor in the resulting type of immune response, as surface exposed peptide on liposomes induced a predominant antibody response,[26] i.e., liposomes associated MUC1 induces either a cellular or humoral immune response depending on the physical association of the peptide and liposome.[26] DNA: Intra-muscular MUC1 cDNA immunization in C57BL/6 mice results in tumor protection in 80% of mice, and humoral and cell-mediated responses, although the humoral response did not correlate with tumor rejection.[15]

Dendritic Cell Fusions

Dendritic cells fused with murine MUC1+ MC38 carcinoma cells can stimulate naive T cells in culture and in vivo induce specific tumor CTLs and reject established metastases in mice.[27] In another study dendritic cells were transduced with adenovirus MUC1 and primary allogeneic mixed lymphocyte reactions, tumor protection and CTLs were demonstrated.[28]

MUC1 in MUC1 Transgenic Mice

The above studies describe the immunogenicity of human MUC1 in mice and have shown that MUC1 as synthetic peptide, fusion protein, whole mucin,—or on tumor cells, is highly immunogenic and can induce both cellular and humoral immunity—depending on the formulation of the antigen. However, the real test of immunogenicity would be to use the same species (e.g., human MUC1 in humans, murine MUC1 in mice) or alternatively, mice transgenic for human MUC1 so that the antigen is not foreign. In several studies it is clear that MUC1 is immunogenic in MUC1 transgenic mice. C57BL/6 transgenic mice were found to be tolerant as stimulation by MUC1 or tumor cells or as peptide failed to demonstrate immunoglobulin class switching to IgG subtypes.[29] MUC1+RMA cells induced MUC1 specific CTLs,[15] although in a separate study immunization of MUC1 transgenic mice with irradiated MUC1+ tumor cells did not induce an effective immune response.[30] However, transgenic mice immunized with fused dendritic cells and MUC1+tumor cells induced both cellular and humoral responses accompanied by the rejection of established metastases.[30] We were able to demonstrate strong antibody and CTL responses in DBA/2 MUC1 transgenic mice immunized with MUC1 using vaccinia MUC1 virus or MUC1-mannan conjugates (Acres et al, manuscript submitted). In all these studies it was of interest to note that there were no signs of autoimmunity, and both T and B cell tolerance appeared to be broken. However, the definition of breaking tolerance may need revision. It is possible that the native mucin exists as glycosylated peptide and that the immune system has never been exposed to non- glycosylated peptides, i.e., it is "foreign"—at least for presentation for B cell responses. As indicated, it is not clear whether glycosylated MUC1 peptides are presented by Class I molecules. However, if immune responses can be generated by non-glycosylated, synthetic MUC1 peptide and CTLs act against MUC1 cells, then this would be a novel way of inducing immune responses.

Induction of CTLs in Mice with Mannan-MUC1 (M-Fp)

The first description of CTLs in mice to MUC1 used oxidized mannan conjugated to MUC1. Many different strains of mice could be immunized with either oxidized mannan MUC1 fusion protein (containing five VNTR repeats) (M-FP) or with mannan MUCI peptides of length 8-30mers. Direct CTLs, without in vitro culture or restimulation, could be found in spleen and lymph nodes after three immunizations.[31,32] The potent CTLs were mirrored with the high CTLp frequency found in the spleen.[31] Indeed, after one injection a measurable CTLp frequency was found which increased greatly after three injections, but thereafter the frequency tended to diminish.[33] The CTLs found were of the 'conventional type' in that they were $CD8^+$ (as they could be blocked by CD8 antibodies or by treating the mice with CD8 antibody prior to harvesting the CTLs). H2 restriction was found in inbred mice of the 5 different H2 haplotypes and confirmed in congenic, recombinant and mutant mice, where only the strains carrying a H2 Class I molecule of the immunizing strain could serve as targets for CTLs.[34-36] Further, the findings with the MUC1 fusion protein (consisting of 5 repeats of MUC1) could be reproduced using 9-20mer synthetic peptides conjugated to KLH.[35] Thus, to produce CTLs to MUC1, it is possible to immunize any mouse strain with MUC1-mannan and harvest the CTLs after 3 injections. The induction of $CD8^+$ CTLs was accompanied by other features of a T1 response, such as the secretion of γIFN, IL-2 and IL-12, but not IL-4.[32,37] It was of interest that immunizing with the fusion protein, in the absence of oxidized mannan, a T2 response was generated, with IL-4 being the predominant cytokine produced, the mice producing no CTLs, γIFN or IL-2 but a high antibody response.[32,37] Thus, the oxidization of MUC1 led to a T1 response; reduction of this with sodium borohydride led to a T2 response (Table 7.1).[32]

MUC1 Peptides Presented by H2 and HLA Class I Molecules (Table 7.2)

The CTL responses described above led to the detection of MUC1 multiple epitopes presented by 5 different H2 alleles (K^b, D^b, D^d, L^d and K^k) and by HLA-A*0201.[35,36] The studies were performed using long (10-20 mer) peptides and overlapping 9mer peptides, and as targets: tumor cells, PHA blasts, single H2 alleles expressed after L cell transfection and RMA-S cells (used as CTL targets after peptide loading and in stabilization assays). The 9mer peptides APDTRPAPG, SAPDTRPAP and TSAPDTRPA (Table 7.2) are presented by $H2K^b$ molecules, but we note that these sequences do not have the appropriate anchors for $H2K^b$ molecules, (F/Y at position 5 (P5) and L/A at P8), although exceptions have been noted.[38,39]

Three peptides APGSTAPPA, PAPGSTAPP and RPAPGSTAP were found to be presented by $H2D^b$ using recombinant mice and with L cells transfected with D^b cDNA (Table 7.2).[35] As with $H2K^b$, $H2D^b$ specificity was confirmed in RMA-S cells, and it was also found that none of the $H2D^b$ binding MUC1 peptides could stabilize RMA-S cells for flow cytometric detection of Class I molecules, indicating their low affinity, although the peptide loaded RMA-S cells could readily be lysed by CTLs (Table 7.2).[35] Like K^b the D^b binding MUC1 peptides do not contain the appropriate anchors set for D^b peptides in that N is absent from P5 (24) (Table 7.2).

Using recombinant mice and L cells transfected with D^d or L^d as targets (Table 7.2),[35] it was demonstrated that $H2D^d$ binding peptide was SAPDTRPAP, the H2Ld binding peptide APDTRPAPG and $H2K^k$ binding peptide PDTRPAPGS. It was of interest that 7 Class I binding epitopes were present in a single 20 mer sequence. The HLA-A*0201/K^b mice were also used for epitope studies: transgenic mice were immunized with M-FP which generated highly active CTLs to MUC1 peptides presented by both the transgenic murine HLA-A*0201/K^b lymphoblast peptide pulsed target cells and by human EBV-B-HLA-A*0201 molecules.[36] The 9mer MUC1 peptide sequences which bound to HLA-A*0201 were three related sequences (APDTRPAPG, SAPDTRPAP, TSAPDTRPA) and also the STAPPAHGV peptide (Table 7.2) this latter epitope having previously been described for HLA-A*0201 and HLA-A11.7 The findings are of interest for tumor immunotherapy, as the CTLs generated in HLA-A*0201 transgenic mice could lyse the HLA-A*0201+ human breast cancer cell line, MCF-7.[36] However, HLA-B27 transgenic mice did not respond to M-FP presented by HLA-B27 (unpublished data), but these mice apparently do not respond to many antigens.

From the epitopes detected in inbred and HLA-A*0201 mice a number of conclusions can be drawn:

i. Many different epitopes are contained within the MUCI VNTR, indeed, the 20 amino acid sequences contain epitopes which bind to at least 6 different Class I molecules to HLA-A*0201 and HLA-A11. This is an unusual finding but not unique, such findings have been previously described for HIV-gp120 and Epstein-Barr virus nuclear antigens.[40,41]

ii The MUC1 epitopes presented differ for some of the Class I molecules, but for many, the sequence SAPDTRPAP or a variant of this (APDTRPAPG or TSAPDTRPA) also binds. In some cases, it appears that the three different epitopes can be presented by one allele, e.g., by K^b and further, the 7 mer APDTRPA can also be presented.[35]

iii. In some cases the peptides which bind are those that would be expected from an analysis of the sequence. For example, the sequence containing APDTRPAPG satisfies the 'rules' for the binding of peptides to L^d molecules, (anchor residues of P at P2 and T at P4),42 nor does the SAPDTRPAP sequence for Dd (preferred anchor residues are G at P2, P at P3 and L/F at P9), although other amino acids have been found at these positions.[35,42] The D^d peptide fits with the anchoring residue at P3 and amino acids in particular at P1 for S, P4

Table 7.2. Epitopes detected by anti-H2 and anti-HLA MUC1 CTLs

	MUC1 VNTR																			
Restriction	**T**	**S**	**A**	**P**	**D**	**T**	**R**	**P**	**A**	**P**	**G**	**S**	**T**	**A**	**P**	**P**	**A**	**H**	**G**	**V**
$H2K^b$	T	S	A	P	D	T	R	P	A											
		S	A	P	D	T	R	P	A	P										
			A	P	D	T	R	P	A	P	G									
$H2D^b$							R	P	A	P	G	S	T	A	P					
								P	A	P	G	S	T	A	P	P				
									A	P	G	S	T	A	P	P	A			
$H2D^d$		S	A	P	D	T	R	P	A	P										
$H2L^d$			A	P	D	T	R	P	A	P	G									
$H2K^k$				P	D	T	R	P	A	P	G	S								
HLA-A*0201	T	S	A	P	D	T	R	P	A											
		S	A	P	D	T	R	P	A	P										
			A	P	D	T	R	P	A	P	G									
												S	T	A	P	P	A	H	G	V
HLA-A11												S	T	A	P	P	A	H	G	V

for D, P5 for T, P7 for P and P9 for A have been shown to associate and bind to H2 Dd. However, it is now apparent that certain peptides which bind to some Class I molecules may not require defined anchors. For example, the sequence APDTRPAPG does not have the consensus sequences set for the binding of peptides to K^b molecules but APDTRPAPG clearly binds satisfactorily to K^b, as it is a suitable target for CTLs: the binding is of low affinity but induces high avidity CTLs.[35] Such findings have also been noted recently for HIV-gag peptides which bind to $H2K^d$ and for other epitopes.[43]

Unusual Features Of MUC1 Peptides Binding to Class 1 Molecules

From the preceding it is clear that MUC1 VNTR peptides have some unusual features and this was further exemplified by antibody binding and modelling studies. Thus far, it is generally accepted that peptides which react in solid or liquid phase with antibodies lose such reactivity when the peptide is buried in the groove of Class I molecules; thus antibodies and T cells react with different epitopes, and anti-peptide antibodies do not block T cell reactivity. The loss in antibody reactivity by peptides in the MHC Class I groove is likely to be due to their loss of secondary structure, and that the amino acids are buried in the groove or hidden by the edges in the groove of H2 molecules. There are exceptions in that several antibodies have been described which react with peptides while bound to Class I but not otherwise.

However, we have recently shown that MUC1 peptides binding to Class I molecules are readily accessible to some anti-MUC1 peptide antibodies, results found by using antibodies which detect 6 different epitopes in the MUC1 20mer VNTR sequence. In these studies, RMA-S cells were pulsed with peptide, washed and analyzed by flow cytometry with anti-MUC1 antibodies. When the $H2K^b$ binding peptide, SAPDTRPAP, was tested using antibodies which react with different parts of this 9mer, the BC2 (APDTR; binding to the N-terminus) and BCP9 (GSTAP; negative control) antibodies were non-reactive but the antibodies BCP8 (DTR; binding to the middle of the sequence) and BCP10 (RPAP; binding to the C-terminus) were both reactive. The same antibodies to the mid and C-terminus were also shown to block CTL killing but not to the N-terminus. By molecular modeling the H2 K^b peptide SAPDTRPAP appears to be anchored at the N-terminus (SA...), PDT looping out, R acting as an anchor and PAP at the C-terminus is free and looping out of the groove.[44] Similar results were obtained with $H2D^b$ peptides, where RMA-S cells were pulsed with the 9mer APGSTAPPA peptide; the antibody BCP9 (GSTAP) was reactive whereas VA1 (APG), BCP8 (DTR) and BCP10 (RPAP) antibodies were not. BCP9 could also block CTL killing. For H2Db by molecular modeling it appears that the N- and C-terminus of the peptide are held down and the central region looping out and accessible to anti-peptide antibodies.[44] For HLA-A*0201 the SAPDTRPAP peptide appears to be buried in the groove of Class I at the N-terminus, and the middle and C-terminus are free and accessible to anti-peptide antibodies. The other HLA-A*0201 binding peptide, STAPPAHGV, binds in a canonical manner with higher affinity, and is not accessible to anti-APPAH antibodies.[44] The antibody binding and modelling studies demonstrated that peptides do not have to be deeply bound within the groove of Class I molecules, indeed low affinity binding peptides can generate effective CTLs. Thus, there is a different mode of MUC1 peptide binding to some Class I molecules. Crystallization studies, currently in progress confirm the findings (Apostolopoulos et al, in preparation).

Methods of Increasing the MUC1 CTL/CTLp Frequency

Although the CTL response to human MUC1 in mice was significant and induced the eradication of tumors, the response to a self antigen was very much less. For example, mice immunized with mouse MUC1 (Xing et al, unpublished observations) had a much lower frequency of CTLp as did monkeys immunized with monkey MUC1 (Vaughan et al, submitted).

A few patients also made CTLs to human MUC1 (see below).[45] We therefore considered methods of increasing the CTLp frequency in mice as a pointer to how this could be done in humans to induce a more effective anti-tumor response. As mannan MUC1 induced potent responses we increased its immunogenicity by a variety of means. Cyclophosphamide could readily increase the CTLp frequency in mice, and 3 injections of M-FP was equivalent to 1 injection M-FP + cyclophosphamide.[46] Adjuvants such as GMDP, MDP and incomplete Freund's adjuvant can also increase the CTLp frequency,[33] as can the cytokines, IL-4+γIFN, IL-2+γIFN, and IL-12 (Lees et al, manuscript submitted). However, one of the most potent modes of immunization is the in vitro targeting of mannose receptor bearing cells (Apostolopoulos et al, manuscript in preparation). Further a number of studies describing dendritic cells pulsed in vitro with peptides and then transferred in vivo give satisfactory CTL responses. We have done similar studies targeting the mannose receptor of murine macrophages and indicate that one immunization with mannan-MUC1 pulsed macrophages can give rise to responses similar to three in vivo immunizations (Apostolopoulos et al, manuscript in preparation). It will be of interest to determine whether the addition of cyclophosphamide, adjuvants, cytokines or macrophages renders MUC1 more immunogenic in a clinical setting, particularly if accompanied by anti-tumor effects.

Immune Responses to MUC1 Outside the VNTR Region

For human MUC1 all of the attention has been focused on peptides within the VNTR region, an obvious target because of its preferential immunogenicity in mice, and, from the studies in humans with T cell responses. However, it has recently been demonstrated that the whole native MUC1 conjugated to mannan generates significant CTL and CTLp responses in mice to various non-VNTR regions in ($H2^b$, $H2^d$ and HLA-A*0201) mice (Pietersz et al, manuscript submitted). This study demonstrated that the whole native MUC1 molecule contains multiple T cell epitopes, thus, making it a more powerful immunogen than the VNTR region alone. Using the whole molecule also increases the ability for more peptides to be presented by Class I molecules to T cells and be targets for MUC1 therapy.

Mimic of MUC1

Recent findings in xenotransplantation have provided an approach to MUC1 tumor immunotherapy. It has been known that natural human IgM and IgG antibodies give rise to hyperacute rejection of pig xenografts by humans, as they react with the Gal(1,3)Gal epitope present on pig cells.[47] In an extension of these studies it was demonstrated that the peptide, DAHWESWL, isolated from a peptide library, blocked the binding of anti- Gal(1,3)Gal antibodies to Gal(1,3)Gal expressed on pig cells and therefore was a peptide mimic of the Gal sugar.[48] In studies to determine the specificity of the DAHWESWL peptide, Gal(1,3)Gal antibodies were shown to bind to MUC1 peptides.[49] It was also shown that the peptide DAHWESWL, could mimic the MUC1 peptide SAPDTRPAP(G), in regards to T cell responses in H2d mice but not H2b nor HLA-A*0201 mice and that tumor protection was also evident after immunizing with the mimic peptide.50 Mutations of the DAHWESWL peptide to a more HLA-A*0201 compatible peptide, DLHWASWV, to include appropriate HLA-A*0201 binding motifs, led to CTLs in HLA-A*0201 mice which lysed human MCF7 breast cancer cells in vitro.[50] A clinical trial is soon to commence with the mimic peptide (HLA-A*0201 mutated mimic, DLHWASWV) to determine if mimics of self MUC1 peptides can induce stronger immune responses in humans than using self MUC1 peptides.

MUC1 in Primates

Dendritic cells are now being examined in cancer immunotherapy for many antigens, including MUC1, and chimpanzee dendritic cells were able to elicit immune responses to ovalbumin and MUC1 peptides. Following an intravenous inoculation of ovalbumin and MUC1, in vitro pulsed dendritic cells and a boost 10 days later,[51] four chimpanzees made an IgG antibody and their T cells proliferated to ovalbumin, but only 1/4 made an antibody response to MUC1 and there were no T cell responses.[51] In another study, autologous Epstein-Barr virus (EBV)-immortalized B cells from two chimpanzees were transfected with MUC1 cDNA, treated with phenyl-N-acetyl-alpha-D-galactosamine to remove sugars and expose protein VNTR epitopes, irradiated, and injected into chimpanzees.[52] Measurable CTLps were found, DTH responses were evident, but no antibodies were generated.[52] Autologous antigen presenting cells, such as dendritic cells, macrophages or EBV-immortalized B cells, expressing tumor-associated antigens could be useful immunogens for the generation of effective immune responses in vivo - perhaps the focus should be on increasing T cells help in these studies. In addition to the above, we have injected rhesus macaque monkeys with human MUC1 conjugated to mannan (M-FP); 4/4 generated antibody responses, 3/4 proliferative T cells and 1/4 a reasonable CTLp frequency (Vaughan et al, manuscript submitted). Human MUC1 is therefore immunogenic in primates for both humoral and cellular immunity; but monkey MUC1 was far less immunogenic.

MUC1 Clinical Trials (Table 7.3)

A Phase I trial with MUC1 synthetic peptides was performed to determine the immunogenicity and toxicity of the protein core peptides. Patients were injected with doses of 100μg-1000μg of synthetic VNTR peptide conjugated to diphtheria toxoid; poor responses were noted, but there were no side effects.[53]

In another study MUC1 synthetic peptide (containing 105 amino acids = 5 VNTR repeats) was used together with BCG to inject 63 patients with adenocarcinoma.[54] The only side effect noted was minor ulceration at the vaccination site; 3/63 patients had a macroscopic DTH response to the MUC1 peptide, but 37/55 biopsies showed intense T cell infiltration.[54] Further 7/22 patients tested had a 2-4 fold increase in MUC1 specific CTLp; no other immune responses such as antibodies or T cell proliferation were detected.[54] The positive microscopic DTH responses in the absence of macroscopic findings in cancer patients is an important observation and indicates the preferential method of assessing DTH responses in such patients. In a separate study, 5μg of a 16 amino acid MUC1 peptide from the VNTR (GVTSAPDTRPAPGSTA) conjugated to KLH and mixed with the DETOX adjuvant was injected with 16 patients with breast cancer;[55] 3/16 generated MUC1 antibody responses, 16/16 KLH antibody responses and 7/11 patients showed some CTL activity at higher E:T ratios (100:1) on Class I matched adenocarcinoma target cell lines.[55] Because of the simple induction of CTLs and protection from MUC1+ tumors in mice using M-FP, we immunized patients with advanced adenocarcinoma with M-FP;[45] 3/10 (HLA-A*0201) patients made CTLs; 4/15 (HLA-A1.B8, DR3) patients generated a T cell proliferative response and 13/25 patients made a strong antibody response.[45] We have completed other trials using M-FP and cyclophosphamide in, (i) breast carcinoma patients injected intramuscularly, (ii) adenocarcinoma patients injected subcutaneously and (iii) colon carcinoma patients injected intraperitoneally (Karanikas et al, manuscript in preparation).

Again strong antibody responses were noted particularly after intraperitoneal injection, and a measurable frequency of CTLps and proliferating T cells. At present, other clinical trials will use native and mutated MUC1 peptides, a mimic peptide, cytokines, in vivo targeted antigen presenting cells and whole mucin conjugated to mannan. Other centres will use MUC1 in liposomes with cDNA, vaccinia virus or peptides with various adjuvants. All of these clinical studies are in the very early phase, Phase I/II stages, using patients with advanced disease and thus far, convincing anti-tumor responses have yet to be seen.

One potential problem with administering a "self" antigen such as MUC1 is the development of auto-immunity normal tissues expressing MUC1 (breast, lung, colon, pancreas and ovary), but in the absence of significant anti-tumor responses. This does not appear to be a problem, and with the one exception of the patient with mastitis, has not been seen in mice or humans. If there were complete disappearances of tumors due to the immunotherapy, some level of autoimmunity, which was not life threatening, could be acceptable. However, autoimmunity is unlikely to be a major problem as; (i) the cell surface MUC1 peptide epitopes are hidden from the immune system (by carbohydrates) but could be presented by Class I antigen; (ii) in the normal state the antigens are unlikely to be accessible as they are present at the secretory pole of the cell; (iii) ovarian cancer patients have received large amounts of radiolabelled 99Yt anti-MUC1 antibody and no side effects on MUC1+ tissues were observed;[56] (iv) MUC1 injected into MUC1 transgenic mice showed no evidence of autoimmunity, nor did monkey MUC1 injected into monkeys, nor mouse MUC1 injected in mice (Vaughan et al and Xing et al, unpublished data) and (vi) clinical trials so far show no evidence of autoimmunity. These considerations are not entirely meaningful at present, as the real answer will occur when the results of clinical trials which induce strong cellular responses are induced. However, as mentioned earlier, a breast cancer survivor had her first pregnancy 5 years after being diagnosed with cancer, and developed severe mastitis with heavy lymphocyte infiltration, high CTLp frequency to MUC1 (1/3,086, 1/673, 1/583 in different experiments compared to 1/106 in normal subjects controls) and both IgG and IgM anti-MUC1 antibodies. This may have been a case of autoimmunity and hopefully, this patient is now protected from a recurrence (she is free of breast cancer after 5 years). It was proposed that the initial cancer primed the patient to MUC1, and that the re-expression of the same MUC1 epitopes in the lactating breast induced a secondary immune response.[6]

Prospects for the Future Using MUC1 Based Immunotherapy

MUC1 peptides are an attractive target for immunotherapy as there is approximately 100-fold more MUC1 in breast cancer cells than normal tissues and human MUC1 is highly immunogenic in both normal and MUC1 transgenic mice leading to both antibody and CTLs. The ease and ability to make antibodies to human MUC1 in patients by immunizing with synthetic peptides suggested that MUC1 peptides, perhaps being of lower affinity, could evade tolerance inducing mechanisms as MUC1 reactive T cells would not be eliminated in the early stage of thymic differentiation. However, it is possible that deglycosylated peptides never occur in the native state, always being heavily glycosylated and thus not a true reflection of tolerance. At present, it is possible to generate both antibodies and CTLs in patients and primates to MUC1 VNTR sequences, and under special circumstances, to non-VNTR sequences. However, while it is early days in clinical testing (Phase I/II studies) no substantial tumor responses have been noted. There are a number of possible reasons for this, such as, the cross reactivity of MUC1 with naturally occurring anti-Gal antibodies which divert the immune response to antibody formation by forming immune complexes;[57] or the refractory nature of patients with advanced breast cancer to immunization. The challenge now, is to create an effective anti-tumor immunization procedure. Studies currently being undertaken to improve the efficacy of MUC1 peptides include in vitro sensitization of DCs (possibly by preferentially targeting the mannose receptor); the use of cytokines (especially IL-12) and examining the prime/boost concept by appropriately priming and boosting with a replicating antigen such as vaccinia or

Table 7.3. Clinical studies using MUC1 based immunogens

MUC1 Formulation	Cellular Response	Humoral Response	Phase	Reference
20mer-DT	Weak	Weak	I/II	53
105mer-BCG	Yes	No	I/II	54
16mer-KLH+DETOX	Yes	Yes	I/II	55
FP-mannan	Yes	Yes	I/II	45

Avipox. Consideration should be given to using VNTR and non-VNTR, parts of MUC1 for immunization and also considering other antigens in breast cancer to produce a cocktail and potential targets include p53, Her2/neu and several other oncogenes.

Abbreviations Used

Ab	antibody
CTL	cytotoxic T-lymphcocytes
DC	dendritic cells
ER	endoplasmic reticulum
FISH	fluorescence in situ hybridization
HPLC	high performance liquid chromatography
HPV	human papilloma virus
hsp	heat shock protein
IFN	interferon
IL	interleukin
kD	kilodalton
LAK	lymphokine activated killer cells
MHC	major histocompatibility complex
NSCLC	non-small cell lung cancer
PBMC	peripheral blood mononuclear cells
PCR	polymerase chain reaction
PEA	polyoma virus enhancer A3
RCC	renal cell carcinoma
SCLC	small cell lung carcinoma
TAA	tumor-associated antigens
TAL	tumor-associated lymphocytes
TCR	T-cell receptor
TIL	tumor-infiltrating lymphocytes
TNF	tumor necrosis factor

References

1. Gendler SJ, Papadimitriou JT, Duhig T et al. A highly immunogenic region of a human polymorphic epithelial mucin expressed by carcinomas is made up of tandem repeats. J Biol Chem 1989; 263:12820-12823.
2. Xing PX, Tjandra JJ, Stacker SA et al. Monoclonal antibodies reactive with mucin expressed in breast cancer. Immunol Cell Biol 1989; 67:183-95.
3. Xing PX, Reynolds K, Tjandra JJ et al. Synthetic peptides reactive with anti-human milk fat globule membrane monoclonal antibodies. Cancer Res 1990; 50:89-96.

4. Jerome KR, Domenech N, Finn OJ. Tumor-specific CTL clones from patients with breast and pancreatic adenocarcinoma recognise EBV-immortalised B cells transfected with polymorphic epithelial mucin cDNA. J Immunol 1993; 151:1654-62.
5. Jerome KR, Kirk AD, Pecher G, Ferguson WW, Finn OJ. A survivor of breast cancer with immunity to MUC-1 mucin, and lactational mastitis. Cancer Immunol Immunother 1997; 43:355-60.
6. Ragupathi G. Carbohydrate antigens as targets for active specific immunotherapy. Cancer Immunol Immunother 1996; 43:152-7.
7. Domenech N, Henderson RA, Finn OJ. Identification of an HLA-A11-restricted epitope from the tandem repeat of the epithelial tumor antigen MUC1. J Immunol 1995; 155:4766-74.
8. Agrawal B, Reddish MA, Longenecker BM. In vitro induction of MUC-1 peptide-specific type 1 T lymphocyte and cytotoxic T lymphocyte responses from healthy multiparous donors. J Immunol 1996; 157:2089-2095.
9. Rughetti A, Turchi V, Ghetti CA et al. Human B-cell immune response to the polymorphic epithelial mucin. Cancer Res 1993; 53:2457-2459.
10. von Mensdorff-Pouilly S, Gourevitch MM, Kenemans P et al. Humoral immune response to polymorphic epithelial mucin (MUC-1) in patients with benign and malignant breast tumors. Eur J Cancer 1996; 32:1325-1331.
11. Kotera Y, Fontenont JD, Pecher G et al. Humoral immunity against a tandem repeat epitope of human mucin MUC-1 in sera from breast, pancreatic and colon cancer patients. Cancer Res 1994; 54:2856-2060.
12. Gourevitch MM, von Mensdorff-Pouilly S, Litvinov SV et al. Polymorphic epithelial mucin (MUC-1)-containing circulating immune complexes in carcinoma patients. Br J Cancer 1995; 72:934-938.
13. Apostolopoulos V, Xing PX, McKenzie IFC. Murine immune response to cells transfected with human MUC1: Immunisation with cellular and synthetic antigens. Cancer Res 1994; 54:5186-5193.
14. Lalani E-N, Berdichevsky F, Boshell M. et al Expression of the gene coding for a human mucin in mouse mammary tumor cells can affect their tumourigenicity. J Biol Chem 1991; 266: 15420-26.
15. Graham RA, Stewart LS, Beverley P et al. MUC1-based immunogens for tumor therapy: development of murine model systems. Tumor Targeting 1995; 1:211-21.
16. Apostolopoulos V, Pietersz GA, Xing PX et al. The immunogenicity of MUC1 peptides and fusion protein. Cancer Letters 1995; 90:21-26.
17. Ding L, Lalani EN, Reddish M. Immunogenicity of synthetic peptides related to the core peptide sequence encoded by the human MUC1 gene: Effect of immunization on the growth of murine mammary adenocarcinoma cells tranfsfected with the human MUC1 gene. Cancer Immunol Immunother 1993; 36:9-17.
18. Zhang S, Graeber LA, Helling F, Ragupathi G, Adluri S, Lloyd KO, Livingston PO. Augmenting the immunogenicity of synthetic MUC1 peptide vaccines in mice. Cancer Res 1996; 56:3315-9.
19. Liu X, Sejbal J, Kotovych G et al. Structurally defined synthetic cancer vaccines: Analysis of structure, glycosylation and recognition of cancer associated mucin, MUC-1 derived peptides. Glycoconj J 1995; 12:607-17.
20. Acres B, Hareuveni M, Balloul JM et al. VV-MUC1 immunization of mice-immune response and protection against the growth of murine tumors bearing the MUC1 antigen. J Immunother 1993; 14:136-143.
21. Hareuveni M, Gautier C, Kieny MP et al. Vaccination against tumor cells expressing breast cancer epithelial tumor antigen. Proc. Natl. Acad. Sci. USA 1990; 87:9498-502.
22. Akagi J, Hodge JW, McLaughlin JP et al. Therapeutic antitumour response after immunization with an admixture of recombinant vaccinia viruses expressing a modified MUC1 gene and the murine T-cell costimulatory molecule B7. J Immunother 1997; 1:38-47.
23. Henderson RA, Konitsky WM, Barratt-Boyes SM et al. Retroviral expression of MUC-1 human tumor antigen with intact repeat structure and capacity to elicit immunity in vivo. J Immunother 1998; 21:247-56.
24. Henderson RA, Nimgaonkar MT, Watkins SC et al. Human dendritic cells genetically engineered to express high levels of the human epithelial tumor antigen mucin (MUC-1). Cancer Res 1996; 56:3763-70.
25. Samuel J, Budzynski WA, Reddish MA et al. Immunogenicity and antitumour activity of a liposomal MUC1 peptide-based vaccine. Int J Cancer 1998; 75:295-302.

26. Guan HH, Budzynski W, Koganty RR et al. Liposomal formulations of synthetic MUC1 peptides: effects of encapsulation versus surface display of peptides on immune responses. Bioconjug Chem 1998; 9:451-458.
27. Gong J, Chen D, Kashiwaba M et al. Induction of antitumour activity by immunization with fusions of dendritic and carcinoma cells. Nat Med 1997; 3:558-61.
28. Gong J, Chen L, Chen D et al. Induction of antigen-specific antitumour immunity with adenovirus-transduced dendritic cells. Gene Ther 1997; 4:1023-8.
29. Rowse GJ, Tempero RM, VanLith ML et al. Tolerance and immunity to MUC1 in a human MUC1 transgenic murine model. Cancer Res 1998; 58:315-321.
30. Gong J, Chen D, Kashiwaba M et al. Reversal of tolerance to human MUC1 antigen in MUC1 transgenic mice immunized with fusions of dendritic and carcinoma cells. Proc Natl Acad Sci USA. 1998; 95:6279-6283.
31. Apostolopoulos V, Pietersz GA, McKenzie IFC. Cell-mediated immune responses to MUC1 fusion protein coupled to mannan. Vaccine 1996;14:930-938.
32. Apostolopoulos V, Pietersz GA, Loveland BE et al. Oxidative/reductive conjugation of mannan to antigen selects for T1 or T2 immune responses. Proc Natl Acad Sci USA 1995; 92:10128-10132.
33. Pietersz G A, Wenjun L, Popovski V et al. Parameters in using mannan-fusion protein (M-FP) to induce cellular immunity. Cancer Immunol Immunother 1998; 45:321-6.
34. Apostolopoulos V, Loveland BE, Pietersz GA et al. CTL in mice immunized with human Mucin 1 are MHC-restricted. J Immunol 1995; 155:5089- 5094.
35. Apostolopoulos V, Haurum JS, McKenzie IFC. MUC1 peptide epitopes associated with 5 different H2 Class I molecules. Eur J Immunol 1997; 27:2579- 2587.
36. Apostolopoulos V, Karanikas V, Haurum J. Induction of HLA-A2 Restricted Cytotoxic T Lymphocytes to the MUC1 Human Breast Cancer Antigen. J Immunol 1997; 159:5211-5218.
37. Lofthouse SA, Apostolopoulos V, Pietersz GA. Induction of T1 (CTL) and/or T2 (antibody) response to a mucin-1 tumor antigen. Vaccine 1997; 15:1586-1593.
38. Mandelboim O, Berke G, Fridkin M et al. CTL induction by a tumor-associated antigen octapeptide derived from a murine lung carcinoma. Nature 1994; 369:67-71.
39. Fremont DH, Stura EA, Matsumara M et al. Crystal structure of an H2Kb-ovalbumin peptide complex reveals the interplay of primary and secondary anchor positions in the major histocompatibility complex binding groove. Proc Natl Acad Sci USA 1995; 92:2479-2483.
40. Mutsunori S, Pendleton D. and Berzovsky JA. Broad recognition of cytotoxic T cell epitopes from the HIV-1 envelope protein with multiple Class I histocompatibility molecules. J Immunol 1995; 148:1657-1667.
41. Thomson S, Khanna R, Gardner J et al. Minimal epitopes expressed in a recombinant polyepitope protein are processed and presented to CD8$^+$ cytotoxic T cells: Implications for vaccine design. Proc Natl Acad Sci USA 1995; 92:5845-5849.
42. Rammensee HG, Friede T, Stevanovic S. MHC ligand and peptide motifs: first listing. Anniversary review. Immunogenetics 1995; 41:178-228.
43. Mata M, Travers PJ, Liu Q. The MHC class I- restricted immune response to HIV-gag in BALB/c mice selects a single epitope that does not have a predictable MHC-binding motif and binds to Kd through interactions between a glutamine at P3 and pocket D. J Immunol 1998; 161:2985-2993.
44. Apostolopoulos V, Chelvanayagam G, Xing PX et al. Anti-MUC antibodies react directly with MUC1 peptides presented by class I H2 and HLA molecules. J Immunol 1998;161:767-75.
45. Karanikas V, Hwang L, Pearson J et al. Antibody and T cell responses of patients with adenocarcinoma immunized with mannan-MUC1 fusion protein. J Clinical Invest 1997; 100:2783-2792.
46. Apostolopoulos V, Popovski V, McKenzie IFC. Cyclophosphamide enhances the CTL precursor frequency in mice immunized with MUC1-mannan fusion protein (M-FP). J Immunother 1998; 21:109-13.
47. Sandrin MS, Vaughan HA, Dabkowski PL et al. Anti-pig antibodies in human serum react predominantly with gal(1,3)gal epitopes. Proc Natl Acad Sci USA 1995; 90:11391-11395.
48. Vaughan HA, Oldenburg KR, Gallop MA et al. Recognition of an octapeptide sequence by multiple Gal(1,3)Gal-binding proteins. Xenotransplantation 1996; 3:18-23.

49. Sandrin MS, Vaughan HA, Xing PX et al. Natural human anti gal(1,3)gal antibodies react with human mucin peptides. Glycoconj J 1997; 14:97-105.
50. Apostolopoulos V, Lofthouse SA, Popovski V et al. Peptide mimics of a tumor antigen induce functional cytotoxic T cells. Nature Biotech 1998; 16:276-280.
51. Barratt-Boyes SM, Kao H, Finn OJ. Chimpanzee dendritic cells derived in vitro from blood monocytes and pulsed with antigen elicit specific immune responses in vivo. J Immunother 1998; 21:142-8.
52. Pecher G, Finn OJ. Induction of cellular immunity in chimpanzees to human tumor-associated antigen mucin by vaccination with MUC-1 cDNA-transfected Epstein-Barr virus-immortalized autologous B cells. Proc Natl Acad Sci USA 1996; 93:1699-704.
53. Xing PX, Apostolopoulos V, Michaels M et al. Phase I study of synthetic MUC1 peptides in cancer. Int J Oncology 1995; 6:1283-1289.
54. Goydos JS, Elder E, Whiteside TL et al. A phase I trial of a synthetic mucin peptide vaccine. Induction of specific immune reactivity in patients with adenocarcinoma. J Surg Res 1996; 63:298-304.
55. Reddish M, MacLean GD, Koganty RR et al. Anti-MUC1 class I restricted CTLs in metastatic breast cancer patients immunized with a synthetic MUC1 peptide. Int J Cancer 1998; 76:817-23.
56. Hird V, Maraveyas A, Snook D et al. Adjuvant therapy of ovarian cancer with radioactive monoclonal antibody. Br J Cancer 1993; 68:403-41.
57. Apostolopoulos V, Osinski C, McKenzie IFC. MUC1 cross-reactive Gal alpha(1,3)Gal antibodies in humans switch immune responses from cellular to humoral. Nat Med 1998; 4:315-2027.

Chapter 8

Cytotoxic T Cell Epitopes and Tissue Distribution of the HER-2/neu Proto-Oncogene: Implications for Vaccine Development

Barbara Seliger, Koji Kono, Y. Rongcun and Rolf Kiessling

Introduction

The development of immunotherapeutic methods to treat cancer is critically dependent on the identification of tumor-associated antigens (TAA). Several immunodominant peptide epitopes, recognized by cytotoxic T-lymphocytes (CTL) lines and clones, have been defined from human melanomas.[1-6] This was mainly accomplished by using genetic methods, such as genomic or cDNA expression libraries,[3-6] but also by employing biochemical approaches using high performance liquid chromatography (HPLC) separated peptide fractions.[2] Tumor-specific CTL, raised from tumor-infiltrating lymphocytes (TIL) of cancer patients by culturing in the presence of interleukin (IL)-2[4] or from peripheral blood mononuclear cells (PBMC) of cancer patients upon repeated stimulation with the autologous tumor,[3,6] have been critical reagents for these studies.

The identification of TAA in melanomas has been paralleled by the demonstration of CTL-derived from solid tumors or ascites of patients with carcinomas, capable of recognizing autologous as well as HLA class I-matched allogeneic tumors.[7-11] One major challenge for treatment of carcinomas is the development of therapeutic tumor vaccinations targeting defined T-cell epitopes. However, the knowledge about TAA in carcinomas is very limited in comparison to melanomas.

HER-2/neu, which is the main topic of this review, is among the few examples of TAA-defined from carcinomas. The HER-2/neu proto-oncogene encodes a 185 kD transmembrane receptor-like glycoprotein with tyrosine-specific kinase activity and shares similarities between both the structure and sequence of the epidermal growth factor receptor.[12] As recently reviewed in detail,[13] as well as discussed in this issue, HER-2/neu has several properties which makes it an attractive candidate for specific immunotherapeutic approaches in a variety of human carcinomas. It is amplified and/or over-expressed in approximately 30% of human ovarian and breast tumors[14] and as we will review below also in several other types of human carcinomas. In addition, over-expression of HER-2/neu may directly contribute to the transformation of human mammary epithelium.[15]

Several ongoing clinical trials are based on HER-2/neu as a target for both antibody- and T-cell based immunotherapy as summarized elsewhere in this volume. We will here focus on the evidence and controversies regarding the potential use of this molecule as a target for CTL-based immunotherapy. Issues relevant in this context are the tumor and tissue distribution of HER-2/neu expression, the existence of multiple HER-2/neu specific CTL epitopes, the natural processing of the antigen and the underlying molecular mechanisms by which HER-2/neu expressing tumors might escape from the host's immune surveillance.

HER-2/neu Expression in Normal Tissue and in Tumors

HER-2/neu Expression in Normal Adult and Fetal Tissues

HER-2/neu is expressed in both fetal and adult tissue derived from all three germ layers. Ectodermal derivatives, such as skin and mammary glands, mesodermal derivatives, i.e., kidney, ureter, vagina, cervix, uterus, and endodermal derivatives, i.e., oropharynx, esophagus, stomach, intestines, pancreas, lungs, prostate and bladder, all express HER-2/neu. Thus, epithelial cells of most organs including the gastrointestinal, respiratory, urinary and reproductive tracts as well as skin reveal a HER-2/neu positive expression profile.[16,17] The HER-2/neu staining pattern of epithelial cells is quite heterogeneous ranging from barely detectable, over weakly to distinctly positive. In comparison to corresponding adult tissue, the level of HER-2/neu expression is more pronounced during fetal development (Table 8.1). Cells of the lamina propria, muscularis and serosa show no immunoreactivity with HER-2/neu specific antibodies neither in fetal nor in adult tissue samples. The determination of the normal expression pattern of HER-2/neu is of importance in terms of its therapeutic value due to its differential high expression levels in tumors.

HER-2/neu Expression in Tumors of Different Origins

Amplification and over-expression of HER-2/neu has been demonstrated in both cell lines and biopsies from human malignancies of distinct histology at a relatively high frequency (Table 8.2). HER-2/neu is ubiquitously expressed in many epithelial tumors including breast and ovarian tumors,[14,18] colon carcinoma,[19] prostatic adenocarcinoma,[20] cervix carcinoma,[21] gastric cancer,[22,23] pancreatic carcinoma,[24] renal cell carcinoma (RCC)[25,26] and non-small cell lung carcinoma (SCLC).[27,28] Its dysregulated expression is often associated with the progress of malignancy, metastatic phenotype and chemoresistance.[29] The relative selectivity of HER-2/neu gene amplicons and/or its over-expression in human tumors of epithelial origin and the association of HER-2/neu with a pathogenic mechanism responsible for the generation of neoplasia in some malignancies, such as breast carcinoma, make the HER-2/neu antigen an attractive target for the development of an immunotherapeutical approach.

Gynecological Tumors

Aberrant HER-2/neu expression, often associated with gene amplification and/or dysregulation, occurs at a distinct frequency in human gynecological adenocarcinomas including those of the ovary, endometrium, cervix and breast. The immunostaining of normal breast tissue displays weak to low levels of HER-2/neu reactivity which is comparable to non-HER-2/neu over-expressing breast cancers. In contrast, breast tissue is negative in any embryonic tissue.[16] Generally, HER-2/neu over-expression is seen in 15 to 40% of invasive breast cancers and in 60 to 80% of ductal carcinomas-in-situ[14,30] as a result of either gene amplification and/or enhanced transcription. Transcriptional upregulation of HER-2/neu leads to a 6- to 8-fold increase of its mRNA level and involves the induction of the transcription factor polyoma virus enhancer A3 (PEA3). In more than 90% of HER-2/neu over-expressing human breast tumor samples, an elevated activity of PEA3 has been shown implicating an important role of PEA3 in the initiation and progression of HER-2/neu positive breast cancer.[31]

Table 8.1. Expression of HER-2/neu oncogene in fetal and adult tissues

Tissues	immunostaining for HER-2/neut Fetal	Adult
breast	N.D.	+
female reproductive		
ovary	-	-*
uterus endometrium/ cervix	+	+/ weak/+
male reproductive		
testis	-	-
prostate	N.D.	weak/+
gastrointestinal		
stomach small and large intestine	+	weak
liver-hepatocytes	+	weak
bile ducts	+	-
pancreatic ducts	+	weak
urinary		
kidney	+/++	weak/+
ureter	+	weak
bladder	+	r-/weak
circulatory (heart, arteries, veins)	-	-
respiratory		
bronchi	+	+
alveoli	-	-
hematologic (liver, spleen, bone marrow)	-	-
musculoskeletal (bone, cartliage, muscle)	-	-
central nervous system (brain, spinal cord, eye, meningis)	-	-
endocrine system (adrenal cortex and medulla, pancreatic islets)	-	-

N.D.: not determined according to Press and coworkers[16]

Both the degree of gene amplification and the level of HER-2/neu over-expression vary widely within and among different breast carcinomas.[32] This heterogeneity provides a potential source for the selection of subclones with increased malignant and metastatic potential, especially in the context of its therapeutic targeting.[33,34] Indeed, HER-2/neu gene amplification and its over- expression has been demonstrated to correlate with poor prognosis in patients with positive lymph-nodes, as measured by both a lower overall and disease free survival, tumor grading, tumor size, nodal involvement and an inverse association with steroid hor-

Table 8.2. Amplification and over-expression of HER-2/neu in human tumors

Tumor type	Over-expression	Amplification	Literature
non-small cell lung cancer	50%.	21% 3-30 copies	27,28
renal cell carcinoma	10-45	none	26,62, Seliger et al,submitted
bladder cancer	30-81#	15% 2-15 copies	122
prostate carcinoma	>20%	44%	20, 60, 61
gastric adenocarcinomas	>20%	20-43%	22,25
esophageal adenocarcinoma	64%	15% 2-3 copies	123
breast cancer with axillary lymph node	90%*	15%-40% 2-20 copies	37
node-positive/ node-negative breast cancer	n.d.	25-29%	10
cervical cancer	20%	n.d.	46
	n.d	14% 5-68 copies.	124
ovarian cancer	20-46%	highly variable, depending on tumor stage	55
endometrial cancer	12-52%	20%	48,50,54
melanoma	10-40%	n.d.	88

The over-expression was mainly determined by either immunohistochemistry, Northern blotting or RT-PCR, whereas amplifactions were determined by Southern blotting or FISH. *% of HER-2/neu amplified tumors # depending on the method employed (Northern blot/immunohistochemistry) n.d. not determined

mone receptors.[35-39] The significance of HER-2/neu over-expression in node-negative patients is still controversial.[40] The recent identification of a truncated HER-2/neu protein, p95 HER-2/neu, which is differentially expressed in HER-2/neu positive tumor tissue samples from lymph-node positive and lymph-node negative patients suggest that this p95 HER-2/neu protein may be employed as an additional marker in breast cancer patients.[41] However, the prognostic significance of this observation has to be validated by a different patient population.

In breast cancer, poor prognosis partially relates to enhanced invasiveness rather than to increased proliferative capacity of the tumor cells.[42] The relationship between HER-2/neu protein over-expression and outcome is complex, and over-expression seemed not to be always associated with aggressive disease. In stage I breast cancer patients, HER-2/neu over-expression has been shown to coincide with a favorable prognosis, which may be due to the presence of tumor-infiltrating lymphocytes (TIL).[43] However, these results have yet to be confirmed.

The hypothesis that HER-2/neu over-expression plays an important role in the pathogenesis of human breast cancer is underlined by studies using mice transgenic with the activated or wild-type neu proto-oncogene under the control of the mouse mammary tumor virus promoter. These transgenic mice develop mammary tumors, but with a distinct kinetic.[44,45]

As evaluated by immunohistochemistry, HER-2/neu over-expression is frequently detected in invasive cervical cancer (20%), preferentially in human papilloma virus (HPV)-16 positive lesions.[46] In comparison to the ectocervical squamous and endocervical columnar epithelium of the adult cervix, HER-2/neu staining is more pronounced in these lesions. HER-2/neu positive tumors display a heterogeneous distribution of immunoreactive tumor cells. The pattern and intensity of HER-2/neu staining is neither influenced by tumor grade nor by histology. In addition, HER-2/neu expression cannot be used as a predictor of survival of cervical cancer patients nor as a marker for high risk patients.[21]

HER-2/neu gene amplification and protein over-expression is an alteration which is associated with the neoplastic phenotype in endometrial neoplasms, but the frequency is controversially discussed.[47-49] In normal cyclic, postmenopausal and hyperplastic endometrial epithelium HER- 2/neu is variably present. Using immunohistochemical staining barely detectable to low levels of HER-2/neu expression is observed, but HER-2/neu is neither amplified nor over-expressed in these tissues.[16] In contrast, 12 to 52% of endometrial cancers show a moderate or high immunostaining pattern for HER-2/neu which correlates with gene amplification in approximately 20% of cases.[48,50] HER-2/neu gene amplification without protein over-expression is only rarely detected within endometrial cancer.[50,51] Since HER-2/neu oncoprotein is found in all clinical stages, it seems to be an early event in the natural history of this disease.[52] The clinical follow-up of patients with endometrial cancer reveals a significant correlation between HER-2/neu gene amplification and/or protein over-expression and poor overall survival.[50,53] Thus, the altered HER-2/neu status is a potential prognostic marker of poor outcome for this disease and patients at a high risk of recurrence might benefit from intensified therapy.[54]

In addition, HER-2/neu over-expression occurs in about 20 to 46% of invasive ovarian carcinomas and in approximately 15% of borderline cancer. The incidence of HER-2/neu gene amplification in the late stage disease is significantly higher than that in the early stage.[53] These data suggest that HER-2/neu amplification may be a prognostic factor for invasive neoplasm and seems to be associated with an unfavorable clinical course. Since HER-2/neu gene amplification is relatively infrequent in borderline tumors, it may represent a late event in ovarian carcinogenesis.[55]

Carcinomas of the Gastrointestinal Tract

In gastric adenocarcinomas, high levels of HER-2/neu gene amplifications are detected in approximately 20 to 43% of all cases.[22,23,25] Using fluorescence in situ hybridization (FISH), amplified HER-2/neu DNA is mainly found in homogeneously staining regions and occasionally in double minute chromosomes and results in over-expression of the protein. The pathological stage and HER-2/neu gene amplification rate are independent parameters in this disease.[22]

HER-2/neu over-expression is common in many pancreatic carcinomas and seems to be related to glandular differentiation and early oncogenesis.[56] In addition, HER-2/neu expression is linked to the hyperplasia grading, in most cases being negative in corresponding normal duct epithelium,[16] weak in flat hyperplasia and moderate to strong in atypically hyperplasia and carcinoma in situ, suggesting that HER-2/neu expression might be used as a marker of hyperplasia, dysplasia and neoplasia.[57]

Similarly to normal and malignant pancreatic tissues, normal colon mucosa, benign lesions and colorectal adenocarcinomas differ in the expression levels and distribution of the HER-2/neu antigen. Staining pattern of normal mucosa is mostly negative whereas a number of benign lesions and adenocarcinomas over-express HER-2/neu protein.[19] In contrast to be-

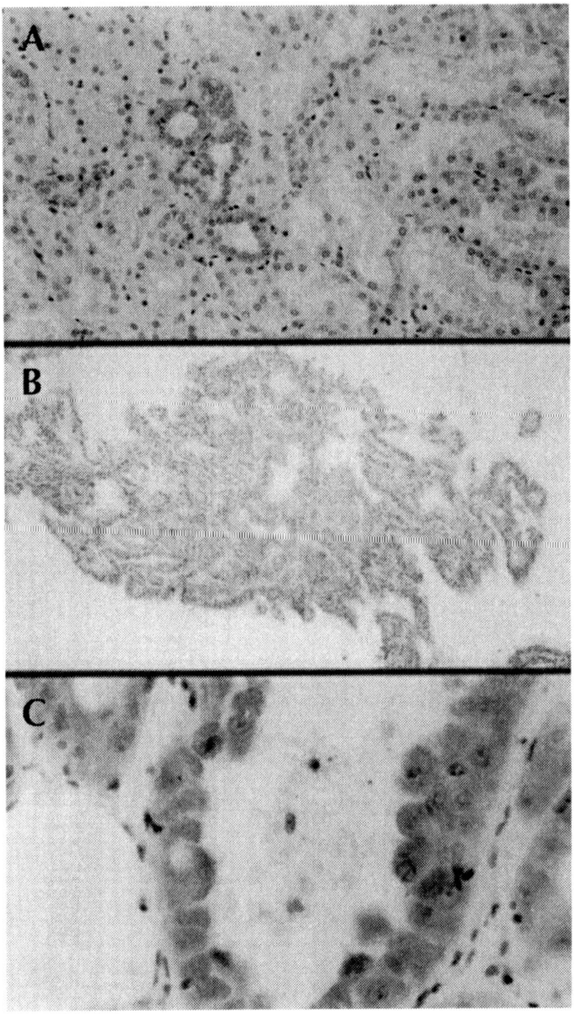

Fig.8.1A. HER-2/neu protein expression in kidney tumors. Immunohistochemical staining of normal kidney epithelium and RCC was performed using standard methods. (A) Corresponding normal kidney epithelium to the tumor shown in Figure 8.B,C. with a very weak positive staining for the cerbB2 antibody (200x) in the epithelium of the distal tubulus system. Chromophilic RCC, grade II (H.E. 100x) with classic papillary growth pattern of eosinophilc tumor cells showing macrophages and edema in the center of some papillae. Immunohistochemical staining for the c-neu antibody (100x) and for the c-neu antibody (400x) in another area of the same tumor as in (B) showing moderate positive staining in the cytoplasm of the tumor cells.

nign lesions, adenocarcinomas are significantly more positive, further demonstrating a correlation between the degree of epithelial abnormality and clinical parameters.

Carcinomas of the Respiratory Tract

In all stages of fetal development and in the neonate, the lung epithelium is immunoreactive for HER-2/neu,[16] whereas adult lungs show a barely detectable HER-2/neu expression profile. Over- expression of HER-2/neu is encountered in a subgroup of non-small cell lung carcinoma (NSCLC), but not in small cell lung carcinoma (SCLC). In adenocarcinomas, high levels of HER-2/neu over-expression is relatively frequent (approximately 50%). Compared to patients with low HER-2/neu expression, patients with HER-2/neu protein over-expression have a significantly higher incidence of early tumor recurrence and a shortened survival rate.[28,58,59] In addition, the analysis of a number of different NSCLC cell lines established from previously untreated patients demonstrate an association of HER-2/neu over-expression with intrinsic multiple drug resistance.[29] These data are further confirmed by enhanced chemoresistance of NSCLC lines upon HER-2/neu gene transfer.[29] In addition, a correlation between the protein over-expression of HER-2/neu coupled to a higher S-phase frequency and shorter doubling time of cells as well as intrinsic chemoresistance has been described.[29]

Carcinomas of the Urinary Tract

Although the prognostic role of the HER-2/neu oncogene has been consistently reported in a number of cancers, studies on the frequency and predictive value of HER-2/neu alterations in prostate adenocarcinoma are limited and with inconsistent results, especially by employing immunohistochemistry.[20,60] In comparison to prostatic adenocarcinomas, prostatic hyperplasia and prostatic intraepithelial neoplasia express higher levels of the HER-2/neu protein.[60] Using FISH, 44% tumor specimens showed HER-2/neu gene amplifications which were associated with tumor grading, non-diploid DNA content and recurrence.[61]

The distribution of HER-2/neu expression in fetal and adult kidney is similar and the intensity of their immunostaining ranges from weak staining in Bowman's capsule and in the proximal tubulus to moderate staining in the collecting ducts (Seliger et al, submitted).[16] However, the level of the HER-2/neu oncoprotein expression rate is very low in normal kidney whereas in renal neoplasm overexpression of HER-2/neu is found in more than 40% of primary epithelial renal tumors and more than 30% of primary renal cell carcinoma (RCC) specimens. However, this is often not due to structural alterations, such as HER-2/neu gene amplification or chromosomal rearrangements.[26,62-64] Based on the histogenetic classification according to Thoenes and coworkers,[65] a distinctive and in the distribution heterogeneous HER-2/neu expression pattern is demonstrated in the different subtypes of kidney tumors with the highest frequency in chromophilic and chromophobic RCC, but not in clear cell RCC (Seliger et al, submitted; Figure 1B,C; Table 3).[26] The corresponding normal kidney epithelium, i.e., the distal tubulus system and collecting ducts, exhibit very low levels of HER-2/neu expression with a focal staining pattern (Figure 1A). The pattern and intensity of HER-2/neu expression does not seem to correlate with tumor stage, grading and progression of this disease (Seliger et al, submitted).[26] Similar to surgically removed RCC lesions, RCC tumor cells display a distinct HER-2/neu expression pattern. In all RCC cell lines analyzed, HER-2/neu protein expression is lower in comparison to HER-2/neu over-expressing ovarian carcinoma cells.

Validity of Methods Employed for the Assessment HER-2/neu Status

Although gene amplification and/or protein over-expression of HER-2/neu may be considered as a prognostic parameter in patients with epithelial tumors, its suitability is still controversial, due to the distinct validity of the different techniques employed for the assessment of the HER-2/neu status in tumors.[66,67] These methods include Southern blotting, slot blot analysis, differential polymerase chain reaction (PCR) and FISH for the detection of gene amplifications and ELISA, Western blotting, immunohistochemistry as well as flow cytometry for the evaluation of HER-2/neu protein over-expression. A number of studies demonstrated that the use of FISH provides a more accurate assessment of the prognostic significance of HER-2/neu gene amplification in tumors.[39,54,61,68] Employing the ELISA technique and differential PCR analysis, a high correlation between the degree of HER-2/neu gene amplification and

Table 8.3. Histopathological data of RCC patients and distinctive HER-2/neu expression pattern in RCC subtypes

Subtype of RCC	Grade	Positive for c-erbB2	Positive for c-neu	Number of tumors
Clear cell	I	3	0	11
	II	3	1	10
	III	2	0	10
Chromophilic	I	3	2	13
	II	20	9	41
	III	0	0	13
Chromophobe	I	4	2	6
	II	10	1	14
Oncocytoma		7	5	10
Total		52	20	128

Renal cell carcinoma (RCC) specimens and normal corresponding kidney epithelium from 128 patients were obtained after radical nephrectomy. Tumor-node-metastasis (TNM) staging and histopathological evaluation was performed according to the classification of Thoenes and coworkers.[65] Informed consent was received from all patients. Formalin fixed, paraffin embedded specimens were subjected to immunohistochemical analyses using the antibodies c-erbB2 and c-neu using standard protocols.

HER-2/neu protein over-expression were obtained. A cellular HER-2/neu expression level higher than or equal to 260 fmol/ mg protein has been defined as a safe cut-off value for an altered HER-2/neu status in given tumor specimens. However, these techniques provide no information about the quantity, quality and heterogeneity of the expression level.[66,67,69,70]

CTL Epitopes Defined from HER-2/neu

Shared Tumor Antigens Defined by Carcinoma Specific HLA-A2 Restricted CTL

MHC class I-restricted lysis by T-cells isolated from several types of human solid tumors, including renal, ovarian and lung carcinomas, was found in early studies.[71] Subsequently, CTL clones were developed from tumor-infiltrating lymphocytes (TIL) of the ascites from patients with ovarian carcinomas, obtained by co-culture of TIL with autologous tumor cells and cloned in the presence of autologous tumor cells.[72] These CTL clones reveal the highest reactivity measuring either cytotoxicity or cytokine release against autologous tumor cells. The fine-specificity analysis of these autologous tumor-specific CTL demonstrated the simultaneous or separate presentation of at least three antigenic epitopes on the tumor cells. Thus, there exists a substantial heterogeneity of antigens recognized by ovarian specific CTL.

The presence of shared tumor antigens among solid tumors was later demonstrated by Peoples and coworkers[10,73] using MHC class I-restricted tumor-specific CTL generated from both tumor- associated lymphocytes (TAL) from malignant ascites and TIL from solid ovarian tumors stimulated by anti-CD3, low dose IL-2, and repeated co-cultivation with tumor cells. The same group also used the approach of acid-eluting HLA-bound peptides from ovarian cancers. Upon HPLC-fractionation of the peptides, they screened for the relevant potential T-cell epitopes by loading the HLA-A2+ T2 cell line with distinct fractions in order to determine whether common TAA exist among HLA-A2+, HER-2/neu+ double positive epithelial cancers. Employing this method, CTL generated from TIL isolated from ovarian, breast and non-small cell lung cancers (NSCLC) recognized at least three identical peptide fractions within each of the respective elution profiles. One of these fractions co-eluted with a HER-2/neu-derived peptide.[73] In our earlier and ongoing studies on the characterization of the fine-specificity of CTL lines produced from TIL of solid tumors or from TAL of ascites fluid of ovarian carcinoma following repeated stimulation with the autologous HLA-A2+ tumor with a low dose of IL-2, cross-reactivity of CTL lines between different HLA-A2+ carcinomas has been consistently found (Rongcun et al, submitted for publication].[74,75] Thus, ovarian-specific CTL recognize shared tumor antigens among various types of carcinomas in an HLA-restricted manner and with HLA-A2 representing one major restriction element.

Evaluation of freshly harvested TIL has demonstrated a restricted use of T-cell receptor (TCR) genes in some instances. Moreover, there is also evidence that HLA-A2-restricted ovarian cancer-specific T-cell populations consist of a limited T-cell repertoire. Selective usage of the Vβ6 or Vβ5 TCR products has been found in TIL lines derived from three out of nine patients with ovarian carcinomas.[76] HLA-A2- restricted, tumor-specific CTL isolated from TIL in ovarian cancer specific for HER-2/neu showed an increase in the proportion of Vβ2, Vβ3 and Vβ6 TCR chains.[11] Their cytotoxic activity could be completely blocked by a combination of monoclonal antibodies (Ab) directed against Vβ2, Vβ3, and Vβ6. However, a final proof of a selective usage of Vβ families in HER-2/neu specific CTL still awaits further studies on a large panel of specific CTL clones.

HER-2/neu as a Shared Tumor Antigen Among Tumor Specific CTL

CTL specific for HER-2/neu are part of the T-cell repertoire within TIL lines derived from ovarian carcinomas as well as other types of tumors. The majority of these HER-2/neu T-cell epitopes are presented in the context of HLA-A2 or -A3 molecules (Table 8.4). Early evidence for the involvement of HER-2/neu in the recognition of tumor- specific CTL was obtained from studies targeting ovarian carcinoma cell lines with high or low expression rate of HER-2/neu: The high HER-2/neu expressing tumor lines displayed a significantly higher sensitivity to CTL killing in comparison to the low level expressing ones.[78] Further support for HER-2/neu being a target for tumor- specific CTL is based on the treatment of tumor cells with interferon (IFN)γ. After this exposure, HER-2/neu expression was significantly decreased in some tumor cells, whereas the expression of HLA class I molecules was significantly increased in all cases. Therefore, despite upregulation of the MHC class I surface antigen density, HER-2/neu specific CTL-mediated lysis was often decreased. To confirm the apparent association between the HER-2/neu expression rate and CTL recognition in tumors HER-2/neu negative, HLA-A2 positive melanoma cells that were insensitive to lysis by tumor-specific CTL were transfected with the HER-2/neu gene. These HLA-A2+, HER-2/neu transfectants were then efficiently recognized by HLA-A2-restricted ovarian tumor-specific CTL, but not by HLA-A2-mismatched CTL.

The group of Eberlein and collaborators have focused on the HER-2/neu specific epitopes $GP2_{654-662}$ and GP2L differing in a point mutation at position 655 (valine to leucine).[8] The GP2L creates a binding motif in GP2 for a mutated HER-2/neu epitope, whereas the GP2

Table 8.4. HER-2/neu specific T-cell epitopes and their specificity

Epitopes	CTL source	Tumor recognition HLA restriction	Reference
P369-377 (KIFGSLAFL)	OVA TAL	Auto and Allo OVA tumor	7
P971-979 (ELVSEFSRM)		(A2+, HER2+) HER-2/neu+	
P369-377 p971-979	OVA TAL and TIL (A2+, HER-2/neu+)	Auto and Allo OVA tumor	74
P 48-56 (HLYQGCQVV) P789-797 (CLTSTVQLV)t	Healthy PBL	Not shown against A2+ tumor, only peptide-coated targets	86
P369-377 P654-662 (IISAVVGIL)	Healthy PBL	Breast, colon and RCC tumor (A2+, HER-2/neu+)	25
P369-377 P773-782 (VMAGVGSPV)	A2.1/CD8 transgenic mice	ePanel of A2+, HER-2/neu+ tumor	82
P369-377 P689-697 (RLLQETELV)	Gastric TAL	Auto and Allo gastric tumor (A2+, HER-2/neu+)	75
P971-980 (ELVSEFSRMA)	OVA TAL	Auto OVA tumor (A2+, HER-2/neu+)	125
P968-981 (RFRELVSEFSRMAR) p971-979 analogue (ELVSEFSRM)	Healthy PBL	Allo OVA tumor (A2+, HER-2/neu+)	126
P654-662	Breast and OVA TAL	Auto and Allo OVA tumor (A2+, HER-2/neu+)	79
P654-662	Lung TIL	Auto and Allo Lung tumor (A2+, HER2+) Allo OVA tumor (A2+, HER-2/neu+)	8
P654-662	Breast and OVA TAL	Auto OVA and Auto Breast tumor (A2+, HER2+) Pancreas tumor (A2, HER-2/neu+)	127
P435-443 (ILHNGAYSL) P5-13 (ALCRWGLLL)	Healthy PBL	A2+, HER-2/neu+ tumors	81
P754-762 (VLRENTSPK)	Healthy PBL	A3+, HER2+ tumors	87

OVA: ovarian tumor; TAL: tumor-associated lymphocytes; TIL: tumor-infiltrating lymphocytes; Auto: autologous; Allo: allogeneic; A2+: HLA-A2 positive; HER2+: HER-2/neu over-expressing; RCC: renal cell carcinoma.

motif represents the wild-type motif. The authors demonstrated the induction of HLA-A2 restricted GP2 and GP2L specific CTL using PBL derived from pancreatic cancer patients. Furthermore, the wild-type GP2 epitope is also recognized by CTL isolated from human HLA-A2+ non-small cell lung cancer (NSCLC). Moreover, these CTL recognize HLA-A2+, HER-2/neu+ autologous and allogeneic NSCLC cell lines as well as HLA-A2+, HER-2/neu+ heterologous ovarian cancer cell lines.[8] In addition, this group found evidence that the GP2 epitope is one of the shared antigens recognized by both HLA-A2-restricted breast and ovarian tumor-specific CTL.[79,80] In contrast, PBL stimulated with the HER-2/neu derived peptide $GP1_{650-658}$ which has a high binding affinity to HLA-A2 were not capable to induce CTL recognizing HER-2/neu+ tumor cells.

Upon acidic elution of HLA-bound peptides from the surface of ovarian cancer cells, one of the peptide fractions analyzed co-elutes with the GP2 epitope.[73] Kawashima and coauthors[81] recently demonstrated that this GP2 epitope ($aa_{654-662}$) is a poor binder to the HLA-A2.1 allele, while two closely related peptides including HER2 [9_{653}] and HER2 [10_{654}] bind well. They speculate that one of the two HLA-A2.1 binding peptides could represent the actual epitope recognized by the CTL targeting the GP2 epitope.

The HER2 [9_{369}] epitope, termed E75,[7] was first defined by synthetic peptide analogues based on the presence of HLA-A2.1 anchor motifs in order to identify potential epitopes of HER-2/neu which might be recognized by antigen-specific tumor-reactive CTL. Nineteen synthetic peptides were selected and tested for recognition by four HLA-A2-restricted CTL lines obtained from PBMC of patients with ovarian cancer. The HER2 [9_{369}] epitope efficiently sensitized T2 cells for lysis by each of the four CTL lines[7] and was specifically recognized by a CD8+ CTL clone isolated from one of these ovarian specific CTL lines. HER2 [9_{369}]-pulsed T2 cells inhibited the specific lysis mediated by the same CTL clone if challenged with either an HLA-A2+, HER-2high ovarian tumor or an HER-2high, HLA-A2 transfected ovarian tumor line, suggesting that this or a structurally similar epitope is specifically recognized by these CTL. The recognition of HER2 [9_{369}] by tumor-specific CTL and the presentation of this epitope on the surface of various tumor targets has also been confirmed by other groups.[25,74,82]

HLA-A2+ patients with metastatic breast, ovarian and colorectal carcinoma, immunized with the HLA-A2, HER-2/neu binding epitope in the presence of Freund's incomplete adjuvant, developed peptide-specific CTL.[83] Although upon peptide immunization a T-cell response to the HER2 [9_{369}] epitope could be generated in three out of four patients, these T-cells failed to lyse HER- 2/neu over-expressing tumor cells. In addition, neither infecting HLA-A2+ cells with recombinant vaccinia virus encoding HER-2/neu nor IFN-γ treatment of HER-2/neu+ cells resulted in antigen-specific cytotoxicity by p369-reactive T-cells. The inability of HER2 [9_{369}] specific T-cells to recognize HER-2/neu expressing tumors in this study may depend on the isolation of T-cells from patients immunized with synthetic peptides, which may favor low-affinity T-cell responses. Alternatively, differences in the CTL induction protocol as compared to those used by other groups which generated HER2/neu specific T-cells may account for this. Thus, further studies are necessary to determine whether immunotherapeutic strategies based on single HER-2/neu derived peptide epitopes, such as HER2 (9_{369}) can successfully be used in immunotherapy against cancer.

Generation of HER-2/neu Specific CTL Through In Vitro Stimulation with Peptides and the "Reverse Immunology" Approach

Former efforts to generate tumor-specific CTL by in vitro peptide priming protocols with the immunodominant epitopes HER2 [9_{369}] (E75) and HER2 [9_{654}] (GP2) have been successful. The group of T. Eberlein was able to generate CTL from ascites of HER-2/neu+ ovarian cancer patients by either stimulating with the HER2 [9_{654}] peptide epitope or with autologous tumor cells. Both modes of stimulation generated specific CTL against autologous ovarian tumors.[84] Likewise, Brossart and co-authors[25] analyzed whether HER-2/neu epitopes were TAA for renal cell carcinoma (RCC) and for colon carcinoma by employing HER-2/neu peptide specific CTL. In their protocol, autologous dendritic cells (DC) generated from peripheral blood monocytes pulsed with the HER-2/neu derived peptides E75 and GP2 were used as antigen presenting cells for the initial CTL priming. An HLA-A2-restricted CTL activity towards peptide-pulsed targets already occurred past two rounds of weekly restimulations. Furthermore, these CTL lysed not only breast cancer cells but also colon carcinoma and RCC cell lines expressing HER-2/neu in a MHC- and antigen-restricted fashion. The cytotoxic activity against tumor cells was blocked by cold HLA-A2-positive targets pulsed with the cognate peptide in cold target inhibition assay and by anti-HLA-A2 directed monoclonal antibody (Ab).

These reports of peptide-induced HER-2/neu specific CTL utilized peptide epitopes already previously identified in tumor-specific CTL from cancer patients. Therefore these epitopes represent a relatively narrow repertoire of identified epitopes which are dominant in the tumor-specific CTL response. The use of a diverse multi-epitope approach based on subdominant epitopes may be of considerable therapeutic value. As an alternative approach towards the identification of potential antigenic determinants which does rely on the generation of tumor-specific CTL from patients the "reverse immunology" approach has been introduced.[85] This method relies on the identification of MHC-binding peptides from known molecules, such as HER-2/neu, and on employing various in vitro priming protocols to generate CTL against the candidate epitopes. In a first application of this technology towards the identification of HER-2/neu derived epitopes, four peptides with HLA-A2.1-binding amino acid motifs were synthesized.[86] Two of the four peptides (HER2 [9_{48}] and HER2 [9_{789}]) were shown to elicit peptide-specific CTL by primary in vitro immunization in a culture system using peripheral blood lymphocytes from a normal individual homozygous for HLA-A2. However, in this study there exists no evidence whether these peptides might be presented on tumor cells. Recently, Kawashima et al[81] used an extensive "reverse immunology" approach to identify a set of CTL epitopes from TAA frequently found on solid epithelial tumors such as breast, lung and gastrointestinal tumors. This study included a large panel of HLA-A2.1 binding peptides from the MAGE2, MAGE3, HER-2/neu and CEA antigens, which were tested for their capacity to elicit in vitro anti-tumor CTL using lymphocytes from normal volunteers and autologous DC as antigen-presenting cells. A total of six new epitopes (MAGE2 [10_{157}], MAGE3 [9_{112}], CEA [9_{691}], CEA [9_{24}], HER2 [9_{435}] and HER2 [9_{5}]) were identified which were all capable of specifically recognizing tumor cell lines expressing HLA-A2.1 and the corresponding TAA. Most of the newly identified epitopes were found to be highly cross-reactive with other common HLA-alleles of the HLA-A2 supertype (A2.2, A2.3, A2.6 and A6802), thus demonstrating their potential in providing broad and non-ethnically biased population coverage.[81] Two new HER-2/neu CTL epitopes were identified in this study (HER2 [9_{435}] and HER2 [9_{5}]) which showed an intermediate binding to HLA-A2.1. This study also confirmed that two previously defined epitopes (HER2 [9_{369}])[7] and HER2 [9_{789}][85] bind with high affinity to HLA-A2.1, although they were only able to generate CTL in a minority (2/29) of the HER2 [9_{369}] and in none (0/3) of the HER2 [9_{789}] stimulated cultures.

Recently, Kawashima and coworkers[87] identified HLA-A3 restricted CTL epitopes from the HER-2/neu and CEA antigens. HLA-A3-binding synthetic peptides from CEA and HER-2/neu were tested for their immunogenicity by performing in vitro primary CTL response assays using peripheral blood mononuclear cells from healthy volunteers. One peptide from HER-2/neu (HER2 [9_{754}]: VLRENTSPK) was shown to induce CTL that were capable of recognizing and eliminating HER-2/neu+ tumor cells. Additional MHC-binding studies to the most common HLA molecules of the HLA-A3 superfamily (HLA-A*1101, -A*3101,

-A*3301, and -A*6801) demonstrated that the HER2[9_{754}] epitope binds to four out of five alleles tested. These results indicate that this novel CTL epitope might be immunogenic in individuals expressing members of the HLA-A3 superfamily.

We have recently used the same approach to identify HER-2/neu epitopes which can elicit tumor-specific responses in T-cells from patients with ovarian, melanoma and renal cell carcinomas (RCC; Seliger et al, submitted for publication, Rongcun et al, submitted for publication). The same set of 22 HLA-A2.1 HER-2/neu derived peptides as selected by Kawashima and coauthors[81] were analyzed for their capacity to elicit peptide and tumor specific CTL responses. Peptide-pulsed DC from the ascites of patients with ovarian carcinomas were used as antigen presenting cells for the stimulation of autologous T-cells isolated from the same ascites. Out of the 22 HER-2/neu derived peptides binding to HLA-A2 with either high (IC < 50) or intermediate (IC < 50 IC < 500) affinity, we were able to define that at least five epitopes, including HER2 [9_{435}], HER2 [9_{665}], HER2 [9_{689}], HER2 [10_{952}], and the known HER2 [9_{369}] epitope[7] which were able to induce specific CTL responses. These epitope specific CTL exert cytotoxic activity to peptide sensitized target cells even when taken from patients with tumors over-expressing HER-2/neu. Most importantly, a HER-2/neu transfected cell line and the autologous tumor cells were also recognized.

Immunohistochemical studies and flow cytometric analyses demonstrate that human melanoma can significantly express the HER-2/neu antigen,[88] but the levels of HER-2/neu expression are lower as compared to ovarian carcinoma lines (Rongcun et al, submitted for publication). In addition, four of the analyzed melanoma lines were specifically killed by both HER-2/neu specific CTL lines and CTL clones specific for the HER2 [9_{369}], the HER2 [9_{435}] and the HER2 [9_{689}] epitopes and their specificity was further confirmed by cold target competition assays. HER-2/neu specific CTL mediated lysis of melanomas may broaden the potential use of immunotherapy targeting the HER-2/neu TAA.

Also renal cell carcinomas (RCC) were shown to be sensitive to HLA-A2-matched allogeneic CTL lines induced by three different peptide epitopes (HER2 [9_{369}], HER2 [9_{435}] and HER2 [9_{689}]) and by two CTL clones specific for the HER2 [9_{435}] and HER2 [9_{689}] epitopes (Seliger et al, submitted for publication). Again, their specificity was further confirmed by cold target inhibition assays. In addition, CTL mediated lysis could be enhanced by pulsing tumor cells with exogenous HER/neu specific peptides. In conclusion, HER/neu is heterogeneously expressed in different subtypes of RCC (Table 3), and HER/neu epitopes presented by RCC can be recognized by HER-2/neu specific CTL as was also shown by others for the HER2(9_{369}) and the HER2 (9_{654}) epitopes.[25] Thus, HER-2/neu might represent a potential antigenic target for the development of RCC-specific vaccines.

MHC Class II HER-2/neu Specific Peptides

In the last decade, most studies have characterized the reactivity of CD8+ CTL against autologous and allogeneic tumor cells. Recently, there is growing information about antigens and epitopes recognized by CD4+ T-cells in tumor patients.[89] In vitro cultured CD4+ T-cells of TIL could recognize antigenic peptides in the context of MHC class II molecules: An HLA-DR4.1-restricted tyrosinase-specific peptide has been identified by Topalian and coworkers,[90,91] which was capable to stimulate secretion of the Th1 cytokine IFN-γ by CD4+ melanoma TIL. CD4+ CTL could also recognize a shared HLA-DR15 presented melanoma associated antigen.[92] In addition, autologous, tumor-specific HLA-DR4 and HLA-DR15 restricted CD4+ T-cells have been identified in sarcoma.[93] An in vivo priming by enhanced presentation of self-peptides due HER-2/neu over-expression has been suggested since breast cancer patients over-expressing HER-2/neu developed anti-HER-2/neu antibodies and T-cells that proliferate in response to both HER-2/neu protein and short peptides.[94] CD4+ T-cells from healthy donors as well as from patients with primary breast and ovarian cancer respond by in vitro proliferation and Th1 cytokine secretion to a number of HER-2/neu peptides.[95] Thus, the characterization of further HER-2/neu epitopes that regulate the Th1 response may have important implications for the development of a multivalent HER-2/neu based tumor vaccine.

HER-2/neu Expression and Immune Escape

Deficiencies of the MHC Class I Antigen Processing Pathway

The major histocompatibiliy complex (MHC) class I molecules present antigenic peptides of 8-11 amino acids to CD8+ cytotoxic T-lymphocytes (CTL).[96,97] In general, the peptide cargo is generated by cytosolic degradation of endogenously synthesized proteins. The proteasome has been defined to be the major cytosolic protease complex that produces such peptide ligands for MHC class I molecules,[98] although some cytosolic MHC class I- binding peptides can be processed by non-proteasomal cytosolic or ER-resident protease.[99] The cleavage preferences of the proteasomes quantitatively and qualitatively control the generation of epitopes. In addition, the IFNγ—inducible PA28 (11S) activator complex enhances the cleavage of short peptide substrates in vitro, and promote the generation of double cleavage products, thus containing significantly higher amounts of potential precursor or mature MHC class I ligands in vitro and in vivo.[100,101]

It is still an open question how peptides are released from the proteasomes and then reach the peptide transporter. Srivastava and coworkers[102] suggested that the cytosolic heat shock proteins (hsp) 70 and 90 serve as chaperones by binding the cytosolic peptides and delivering them to the peptide transporters. Most of the peptides are then shuttled from the cytosol compartment into the lumen of the endoplasmic reticulum (ER) by the ATP-dependent heterodimeric transporter-associated with antigen processing complex TAP1 and TAP2. ATP hydrolysis is required for peptide translocation, but not for the initial binding of peptide to TAP. TAP1 forms a complex with a subset of MHC class I molecules.[103] The TAP interaction is influenced by both MHC alleles and structural alterations of the MHC class I molecules.[104] TAP can act as a chaperone by supporting correct loading of peptides onto MHC class I molecules and seems to function as a major checkpoint for the MHC class I maturation process. The sequence specificity and peptide length preference of TAP roughly meet the requirements sufficient for the binding to MHC class I molecules.

Assembly of MHC class I-β_2-m dimers in the ER involves a variety of chaperones. In the ER, the MHC molecules are consecutively stabilized by calnexin and calreticulin. In addition, other ER resident chaperones such as the 48 kD tapasin,[105] facilitate the delivery of peptide ligands onto MHC class I molecules. Upon peptide binding, the trimolecular MHC class I/ peptide complex is released from the multimeric assembly complex at least consisting of TAP, tapasin and calnexin, and then exported via the trans-Golgi apparatus to the cell surface.

Reduced or lack of MHC class I surface expression has been demonstrated in a variety of human tumors in vitro and in vivo as well as in murine models of oncogenic transformation and is often associated with the metastatic potential of tumor cells.[106-109] There exist different mechanisms leading to deficient MHC class I surface expression including structural alterations of the MHC class I heavy and light chain molecules and or other components of the antigen processing machinery, such as peptide transporters TAP and proteasome subunits, as well as dysregulation of the various elements of the MHC class I antigen processing and presentation pathway. The impaired expression and function of the molecules in tumor cells results in resistance to MHC class I restricted T-cell recognition.[110.]

Escape from HER-2/neu Specific CTL; Role of Defect MHC Class I Expression

In order to study the underlying mechanisms by which tumor cells evade from the T-cell-mediated immune surveillance, tumor escape variants from progressively growing HER-2/neu over-expressing ovarian cancer cells were established.[74] Like in melanoma,[111] CTL-resistant tumor variants could be efficiently selected. In most escape variants, a correlation between the level of HER-2/neu specific CTL recognition and MHC class I surface expression was found. One tumor variant was however not lysed by these CTL despite high levels of MHC class I surface expression. These data suggest that two distinct mechanisms, one dependent and another unrelated to MHC class I surface expression, can be involved in the development of immune escape variants.[74]

Furthermore, it has recently been shown that HER-2/neu over-expression can be associated with either loss or down-regulation of MHC class I surface expression in vivo. Transgenic animals expressing the rat HER-2/neu proto-oncogene under the tissue specific control of the mammary tumor virus (MMTV) long terminal repeats developed spontaneous mammary carcinomas with similar histology as human breast carcinomas.[112] The developing tumors expressed high levels of HER-2/neu, but distinct amounts of MHC class I surface antigens. When cultured in vitro, the impaired constitutive H-2 antigen expression level in these carcinoma cells could be significantly induced by IFN-γ treatment.[113] These data suggest (1) that HER-2/neu expressing tumors cells can develop an immune escape phenotype characterized by downregulation of MHC class I surface expression and (2) that MHC class I deficiencies seem to be due to dysregulation rather than structural alterations. In addition preliminary results (Seliger et al, unpublished observations) confirmed the observation: HER-2/neu over-expression in murine fibroblasts results in a strong suppression of H-2 surface antigen expression which seems to be linked to either inhibition of the peptide transporter and/or proteasome subunits.

Tumor variants resistant to lysis by tumor-specific CTL resulting from loss of CTL epitopes have been established from murine mastocytoma and human melanomas by repeated in vitro exposure of tumor cells to tumor-specific CTL clones.[114,115] The risk of selection for antigenic loss variants during immunotherapy has been demonstrated by immunizing with CTL- derived epitopes from melanocyte differentiation antigens, such as Melan-A/MART-1, where antigen-loss variants have been shown to develop in vivo during tumor progression in melanoma patients.[116] One of the attractive features of HER-2/neu based cancer vaccines however is that the risk of inducing CTL resistance due to antigenic loss should be minimal, since HER-2/neu may be directly involved in the pathogenic process.[118] We have tried to generate HER-2/neu loss variants by co-culturing ovarian tumor cells that over-express HER-2/neu with autologous CTL clones specific for this proto-oncogene, but failed to detect loss of HER-2/neu as a result of CTL selection (Kono et al, unpublished observations]. This is also supported by the observations of a relatively stable HER-2/neu expression rate in carcinomas over time both through the clinical course of the disease and at metastatic sites.[117] The difficulty in selecting for HER-2/neu loss tumor variants and the fact that its expression rate in human tumors is rarely heterogeneous is encouraging for the potential use of HER-2/neu based new therapies targeted at this molecule.

Escape from Innate Immunity by Tumors Over-Expressing HER-2/neu

High expression of HER-2/neu may induce malignant transformation and thus an increased inherent proliferative potential.[118] In addition, increased resistance towards host-resistance factors by HER-2/neu over-expressing tumors may also play a role. Thus, an intriguing set of observations support the view that HER-2/neu over-expressing tumor cells are more resistant to innate immunity mediated by tumor necrosis factor (TNF)-α and lymphokine-activated killer (LAK) cells. Over-expression of HER-2/neu was shown to induce resistance of NIH3T3 cells towards the cytotoxic effects of recombinant TNF-α or macrophages. A correlation was also found between HER-2/neu expression rates and the TNF resistance seen in continuously passaged breast and ovarian carcinomas.[119-121] Of particular interest, ovarian and breast cancer lines displaying amplified HER-2/neu genes and protein over-expression along with TNF resistance were also resistant to LAK cells. In contrast, seven out of eight non-expressing lines showed sensitivity for TNF and all eight were sensitive to LAK-mediated lysis. These data indicate that expression of HER-2/neu may impart a proliferative advantage in tumor cells due to an induction of resistance towards several different cytotoxic processes.[119] The mechanism remains to be elucidated, and may involve some resistance inducing factor which acts at a molecule shared between the pathways of TNF- and LAK-mediated lysis. Clearly, however, cytotoxicity mediated by specific CTL is exempt from the resistance imparted by HER-2/neu expression, as discussed above.

Concluding Remarks

The growing number of reports documenting HER-2/neu over-expression in tumors of distinct histology implies that HER-2/neu is an attractive target for T-cell based immunotherapy. Using tumor associated lymphocytes from tumor patients or peripheral blood mononuclear cells from healthy volunteers or cancer patients, several HLA-A2- and HLA-A3 restricted, HER-2/neu specific peptides have been identified. The HER-2/neu specific CTL lysed autologous as well as allogeneic tumors, suggesting that HER-2/neu antigens are often shared between several distinct tumor types.

The observations (1) that HER-2/neu over-expression might downregulate MHC class I surface expression thereby resulting in immune evasion and (2) that human tumors are often poorly immunogenic emphasizes the need for the development of approaches to induce and augment an HER-2/neu specific immune response. Recently, a dominant HLA-DR4 restricted HER-2/neu specific epitope for CD4+ T- cells has been identified which might play an important role for activation and regulation of T-cell differentiation.[95] It might be also beneficial for CTL activation and expansion. In this context, the characterization of further HER-2/neu epitopes that regulate Th1 responses, which can in turn control the spread of Th1/Th2 responses by other HER-2/neu peptides may have important implications for CTL induction and also for the understanding of the regulation of human tumor immunity. Thus, HER-2/neu specific vaccination strategies might be improved by increasing the arsenal of HER- 2/neu specific peptides presented by both MHC class I and class II molecules.

References

1. Traversari C, van der Bruggen P, Luescher IF et al. A nonapeptide encoded by human gene MAGE-1 is recognized on HLA-A1 by cytolytic T lymphocytes directed against tumor antigen MZ2-E. J Exp Med 1992; 176:1453-1457.
2. Cox AL, Skipper J, Chen Y et al. Identification of a peptide recognized by five melanoma-specific human cytotoxic T cell lines. Science 1994; 264:716-719.
3. Coulie PG, Brichard V, Van Pel A et al. A new gene coding for a differentiation antigen recognized by autologous cytolytic T lymphocytes on HLA-A2 melanomas. J Exp Med 1994; 180:35-42.
4. Kawakami Y, Eliyahu S, Delgado CH et al. Identification of a human melanoma antigen recognized by tumor-infiltrating lymphocytes associated with in vivo tumor rejection. Proc Nat Acad Sci USA 1994; 91:6458-6462.
5. Kawakami Y, Eliyahu S, Sakaguchi K et al. Identification of the immunodominant peptides of the MART-1 human melanoma antigen recognized by the majority of HLA-A2-restricted tumor infiltrating lymphocytes. J Exp Med 1994; 180:347-352.
6. Brichard V, Van Pel A, Wolfel T et al. The tyrosinase gene codes for an antigen recognized by autologous cytolytic T lymphocytes on HLA-A2 melanomas. J Exp Med 1993; 178:489-495.

7. Fisk B, Blevins TL, Wharton JT et al. Identification of an immunodominant peptide of HER-2/neu protooncogene recognized by ovarian tumor-specific cytotoxic T lymphocyte lines. J Exp Med 1995; 181:2109-17
8. Yoshino I, Goedegebuure PS, Peoples GE et al. HER2/neu-derived peptides are shared antigens among human non-small cell lung cancer and ovarian cancer. Cancer Res 1994; 54:3387-3390.
9. Ioannides CG, Freedman RS, Platsoucas CD et al. Cytotoxic T cell clones isolated from ovarian tumor-infiltrating lymphocytes recognize multiple antigenic epitopes on autologous tumor cells. J Immunol1991; 146:1700-1707.
10. Peoples GE, Goedegebuure PS, Andrews JV et al. HLA-A2 presents shared tumor- associated antigens derived from endogenous proteins in ovarian cancer. J Immunol 1993; 151:5481-5491.
11. Peoples GE, Yoshino I, Douville CC et al. TCR V beta 3+ and V beta 6+ CTL recognize tumor-associated antigens related to HER2/neu expression in HLA-A2+ ovarian cancers. J Immunol 1994; 152:4993-4999.
12. Coussens L, Yang-Feng TL, Liao YC et al. Tyrosine kinase receptor with extensive homology to EGF receptor shares chromosomal location with neu oncogene. Science 1985; 230:1132-1139.
13. Disis ML, Cheever MA. HER-2/neu oncogenic protein: issues in vaccine development. Crit Rev Immunol 1998; 18:37-45.
14. Pierce JH, Arnstein P, DiMarco E et al. Oncogenic potential of erbB-2 in human mammary epithelial cells. Oncogene 1991; 6:1189-1194.
15. Quirke P, Pickles A, Tuzi NL et al. Pattern of expression of c-erbB-2 oncoprotein in human fetuses. Br J Cancer 1989; 60:64-69.
16. Press MF, Cordon-Cardo C, Slamon DJ. Expression of the HER-2/neu proto-oncogene in normal human adult and fetal tissues. Oncogene 1990; 5:953-962.
17. Cohen JA, Weiner DB, More KF et al. Expression pattern of the neu (NGL) gene- encoded growth factor receptor protein (p185neu) in normal and transformed epithelial tissues of the digestive tract. Oncogene 1989; 4:81-88.
18. Slamon DJ, Godolphin W, Jones LA et al. Studies of the HER-2/neu proto-oncogene in human breast and ovarian cancer. Science 1989; 244:707-712.
19. Kapitanovic S, Radosevic S, Kapitanovic M et al. The expression of p185(HER-2/neu) correlates with the stage of disease and survival in colorectal cancer. Gastroenterol 1997; 112:1103-1113.
20. Latil A, Baron JC, Cussenot O et al. Oncogene amplifications in early-stage human prostate carcinomas. Int J Cancer 1994; 59:637-638.
21. Ndubisi B, Sanz S, Lu L et al. The prognostic value of HER-2/neu oncogene in cervical cancer. Ann Clin Lab Sci 1997; 27:396-401.
22. Brien TP, Depowski PL, Sheehan C et al. Prognostic factors in gastric cancer. Mod Pathol 1998; 11:870-877.
23. Ishikawa T, Kobayashi M, Mai M, Suzuki T, Ooi A. Amplification of the c-erbB-2 (HER-2/neu) gene in gastric cancer cells. Detection by fluorescence in situ hybridization. Am J Pathol 1997;151:761-768.
24. Yamanaka Y. The immunohistochemical expressions of epidermal growth factors, epidermal growth factor receptors and c-erbB-2 oncoprotein in human pancreatic cancer. Nippon Ika Daigaku Zasshi 1992; 59:51-61.
25. Brossart P, Stuhler G, Flad T et al. Her-2/neu-derived peptides are tumor-associated antigens expressed by human renal cell and colon carcinoma lines and are recognized by in vitro induced specific cytotoxic T lymphocytes. Cancer Res 1998; 58:732-736.
26. Rotter M, Block T, Busch R et al. Expression of HER-2/neu in renal-cell carcinoma. Correlation with histologic subtypes and differentiation. Int J Cancer 1992; 52:213-217.
27. Weiner DB, Nordberg J, Robinson R et al. Expression of the neu gene-encoded protein (P185neu) in human non-small cell carcinomas of the lung. Cancer Res 1990; 50:421-425.
28. Hsieh CC, Chow KC, Fahn HJ et al. Prognostic significance of HER-2/neu overexpression in stage I adenocarcinoma of lung. Ann Thorac Surg 1998; 66:1159-1163; discussion 1163-1164.
29. Tsai CM, Chang KT, Wu LH et al. Correlations between intrinsic chemoresistance and HER-2/neu gene expression, p53 gene mutations, and cell proliferation characteristics in non-small cell lung cancer cell lines. Cancer Res 1996; 56:206-209.

30. van de Vijver MJ, Mooi WJ, Wisman P et al. Immunohistochemical detection of the neu protein in tissue sections of human breast tumors with amplified neu DNA. Oncogene 1988; 2:175-178.
31. Benz CC, O'Hagan RC, Richter B et al. HER2/neu and the Ets transcription activator PEA3 are coordinately upregulated in human breast cancer. Oncogene 1997; 15:1513-1525.
32. Shackney SE, Pollice AA, Smith CA et al. Intracellular coexpression of epidermal growth factor receptor, Her- 2/neu, and p21ras in human breast cancers: evidence for the existence of distinctive patterns of genetic evolution that are common to tumors from different patients. Clin Cancer Res 1998;4: 913-928.
33. Symmans WF, Liu J, Knowles DM et al. Breast cancer heterogeneity: evaluation of clonality in primary and metastatic lesions. Hum Pathol 1995; 26:210-216.
34. Cuny M, Simony-Lafontaine J, Rouanet P et al. Quantification of ERBB2 protein expression in breast cancer: three levels of expression defined by their clinico-pathological correlations. Oncol Res 1994; 6:169-176.
35. Seshadri R, Firgaira FA, Horsfall DJ et al.Clinical significance of HER-2/neu oncogene amplification in primary breast cancer. The South Australian Breast Cancer Study Group. J Clin Oncol 1993;11: 1936-1942.
36. Gullick WJ. The role of the epidermal growth factor receptor and the c-erbB-2 protein in breast cancer. Int J Cancer Suppl 1990; 5:55-61.
37. Slamon DJ, Clark GM, Wong SG et al. Human breast cancer: correlation of relapse and survival with amplification of the HER-2/neu oncogene. Science 1987; 235:177-182.
38. Gusterson BA, Gelber RD, Goldhirsch A et al. Prognostic importance of c-erbB-2 expression in breast cancer. International (Ludwig) Breast Cancer Study Group [see comments]. J Clin Oncol 1992; 10:1049-1056.
39. Press MF, Bernstein L, Thomas PA et al. HER-2/neu gene amplification characterized by fluorescence in situ hybridization: poor prognosis in node-negative breast carcinomas. J Clin Oncol 1997; 15:2894-904.
40. Press MF, Pike MC, Chazin VR et al. Her-2/neu expression in node-negative breast cancer: direct tissue quantitation by computerized image analysis and association of overexpression with increased risk of recurrent disease. Cancer Res 1993; 53:4960-4970.
41. Christianson TA, Doherty JK, Lin YJ et al. NH2-terminally truncated HER-2/neu protein: relationship with shedding of the extracellular domain and with prognostic factors in breast cancer. Cancer Res 1998; 58:5123-5129.
42. Xu FJ, Stack S, Boyer C et al. Heregulin and agonistic anti-p185(c-erbB2) antibodies inhibit proliferation but increase invasiveness of breast cancer cells that overexpress p185(c-erbB2): increased invasiveness may contribute to poor prognosis [In Process Citation]. Clin Cancer Res 1997; 3:1629-1634.
43. Rilke F, Colnaghi MI, Cascinelli N et al. Prognostic significance of HER-2/neu expression in breast cancer and its relationship to other prognostic factors. Int J Cancer 1991; 49:44-49.
44. Muller WJ, Ho J, Siegel PM. Oncogenic activation of Neu/ErbB-2 in a transgenic mouse model for breast cancer. Biochem Soc Symp 1998; 63:149-157.
45. Bouchard L, Lamarre L, Tremblay PJ et al. Stochastic appearance of mammary tumors in transgenic mice carrying the MMTV/c-neu oncogene. Cell 1989; 57:931-936.
46. Roland PY, Stoler MH, Broker TR et al. The differential expression of the HER-2/neu oncogene among high-risk human papillomavirus-infected glandular lesions of the uterine cervix. Am J Obstet Gynecol 1997; 177:133-138.
47. Wang D, Konishi I, Koshiyama M et al. Expression of c-erbB-2 protein and epidermal growth receptor in endometrial carcinomas. Correlation with clinicopathologic and sex steroid receptor status. Cancer 1993; 72:2628-2637.
48. Rasty G, Murray R, Lu L et al. Expression of HER-2/neu oncogene in normal, hyperplastic, and malignant endometrium. Ann Clin Lab Sci 1998; 28:138-143.
49. Berchuck A, Rodriguez G, Kinney RB et al. Overexpression of HER-2/neu in endometrial cancer is associated with advanced stage disease. Am J Obstet Gynecol 1991; 164:15-21.
50. Saffari B, Jones LA, el-Naggar A et al. Amplification and overexpression of HER-2/neu (c-erbB2) in endometrial cancers: correlation with overall survival. Cancer Res 1995; 55:5693-5698.

51. Press MF, Pike MC, Hung G et al. Amplification and overexpression of HER-2/neu in carcinomas of the salivary gland: correlation with poor prognosis. Cancer Res 1994; 54:5675-5682.
52. Kohlberger P, Loesch A, Koelbl H et al. Prognostic value of immunohistochemically detected HER-2/neu oncoprotein in endometrial cancer. Cancer Lett 1996; 98:151-155.
53. Hetzel DJ, Wilson TO, Keeney GL et al. HER-2/neu expression: a major prognostic factor in endometrial cancer. Gynecol Oncol 1992; 47:179-185.
54. Riben MW, Malfetano JH, Nazeer T et al. Identification of HER-2/neu oncogene amplification by fluorescence in situ hybridization in stage I endometrial carcinoma. Mod Pathol 1997; 10:823-831.
55. Wong YF, Cheung TH, Lam SK et al. Prevalence and significance of HER-2/neu amplification in epithelial ovarian cancer. Gynecol Obstet Invest 1995; 40:209-212.
56. Dugan MC, Dergham ST, Kucway R et al. HER-2/neu expression in pancreatic adenocarcinoma: relation to tumor differentiation and survival. Pancreas 1997; 14:229-236.
57. Tomaszewska R, Okon K, Nowak K, et al. HER-2/Neu expression as a progression marker in pancreatic intraepithelial neoplasia. Pol J Pathol 1998; 49:83-92.
58. Tsai CM, Chang KT, Perng RP et al. Correlation of intrinsic chemoresistance of non- small-cell lung cancer cell lines with HER-2/neu gene expression but not with ras gene mutations. J Natl Cancer Inst 1993; 85:897-901.
59. Kerns BJ, Jordan PA, Huper G et al. Assessment of c-erbB-2 amplification by immunohistochemistry in paraffin-embedded breast cancer. Mod Pathol 1993; 6:673-678.
60. Ibrahim GK, MacDonald JA, Kerns BJ et al. Differential immunoreactivity of HER-2/neu oncoprotein in prostatic tissues. Surg Oncol 1992; 1:151-155.
61. Kallakury BV, Sheehan CE, Ambros RA et al. Correlation of p34cdc2 cyclin-dependent kinase overexpression, CD44s downregulation, and HER-2/neu oncogene amplification with recurrence in prostatic adenocarcinomas. J Clin Oncol 1998; 16:1302-1309.
62. Weidner U, Peter S, Strohmeyer T et al. Inverse relationship of epidermal growth factor receptor and HER2/neu gene expression in human renal cell carcinoma. Cancer Res 1990; 50:4504-4509.
63. Freeman MR, Washecka R, Chung LW. Aberrant expression of epidermal growth factor receptor and HER-2 (erbB- 2) messenger RNAs in human renal cancers. Cancer Res 1989; 49:6221-6225.
64. Danova M, Giordano M, Torelli F et al. HER-2/neu oncogene expression and DNA ploidy in normal human kidney and renal cell carcinoma. Eur J Histochem 1992; 36:279-288.
65. Thoenes W, Störkel S, Rumpelt HJ Histopathology and classification of renal cell tumors (adenomas, oncocytomas and carcinomas): the basic cytological histopathological elements and their use for diagnostics. Pathol Res Pract 1986; 181:125-143.
66. Szollosi J, Balazs M, Feuerstein BG ERBB-2 (HER2/neu) gene copy number, p185HER- 2 overexpression, and intratumor heterogeneity in human breast cancer. Cancer Res 1995; 55:5400-5407.
67. Valeron PF, Chirino R, Fernandez L et al. Validation of a differential PCR and an ELISA procedure in studying HER- 2/neu status in breast cancer. Int J Cancer 1996; 65:129-133.
68. Persons DL, Borelli KA, Hsu PH. Quantitation of HER-2/neu and c-myc gene amplification in breast carcinoma using fluorescence in situ hybridization. Mod Pathol 1997; 10:720-727.
69. Zhou DJ, Ahuja H, Cline MJ. Proto-oncogene abnormalities in human breast cancer: c- ERBB-2 amplification does not correlate with recurrence of disease. Oncogene 1989; 4:105-108.
70 Dawkins HJ, Robbins PD, Sarna M et al. c-erbB-2 amplification and overexpression in breast cancer: evaluation and comparison of Southern blot, slot blot, ELISA and immunohistochemistry. Pathol 1993; 25:124-132.
71. Vanky F, Roberts T, Klein E et al. Auto-tumor immunity in patients with solid tumors: participation of CD3 complex and MHC class I antigens in the lytic interaction. Immunol Lett 1987; 16:21-26.
72. Ioannides CG, Platsoucas CD, Rashed S et al. Tumor cytolysis by lymphocytes infiltrating ovarian malignant ascites. Cancer Res 1991; 51:4257-4265.
73. Peoples GE, Smith RC, Linehan DC et al. Shared T cell epitopes in epithelial tumors. Cell Immunol 1995; 164:279-286.
74. Kono K, Halapi E, Hising C et al. Mechanisms of escape from CD8+ T-cell clones specific for the HER-2/neu proto-oncogene expressed in ovarian carcinomas: related and unrelated to decreased MHC class 1 expression. Int J Cancer 1997; 70:112-119.

75. Kono K, Rongcun Y, Charo J et al. Identification of HER2/neu-derived peptide epitopes recognized by gastric cancer-specific cytotoxic T lymphocytes. Int J Cancer 1998; 78:202-208.
76. Halapi E, Yamamoto Y, Juhlin C et al. Restricted T cell receptor V-beta and J-beta usage in T cells from interleukin-2-cultured lymphocytes of ovarian and renal carcinomas. Cancer Immunol Immunother 1993; 36:191-197.
77. Peoples GE, Yoshino I, Douville CC et al. TCR V beta 3+ and V beta 6+ CTL recognize tumor-associated antigens related to HER2/neu expression in HLA-A2+ ovarian cancers. J Immunol 1994; 152:4993-4999.
78. Yoshino I, Peoples GE, Goedegebuure PS et al. Association of HER2/neu expression with sensitivity to tumor-specific CTL in human ovarian cancer. J Immunol 1994; 152: 2393-2400.
79. Peoples GE, Goedegebuure PS, Smith R et al. Breast and ovarian cancer-specific cytotoxic T lymphocytes recognize the same HER2/neu-derived peptide. Proc Natl Acad Sci USA 1995; 92:432-436.
80. Linehan DC, Goedegebuure PS, Peoples GE Tumor-specific and HLA-A2-restricted cytolysis by tumor-associated lymphocytes in human metastatic breast cancer. J Immunol 1995; 155:4486-4491.
81. Kawashima I, Hudson SJ, Tsai V et al. The multi-epitope approach for immunotherapy for cancer: identification of several CTL epitopes from various tumor-associated antigens expressed on solid epithelial tumors. Hum Immunol 1998; 59:1-14.
82. Lustgarten J, Theobald M, Labadie C et al. Identification of Her-2/Neu CTL epitopes using double transgenic mice expressing HLA-A2.1 and human CD.8. Hum Immunol 1997; 52:109-118.
83. Zaks TZ, Rosenberg SA. Immunization with a peptide epitope (p369-377) from HER- 2/neu leads to peptide-specific cytotoxic T lymphocytes that fail to recognize HER- 2/neu+ tumors. Cancer Res 1998; 58:4902-4908.
84. Linehan DC, Peoples GE, Hess DT et al. In vitro stimulation of ovarian tumour- associated lymphocytes with a peptide derived from HER-2/neu induces cytotoxicity against autologous tumour. Surg Oncol 1995; 4:41-9.
85. Celis E, Tsai V, Crimi C et al. Induction of anti-tumor cytotoxic T lymphocytes in normal humans using primary cultures and synthetic peptide epitopes. Proc Natl Acad Sci USA 1994; 91:2105-2109.
86. Disis ML, Smith JW, Murphy AE et al. In vitro generation of human cytolytic T-cells specific for peptides derived from the HER-2/neu protooncogene protein. Cancer Res 1994; 54:1071-1076.
87. Kawashima I, Tsai V, Southwood S et al. Identification of HLA-A3-restricted cytotoxic T lymphocyte epitopes form carcinoembryonic antigen and HER-2/neu by primary in vitro immunization with peptide-pulsed dendritic cells. Cancer Res 1999; 59: 431-435.
88. Bodey B, Bodey B, Jr., Groger AM et al. Clinical and prognostic significance of the expression of the c-erbB-2 and c-erbB-3 oncoproteins in primary and metastatic malignant melanomas and breast carcinomas. Anticancer Res 1997; 17:1319-1330.
89. Dadmarz R, Sgagias MK, Rosenberg SA et al. CD4+ T lymphocytres infiltrating human breast cancer recognize autologous tumor in an MHC class II restricted fashion. Cancer Immuno Immunother 1995; 40:1-9.
90. Topalian SL, Gonzales MJ, Parkhurst M et al. Melanoma specific CD4+ T cells recognize nonmutated HLA-DR restricted tyrosinase epitopes. J Exp Med 1996; 183:1965-1971.
91. Topalian SL, Rivoltini L, Mancini M et al. Human CD4+ T cells specifically recognize a shared melanoma-associated antigen encoded by the tyrosinase gene. Proc Natl Acad Sci USA 1994; 91:9461-9465.
92. Takahashi T, Chapman PB, Yang SY et al. Reactivity of autologous CD4+ T lymphocytes against human melanoma. J Immunol. 1995; 154: 772-779.
93. Heike M, Schlaak J, Schulze-Bergkamen H et al. Specificities and functions of CD4+ HLA-class II-restricted T cell clones against a human a human sarcoma: evidence for several recognized antigens. J Immunol 1996; 156:2205-2213.
94. Disis ML, Calenoff E, McLaughlin G et al. Existent T cell and antibody immunity to HER-2/neu protein in patients with breast cancer Cancer Res 1994; 54: 16-20.
95. Tuttle TM, Anderson BW, Thompson WE et al. Proliferative and cytokine responses to class II HER-2/neu associated peptides in breast cancer patients. Clin Cancer Res 1998; 4: 2015-2024.
96. Zinkernagel RM, Doherty PC. Restriction of in vitro T cell-mediated cytotoxicity in lymphocytic choriomeningitis within a syngeneic or semiallogeneic system. Nature 1974; 248:701-702.
97. Rammensee HG, Falk K, Rotzschke O. Peptides naturally presented by MHC class I molecules. Annu Rev Immunol 1993; 11:213-244.

98. Rock KL, Gramm C, Rothstein L et al. Inhibitors of the proteasome block the degradation of most cell proteins and the generation of peptides presented on MHC class I molecules. Cell 1994; 78:761-771.
99. Yang MX, Cederbaum AI. Role of the proteasome complex in degradation of human CYP2E1 in transfected HepG2 cells. Biochem Biophys Res Commun 1996; 226:711-716.
100. Dick TP, Ruppert T, Groettrup M et al. Coordinated dual cleavages induced by the proteasome regulator PA28 lead to dominant MHC ligands. Cell 1996; 86:253-262.
101. Groettrup M, Soza A, Eggers M et al. A role for the proteasome regulator PA28alpha in antigen presentation. Nature 1996; 381:166-168.
102. Srivastava PK, Udono H. Heat shock protein-peptide complexes in cancer immunotherapy. Curr Opin Immunol 1994; 6:728-732.
103. Carreno BM, Solheim JC, Harris M TAP associates with a unique class I conformation, whereas calnexin associates with multiple class I forms in mouse and man. J Immunol 1995; 155:4726-4733.
104. Peace-Brewer AL, Tussey LG, Matsui M et al. A point mutation in HLA-A*0201 results in failure to bind the TAP complex and to present virus-derived peptides to CTL. Immunity 1996; 4:505-514.
105. Sadavsan B, Lehner PJ, Ortmann B et al. Roles for calreticulin and a novel glycoprotein, tapasin, in the interaction of MHC class I molecules with TAP. Immunity 1996; 5:103-114.
106. Seliger B, Maeurer MJ, Ferrone S. TAP off—tumors on. Immunol Today 1997; 18:292- 299.
107. Ferrone S, Marincola FM. Loss of HLA class I antigens by melanoma cells: molecular mechanisms, functional significance and clinical relevance. Immunol Today 1995; 16:487-494.
108. Garrido F, Ruiz-Cabello F, Cabrera T et al. Implications for immunosurveillance of altered HLA class I phenotypes in human tumours. Immunol Today 1997; 18:89-95.
109. Kaklamanis L, Leek R, Koukourakis M, Gatter KC et al. Loss of transporter in antigen processing 1 transport protein and major histocompatibility complex class I molecules in metastatic versus primary breast cancer. Cancer Res 1995; 55:5191-5194.
110. Restifo NP, Esquivel F, Kawakami Y et al. Identification of human cancers deficient in antigen processing. J Exp Med 1993; 177:265-272.
111. Boon T, Van Pel A, De Plaen E et al. Genes coding for T-cell-defined tum transplantation antigens: point mutations, antigenic peptides, and subgenic expression. Cold Spring Harb Symp Quant Biol 1989; 54:587-596.
112. Guy CT, Webster MA, Schaller M et al. Expression of the neu protooncogene in the mammary epithelium of transgenic mice induces metastatic disease. Proc Natl Acad Sci USA 1992; 89:10578-10582.
113. Lollini PL, Nicoletti G, Landuzzi L et al. Down regulation of major histocompatibility complex class I expression in mammary carcinoma of HER-2/neu transgenic mice. Int J Cancer 1998; 77:937-941.
114. Uyttenhove C, Maryanski J, Boon T. Escape of mouse mastocytoma P815 after nearly complete rejection is due to antigen-loss variants rather than immunosuppression. J Exp Med 1983; 157:1040-1052.
115. Van den Eynde B, Hainaut P, Herin M et al. Presence on a human melanoma of multiple antigens recognized by autologous CTL. Int J Cancer 1989; 44:634-640.
116. Jaeger E, Ringhoffer M, Karbach J et al.Inverse relationship of melanocyte differentiation antigen expression in melanoma tissues and CD8+ cytotoxic-T-cell responses: evidence for immunoselection of antigen-loss variants in vivo. Int J Cancer 1996; 66:470-476.
117. Niehans GA, Singleton TP, Dykoski D et al. Stability of HER-2/neu expression over time and at multiple metastatic sites. J Natl Cancer Inst 1993; 85:1230-1235.
118. Di Fiore PP, Pierce JH, Kraus MH et al. erbB-2 is a potent oncogene when overexpressed in NIH/3T3 cells. Science 1987; 237:178-182.
119. Lichtenstein A, Berenson J, Gera JF et al. Resistance of human ovarian cancer cells to tumor necrosis factor and lymphokine-activated killer cells: Correlation with expression of HER2/neu oncogenes. Cancer Res 1990; 50:7364-7370.
120. Shepard HM, Lewis GD. Resistance of tumor cells to tumor necrosis factor. J Clin Immunol 1988; 8:333-341.

121. Hudziak RM, Lewis GD, Shalaby MR et al. Amplified expression of the HER2/ERBB2 oncogene induces resistance to tumor necrosis factor alpha in NIH 3T3 cells. Proc Natl Acad Sci USA 1988; 85:5102-5106.
122. Wood DP, Wartinger DD, Reuter V et al. DNA, RNA and immunohistochemical characterization of the HER-2/neu oncogene in transitional cell carcinoma of the bladder. J Urol 1991; 146: 1398-1401.
123. Al-Kasspooles M, Moore JH, Orringer MB et al. Amplification and over-expression of the EGF-R and erbB2 genes in human esophageal adenocarcinomas. Int J Cancer 1993; 54: 213-219.
124. Mitra AB, Murty VVVS, Pratap M et al. ErbB2 (HER2/neu) oncogene is frequently amplified in squamous cell carcinoma of the uterine cervix. Cancer Res 1994; 54:637-639.
125. Ioannides CG, Fisk B, Fan D et al. Cytotoxic T cells isolated from ovarian malignant ascites recognize a peptide derived from the HER-2/neu proto-oncogene. Cell Immunol 1993; 151:225-234.
126. Fisk B, Chesak B, Pollack MS et al. Oligopeptide induction of a cytotoxic T lymphocyte response to HER- 2/neu proto-oncogene in vitro. Cell Immunol 1994; 157:415-427.
127. Peiper M, Goedegebuure PS, Linehan DC et al. The HER2/neu-derived peptide p654-662 is a tumor-associated antigen in human pancreatic cancer recognized by cytotoxic T lymphocytes. Eur J Immunol 1997; 27:1115-1123.

Chapter 9

Clinical Trials of HER-2/neu Peptide-Based Vaccines

Mary L. Disis and Martin A. Cheever

Introduction

Cancer vaccines are not used routinely in the clinical practice of most oncologists, despite decades of study. Several advances in basic immunology over the last few years have forced a re-evaluation of cancer vaccine development. The most important finding has been that human tumors are immunogenic and the definition of some cancer related proteins that are tumor antigens.

Abundant or overexpressed oncogenic proteins can be immunogenic. Studies by several groups have identified "self" antigens, expressed on tumor cells, as tumor antigens.[1, 2] These proteins are not mutated in any way, but are clearly immunogenic in patients with cancer and have been shown to generate both antibody, T helper, and cytolytic T cell (CTL) responses.[1, 3.] Many of these proteins are present at much higher concentrations in malignant cells than in the normal cells with which they are associated.[3] Gene amplification results in overexpression of normal cellular proteins in cancer cells and is an etiologic factor in the malignant transformation of many solid tumors. Overexpressed oncoproteins are distinct from their normal counterparts only by virtue of their greater concentration in cancer cells. Intuitively, these proteins would not be considered potential tumor antigens as patients should be tolerant to self-proteins. The finding, that many tumor antigens are self-proteins, has resulted in a "paradigm shift".[4.] The new paradigm includes self-proteins as tumor antigens and tolerance induction as a possible mechanism of immune escape. Although issues of tolerance and the potential for precipitation of autoimmune disease by augmenting immune responses to these self cancer proteins is a concern, overexpressed oncoproteins offer the advantage of containing epitopes capable of interacting with any major histocompatibility molecule (MHC). The problem facing tumor immunologists studying immunity to self-tumor antigens is how to invoke immunity to self for cancer therapeutics.

HER2 is an Overexpressed Growth Factor Receptor

The enthusiasm for targeting immunotherapy against HER2 lays mainly in biological characteristics of the molecule. HER2 is a member of the epidermal growth factor receptor family and functions as a growth factor receptor.[5, 6] This transmembrane protein consists of an extracellular domain (ECD) which binds ligand and an intracellular cytoplasmic domain (ICD) with signaling activity. In humans, the HER2 protein is expressed during fetal development.[7] In adults, the protein is weakly detectable in the epithelial cells of many normal tissues by immunohistochemical staining. The HER2 gene is present, in normal cells, as a single copy.[7] Amplification of the gene and/or overexpression of the associated protein has been identified in many human cancers; breast, ovary, uterus, stomach, and adenocarcinoma of the lung[8-12] to name a few. The clinical consequences of the oncogenic protein overexpression have been best studied in breast and ovarian cancer. HER2 protein overexpression occurs in 20-40% of intraductal carcinomas of the breast and 30% of ovarian cancers. It is associated with a poor prognosis in both diseases.[13] In node positive breast cancer, HER2 protein overexpression has been linked to more aggressive disease and a poorer prognosis.

The HER2 oncogenic protein, therefore, is a good vaccine target from both a biologic and immunologic standpoint. Biologically, HER2 is associated with more aggressive disease. Immunologically, portions of this transmembrane protein are likely to be available to both the class I and II antigen processing pathways. As a self-protein HER2 should contain epitopes appropriate for binding to most, if not all class I and class II MHC molecules and be potentially recognizable by all patients. Finally, the marked overexpression of HER2 by malignant cells may allow the HER2 specific T cell response to be selective with minimal toxicity.

Some Patients with HER2 Overexpressing Cancers have a Pre-Existent Immune Response Directed against HER2

Immune responses directed against the HER2 protein have been discovered to exist in some patients with HER2 positive cancer. The existent immune responses occur as a consequence of overexpression of HER2 by autologous cancer cells. The co-existence of cancer specific immunity in the face of a growing tumor is a conundrum for tumor immunologists, yet the phenomenon is not without precedent. For example, most patients who die of an infectious disease, die with some level of immunity directed against the infectious agent, but the immune response was ineffective. The identification of human tumor antigens, such as HER2, now allows some definition of why cancer and immunity against cancer co-exist. Existent antibody, helper T cell, and cytotoxic T cell immunity to HER2 have been detected in humans. [14-17] In the majority of patients the HER2 specific antibody response is of low magnitude. One recent study of HER2 specific antibody immunity [18] identified a minority of stage I or II patients who had high titer HER2 specific antibodies. Much lower levels of HER2 antibody were detected in patients with advanced disease. Advanced stage patients, however, had levels of antibody immunity specific for candida antigen similar to that seen in a population not affected by cancer. Thus, the blunted antibody immune response to the HER2 protein, in these patients, could not be explained entirely by the immunosuppression associated with advanced malignancy. The high titer HER2 specific immunity, found in these few patients with early stage disease, may indicate the development of an immune response to HER2 protein expressed by growing tumor limits further growth and metastasis. It has not yet been proven that existent immunity to HER2 predicts for an improved survival, but existent immunity does predict that vaccines will be able to induce and boost immunity to HER2.

Another study of the HER2 specific antibody response in patients with HER2 overexpressing colon cancer validated that antibodies were present, but were low level. Sera were obtained from 57 patients with colorectal cancer at or near the time of diagnosis and analyzed for the presence of human antibodies directed against the HER2 oncoprotein.[19] Antibodies at titers of 1:100 or more were detected in eight of 57 (14%) colorectal cancer patients compared to 0 of 200 (0%) of normal controls (p=<0.01). Antibody titers ranged from 1:100 to 1:800. Endogenous HER2 specific antibody responses, titer Æ 1:100, correlated to HER2 protein overexpression in the tumor. Six of 13 (46%) patients with HER2 positive tumors had HER2 specific antibodies as compared to two of 44 (5%) patients with HER2 negative tumors (p=<0.01). We compared the magnitude of the HER2 specific IgG antibody response with that of a vaccinated antigen, tetanus toxoid in these patients. The neu antibodies present in the colon cancer population studied were 1000 fold less in concentration than tetanus toxoid specific antibodies in a population of volunteer blood donors. Three of the colon cancer patients had both tetanus toxoid and HER2 specific antibody responses. Although their tetanus toxoid antibody level was similar to normal, the HER2 specific antibody level was still 1000 fold less. These patients could be immunized with a tetanus toxoid vaccine, but immunization to HER2 by virtue of their tumor overexpressing the protein was inefficient.

T helper responses to HER2 recombinant protein and peptides have been performed for over 60 patients with breast cancer using a modified limiting dilution analysis.[20.] These experiments were initiated to evaluate theoretical helper epitopes for their ability to stimulate a T cell response. Overall, a minority of patients, <10%, demonstrated T cell responses which could be defined as a stimulation index (S.I.) greater than 2. The majority of samples analyzed, >90%, demonstrated proliferative T cell responses which could only be described as a significant number of positive wells in a modified limiting dilution analysis. The overall calculated S.I. of these 24 well replicates, however, was less 2. Similarly, studies of CTL responses directed to HER2 protein bearing tumor cells and Class I binding peptides define CTL only after multiple restimulations in vitro with HER2 protein expressing cells or peptide loaded cells to augment the weak immune response.[16, 17.] These studies are an indication the HER2 specific T cell response present in patients with cancers of low magnitude.

Investigations defining HER2 specific immunity in patients with cancer, by our laboratory and others, indicate that high levels of immunity can exist in some patients, but most patients have low or non-existent HER2 specific immunity. Therefore, strategies aimed at boosting the immunity already present may be very effective in inducing an anti-tumor effect. Given that pre-existent immunity to HER2 occurs in some cancer patients, and can most likely be augmented, the current therapeutic questions are: what types of immune responses will be most effective, how best to immunize, and whether vaccine induced immunity can mediate a therapeutic effect.

HER2 Specific Immunity May Have an Anti-Cancer Effect

Interest in HER2 directed immune therapy has been fueled by several recent studies in animal models and in humans establishing whether HER2 specific cancer treatment can be effective. HER specific antibodies may mediate an anti-tumor response. In animal studies, a monoclonal antibody specific for the ECD of rat neu, a homologue to human HER2, was infused into mice transgenic for rat neu; animals destined to develop and die of breast cancer. This antibody to neu dramatically prevented the development of breast cancer, [21] as well as mediating regression of established tumors. The anti- tumor effect of neu specific antibody infusion was dose dependent. In initial human clinical trials, infusion of another HER2 specific monoclonal antibody induced tumor regression in some patients with advanced cancer.[22] More recent studies demonstrate passive HER2 antibody infusion may have a marked therapeutic effect in patients with HER2 overexpressing tumors.[23, 24.] An initial study in 213 evaluable women with metastatic HER overexpressing breast cancer treated with Herceptin, a humanized antibody targeting the HER2 ECD, revealed on overall response rate of 21%. [24.] Similarly, results from a randomized Phase III trial of the addition of Herceptin to initial chemotherapy in patients with metastatic HER2 overexpressing breast cancer was reported.[23.] In this study 234 patients randomized to receive chemotherapy alone and 235 patients randomized to receive chemotherapy and herceptin. Those patients who received the HER2 humanized antibody had a markedly improved response rate, 62%, as compared to 36% tumor response in patients who received standard treatment only. These data indicate that an antibody response, elicited by immunization, and directed against the HER2 ECD may have an anti-tumor effect.

Cytotoxic T cells specific for HER2 have been shown to lyse HER2 expressing tumor cells in vitro.[16,17] The role CTL can play in tumor eradication, while not yet defined specifically for

HER2, has been defined for other antigens in other solid tumors such as melanoma and renal cell carcinoma. The role of T helper cells is critical in the function and expansion of CTL. A mouse leukemia model, FBL-3, has been instructive in determining how to generate T cells which will eradicate tumor and defining the role of T cells in tumor eradication.[25] Experiments have also shown that infusion $CD4^+$ helper cells alone is not curative; and, that for $CD8^+$ cells to be curative, it is necessary to concurrently inject $CD4^+$ cells or administer exogenous IL-2.[26] Indeed, Phase I studies infusing CMV specific $CD8^+$ T cell clones, demonstrated adoptive transfer of cells could safely reconstitute protective levels of $CD8^+$ immunity early post-transplantation, however, the magnitude of CMV specific CTL gradually declined in patients who did not have endogenous recovery of $CD4^+$ Th.[27,28] As with passive infusion of antibodies specific for an antigen, passive infusion of T cells specific for a tumor antigen may also mediate an anti-tumor response. T cells, infiltrating these solid tumors, have been isolated, expanded ex vivo, and given back to the patient in large numbers. Infusion of TIL has resulted in anti-tumor responses and even some complete responses.[29] Unfortunately, clinical responses after TIL infusions are not common occurring in about 10-20% of treated patients. Recently, advances in molecular biology techniques have allowed the identification of many tumor antigens, particularly in melanoma.[30] The recognition of which proteins in melanoma are immunogenic has allowed a reanalysis of TIL for their antigen specificity. Kawakami et al reported that adoptive transfer of TIL which were reactive to the melanoma antigen gp100 was associated with a clinical anti-tumor response.[31] T cell immunity, both T help and CTL, clearly plays a major role in the potential therapeutic effect of any cancer vaccine.

Peptide-Based Vaccines are an Effective Method for Immunizing against a "Self" Antigen in a Pre-Clinical Model

Which method of vaccination would be most effective in eliciting HER2 specific immunity? Tolerance is an issue when evaluating the ability to generate immune responses to self. Animal models, therefore, must be autologous, e.g., generate immunity to rat neu in the rat. Rat neu has been well characterized. In addition, previous investigators had shown that rats could not be immunized with rat neu protein when rat ECD was constructed in a vaccinia vector.[32] Studies from our laboratory confirmed those findings by vaccinating rats with purified rat neu in a classic immunization regimen.[33] Rats were presumed tolerant to rat neu. We questioned whether tolerance to HER2 could prospectively be circumvented by immunization with peptide based vaccines. Immunization to foreign proteins normally elicits immunity to only a subset of potential epitopes, dominant epitopes, whereas other potentially immunogenic epitopes, subdominant epitopes, are ignored. Immunization to self proteins usually fails to elicit immunity. It has been proposed that autologous T cells recognize and become tolerant to the dominant epitopes of self proteins, but "ignore" the subdominant epitopes.[34,35] Immunity to subdominant epitopes can be generated by immunization to proteins truncated to not contain the dominant epitopes or to peptides representing the subdominant epitopes alone. To test the ability of peptide based vaccines to circumvent tolerance to HER2 protein, we immunized rats with rat neu peptide epitopes predicted to interact with Class II by computer modeling.[33] Rats were immunized with rat neu peptides designed for eliciting $CD4^+$ T cell responses. In addition, some animals were immunized with purified rat neu in a similar manner to compare the immunogenicity of intact protein vs. peptides. Antibody and T cell responses specific for both the immunizing peptides and protein were generated in the peptide immunized animals. No antibody or T cell responses were observed in animals immunized with intact rat neu protein.[33] A neu peptide based vaccine proved to be the most effective construct in generating neu specific immunity in the rat.

Protective immunity generated by vaccination depends on the ability of the vaccine to elicit the appropriate immune response to eradicate the antigen-bearing cell. The use of peptide vaccines will allow the expansion of T cell populations specific for defined immunogenic epitopes thus allowing a more specific and, perhaps, a more effective immune response. Unfortunately, peptides are weak immunogens. A major obstacle in the development of effective peptide based vaccines has been the identification of an appropriate adjuvant to use for human injections. Subunit vaccines require multiple injections to generate significant immune responses and these responses are not long lasting. Immunization studies, in the rat, validate that HER2 peptide immunization can elicit $CD4^+$ T cell responses and antibody responses. Immunization studies in the mouse show that HER2 peptide specific immunization can elicit CTL which result in rejection of tumors which express HER2.[36] The peptides were injected with complete Freund's adjuvant (CFA). CFA is toxic and, thus, problematic for human studies. Most human vaccine studies have involved immunization to foreign antigens rather than self peptides and most vaccine studies have involved immunization with whole organisms or with whole proteins. Too little information is available concerning the optimal methods for immunization to either peptides or self epitopes. There are no standard regimens or adjuvants. Studies, immunizing multiple sclerosis patients with self peptides derived from the amino acid sequence of the T cell receptors most associated with multiple sclerosis, have shown that peptide specific T cells can be generated by the use of intradermal (i.d.) injections without adjuvant.[37] This has not been the case in our rat model using rat neu peptides.

GM-CSF is an attractive adjuvant for priming and boosting to HER2 peptides. Other investigators have shown that tumors engineered to secrete GM-CSF can be immunogenic in animal models.[38] In addition, GM-CSF plays a role in the maturation and function of antigen presenting cells such as dendritic cells (DC) and macrophages.[39] Dendritic cells are specialized antigen-presenting cells that are believed to be responsible for stimulating naive T cell responses. They have also been shown in experimental models to augment secondary immune responses better that other antigen presenting cells. The dermis is a site for skin DC (Langerhans cells) that are important in initiating early immune responses by migrating to draining lymph nodes after being exposed to antigen and presenting the antigen to T cells. Although GM-CSF has been used extensively in humans as a myeloid growth factor, little is known about the use of GM-CSF in vivo as a vaccine adjuvant. To test the effect of GM-CSF as an adjuvant, we immunized rats with a foreign protein, tetanus toxoid, in a sub-optimal fashion (low dose and single injection). GM-CSF was mixed with the antigen and both were injected intradermally on days 1 and 2. GM-CSF was injected for 5 days. The immunizations were administered at the same site each day for potential stimulation of dermal DC.[40] The remaining three days animals received either GM-CSF or PBS intradermally. After 2 weeks, the rats were evaluated for their response to tetanus toxoid. All animals immunized with GM-CSF as adjuvant developed significant IgG antibody responses against tetanus. None of the animals immunized with tetanus toxoid in saline developed an antibody response. A similar strategy was used immunizing animals with peptides derived from the rat neu protein structure and known to generate an immune response in rats when CFA was used as an adjuvant. Rats immunized with the peptides and GM-CSF developed evidence of a delayed type hypersensitivity (DTH) response to the peptides after 1 immunization whereas animals who received peptides in saline did not.[41] This immunization regimen has been tailored to a single injection of GM-CSF with a group of defined ECD and ICD immunogenic peptides, and animals received GM-CSF/peptide injections once a month for three months. All animals immunized in this fashion develop significant DTH responses comparable to those animals immunized with the peptides in CFA during the same time period. Histopathologic examination of tissues expressing basal levels of rat neu protein, in animals who developed immune responses, showed no evidence of microscopic autoimmune damage. Thus, a single inoculation of GM-CSF with immunizing antigen intradermally, administered each month, is an effective way to augment or generate immune responses to peptide antigens.

GM-CSF has been used as a human vaccine adjuvant. In a phase I study of GM-CSF as an adjuvant for hepatitis B vaccine in humans, doses of 20-80 ug was administered just prior to and at the same site as the vaccination. Of 81 subjects treated with GM-CSF, 15 developed anti-HBs antibody after only one injection (10 had protective titers) while only one of 27 patients vaccinated without GM-CSF produced a weak and transient antibody response.[42] In a pilot study, 15 hemodialysis patients who failed to respond to hepatitis B vaccine were given a single dose of GM-CSF 24 hours before re-vaccination with one of the six patients who received 0.5 ug/kg, four of five who received 5 ug/kg and two of four who received 10 ug/kg responding with a protective level of anti-HBs antibody.[43].

Design of HER2 Peptide-Based Vaccines for Human Trials used Computer Modeling and Extensive In Vitro Testing of Potential Immunogenic Epitopes

Peptide vaccination was an attractive method to test immunization to HER2. Computer modeling systems had been effective in determining which epitopes in rat neu could potentially be immunogenic. We used the same strategy to determine immunogenic epitopes of human HER2. The peptides were selected using a computer protein sequence analysis package, T Sites, which used two searching algorithms. The first was the AMPHI algorithm for identifying motifs according to charge and polarity patterns. The second was the Rothbard and Taylor algorithm for identifying motifs according to charge and polarity patterns.[44] Each of the searching algorithms had empirically been successful in identifying a substantial proportion (50-70%) of helper T cell epitopes in foreign proteins.[45,46] HER2 peptides predicted by both algorithms were constructed and were 15-18 amino acids in length. PBMC, obtained from breast cancer patients with HER2 overexpressing tumors were analyzed for a proliferative T cell response to HER2 potential immunogenic peptides. Seven of 26 peptides tested demonstrated the ability to elicit T cell responses, in vitro, in at least 20% of breast cancer patients evaluated. The ability of a single peptide to generate immune responses in multiple individuals of diverse MHC backgrounds is not unique. Universal epitopes for class II have been defined for tetanus toxoid.[47,48] In addition, one of the peptides our group identified in screening studies has been further characterized.[49] PBMC from 10 breast cancer patients could respond to p777-789. Responses were associated with HLA-DR4.

Our initial clinical vaccination strategies have concentrated on eliciting a $CD4^+$ T helper response for several reasons. First, the importance of the $CD4^+$ helper T cells in mediating an anti-tumor response is increasingly being emphasized. The pre-existent HER2 specific immune responses detected in patients is, for the most part, low level. A vigorous T helper response may serve to augment the production of HER2 antibodies and/or HER specific CTL, both of which could be therapeutic. In addition, $CD4^+$ T cells play a major role in the maintenance of immunologic memory. Our initial clinical studies would attempt to generate a higher magnitude HER2 T helper immune response. We hypothesized the two domains of the HER2 protein may behave differently immunologically. The ICD may be more immunogenic. As an intercellular protein it is not readily available for immune recognition. The ECD protein is available as an extracellular protein and in many patients may actually become truncated and circulate in the sera as a soluble protein.[50]. For this reason we formulated 2 vaccines for use in humans containing helper peptides derived from the different domains, the ECD and the ICD. Three peptides were incorporated in each vaccine as a compromise to include a sufficient number of peptides that some response may be elicited in individual patients tested, but not so many peptides included that an epitope which may inhibit the development of a HER2 specific immune response, e.g., an "immunodominant" epitope, was present.

In addition, we designed a vaccine formulation for specifically eliciting HER2 specific CTL in patients who are HLA-A2. However, our approach to vaccine design was different from immunizing with HLA-A2 binding peptides. The best way to immunize patients to generate a predominant CTL response has not yet been defined. In many models the generation of CTL specific for an antigen is difficult without requisite T cell help.[51] The recent definition of motifs that predict CTL epitopes for a particular HLA molecule have generated much interest in the use of peptides for immunization.[52] In fact, there has been a recent report of immunizing patients with an HLA-A2 binding epitope specific for HER2.[53] Immunization resulted in the generation of HER2 peptide specific IFN-γ producing T cells which would not lyse HER2 overexpressing tumors. Supplying T cell help along with CTL peptide immunization has been an area of intense investigation. One ploy has been to immunize patients with CTL peptides linked to a strong helper epitope, such as that derived from tetanus toxoid. Several studies have attempted immunization to malarial antigens using such constructs.[54,55] One of the major problems in these studies, was the generation of vigorous tetanus responses that overwhelmed the weak malarial specific response. Other studies, in mice, have advocated linking CTL peptides to helper peptides defined for that antigen, even constructing helper peptides that encompassed the CTL epitopes and thereby supplying the CTL epitope and Th epitope within the same peptide.[56-58] This is the strategy we are using with the HLA-A2 CTL vaccine formulation. The vaccine consists of three helper epitopes that encompass, within their natural sequence, HLA-A2 binding peptides.

Preliminary Results of a Phase I Study of HER2 Peptide-Based Vaccines Indicate Immune Responses Directed against the HER2 Protein can be Generated

Our current trial was most recently updated in May 1998.[59] The trial tests the three helper T cell vaccines, each composed of three peptides, 15-18 amino acids in length. One vaccine targets the ICD, ECD, and the final includes helper peptides that encompass HLA-A2 binding peptides. We use GM-CSF as an adjuvant. The immunizations are given i.d., monthly for six months. The primary endpoint of the study is safety and feasibility and the secondary endpoint is to determine whether we can generate significant detectable immunity to the HER2 protein. Ongoing studies are evaluating the HER2 specific antibody, CTL, and T helper response. Immunologic results at this time are reported for T helper responses.

The population enrolled consists of advanced stage III and IV patients who have breast, ovarian, or lung cancers that overexpress HER2. As concurrent treatment, only hormones or local radiation is allowed. The patients are tested for functional immune systems by DTH to seven recall antigens and only those who are not anergic are entered. As of May 1998, 51 patients have been enrolled on study. Thirty seven have completed greater than two vaccinations and 22 have completed all six. Nine patients are in follow-up. To date, the majority of patients enrolled have breast cancer. Of the 40 patients with breast cancer, 10 have localized disease whereas 30 have recurrent or metastatic disease. Patients are either without evidence of disease at the time of immunization or in a minimal disease state following optimal treatment. Ten patients have ovarian cancer and one patient has non-small cell lung cancer. The age range of enrolled patients is 24-86 years.

The primary endpoints of the study are feasibility and safety. The feasibility of completing all six immunizations was dependent, for the most part, on the disease status of the patient at the time of entry. Only patients with minimal or no evidence of disease were able to complete all six vaccines. Of the 22 patients who have completed all 6 immunizations only two of those had measurable disease at the time of entry, one with chest wall and one with bone metastasis. Sixteen patients have had to withdraw prior to completing the full schedule, 13 from progressive disease. Eleven of those 13 had measurable disease at the time of entry. These patients were able to receive a mean of only three vaccines. Three patients withdrew for reasons that were not disease related.

The primary endpoint of the study is to determine any toxicity associated with the vaccine. Despite the generation of HER2 specific T cell immunity similar to the level seen for tetanus toxoid after a tetanus toxoid vaccination, observed toxicity associated with the vaccine, at this time, is minimal. Two of 40 patients developed grade 2 urticaria and one patient experienced grade 2 myalgia. Both patients with urticaria were treated and responsive to oral diphenhydramine, however, one withdrew from study because of the side effect. No evidence of autoimmune phenomenon to organs expressing basal levels of HER2 protein, such as liver, gut, skin, or lung have been observed despite detecting immunity directed against the HER2 protein after immunization.

The secondary endpoint of the study is to determine whether we could generate measurable immunity to HER2. Of the 22 patients who have completed 6 vaccines, the majority developed peptide and protein specific immunity. The detection of protein specific T cells after peptide immunization implies that one or more of the immunizing peptides represent natural epitopes of the HER2 protein. Epitope spreading was also observed in the majority of patients. This phenomenon, first described in autoimmune disease. Reference 60 suggests that the immune repertoire to HER2 evolves during the course of vaccination. The development of a T cell response with a variety of specificities during immunization indicates that naturally expressed HER2 protein is being processed and presented in the MHC. In those patients who have been followed longer than two months after the last immunization, eight of nine patients have persistent immunity three-seven months after their last vaccine. There has been no correlation, to date, of an immune response to any particular peptide with any particular HLA class I or II type.

As an adjunct to the measure of the peripheral blood T cell response, we also evaluated the DTH response to the individual immunizing peptides at a site distant from the vaccine site. The DTH testing to peptides in the immunizing mix was performed 30 days after the last immunization. To date, 60 peptide DTH skin tests have been placed. A peptide specific DTH $\geq$ 10 mm was highly correlative to a peripheral blood peptide specific T cell response $\geq$ S.I. 2.0 ($p=0.04$, Odds Ratio 11.0).[61]

Interestingly, GM-CSF, used as a vaccine adjuvant, is in itself, immunogenic.[62] An antibody response to GM-CSF occurred in the majority of patients (72%). By 48-hour DTH testing, 17% of patients were shown to have a cellular immune response to the adjuvant GM-CSF alone. Likewise, thymidine incorporation assays also demonstrated a peripheral blood T cell response (S.I.>2) to GM-CSF in at least 17% of the patients. The generation of GM-CSF-specific T cell immune responses elicited in this fashion is an important observation because GM-CSF is being used as a vaccine adjuvant in various vaccine strategies. GM-CSF specific immune responses may incorrectly be interpreted as antigen specific immunity, particularly when local DTH responses to vaccination are the primary means of immunologic evaluation.

Conclusions and Future Directions

In conclusion, using this strategy, immunizing patients with HER2 peptides and GM-CSF resulted in the majority of patients developing HER2 protein specific responses, which we hypothesize to be an in vitro surrogate for the potential to recognize HER2 protein on tumor cells. The development of HER2 protein immunity correlated significantly with the development of epitope spreading. The immunizations themselves were well tolerated and associated with minimal toxicity and the immune responses generated persist after vaccinations have stopped, although follow-up is limited at this time. Finally, we believe these data demonstrate an effective vaccine strategy for generating immunity to a specific tumor antigen, HER2. We are currently applying these principals to other antigens in the laboratory as well as evaluating the role of vaccines in preventing cancer relapse.

Acknowledgments

We would like to express our gratitude to the patients who participated in this study and to the oncologists who referred their patients to us. We are grateful for the expert medical care and protocol coordination by Kathy Schiffman, PA-C and Donna Davis, R.N. We acknowledge the excellent technical support of Paul Crosby and Faith Shiota. We thank Kevin Whitham for assistance in the manuscript preparation.

This work is supported for M.L.D. by grants from the NIH, NCI (K08 CA61834, R29 CA68255, and R01 CA75163) and the Cancer Research Treatment Foundation. Patient care was conducted through the Clinical Research Center Facility at the University of Washington that is supported through NIH grant MO1-RR-00037.

References

1. Pardoll DM. A new look for the 1990s. Nature 1994; 369:357-358.
2. Houghton AN. Cancer antigens: immune recognition of self and altered self [comment]. J Exp Med 1994; 180(1):1-4.
3. Disis ML, Cheever MA. Oncogenic proteins as tumor antigens. In: Alt F, Marrack P, eds. Current Opinion in Immunology. London: Current Biology Ltd., 1996:637-642.
4. Nanda NK, Sercarz EE. Induction of anti-self-immunity to cure cancer. Cell 1995; 82(1):13-17.
5. Bargmann C, Hung M, Weinberg R. The *neu* oncogene encodes an epidermal growth factor receptor-related protein. Nature 1986; 319:226-230.
6. Coussens L, Yang-Feng TL, Chen YLE et al. Tyrosine kinase receptor with extensive homology to EGF receptor shares chromosomal location with *neu* oncogene. Science 1985; 230:1132-1139.
7. Press M, Cordon-Cardo C, Slamon D. Expression of HER-2/*neu* proto-oncogene in normal human adult and fetal tissues. Oncogene 1990; 5:953-962.
8. Berchuck A, Kamel A, Whitaker R et al. Overexpression of HER-2/neu is associated with poor survival in advanced epithelial ovarian cancer. Cancer Res 1990; 50:4087-4091.
9. Berchuck A, Rodriguez G, Kinney RB et al. Overexpression of HER-2/neu in endometrial cancer is associated with advanced stage disease. Am J Obstet Gynecol 1991; 164:15-21.
10. Kern JA, Schwartz DA, Nordberg JE et al. $p185^{neu}$ expression in human lung adenocarcinomas predicts shortened survival. Ca Res 1990; 50:5184-5191.
11. Slamon D, Clark G, Wong S et al. Human breast cancer: correlation of relapse and survival with amplification of the HER-2/neu oncogene. Science 1987; 235:177-182.
12. Yonemura Y, Ninomiya I, Yamaguchi A et al. Evaluation of immunoreactivity for erbB-2 protein as a marker of poor short term prognisis in gastric cancer. Ca Res 1991; 51:1034-1038.
13. Slamon D, Godolphin W, Jones L et al. Studies of the HER-2/*neu* proto-oncogene in human breast and ovarian cancer. Science 1989; 244:707-712.
14. Pupa SM, Menard S, Andreola S et al. Antibody response against the c-erbB-2 oncoprotein in breast carcinoma patients. Cancer Res 1993; 53(24):5864-5866.
15. Disis ML, Calenoff E, McLaughlin G et al. Existent T cell and antibody immunity to HER- 2/*neu* protein in patients with breast cancer. Cancer Res 1994; 54:16-20.
16. Fisk B, Blevins TL, Wharton JT et al. Identification of an immunodominant peptide of HER- 2/neu protooncogene recognized by ovarian tumor-specific cytotoxic T lymphocyte lines. J Exp Med 1995; 181(6):2109-2117.
17. Peoples GE, Goedegebuure PS, Smith R et al. Breast and ovarian cancer-specific cytotoxic T lymphocytes recognize the same HER2/neu-derived peptide. Proc Natl Acad Sci, USA 1995; 92:432-436.
18. Disis ML, Pupa SM, Gralow JR et al. High titer HER-2/neu protein specific antibody immunity can be detected in patients with early stage breast cancer. J Clin Oncol 1997; 15(11):3363-3367.

19. Ward RL, Hawkins NJ, Coomber D et al. Characterization of Human HER-2/neu antibodies in patients with colon cancer. Submitted.
20. Disis M, Cheever M. HER-2/neu oncogenic protein: Issues in vaccine development. Crit Rev Immunol 1998; 18:37-45.
21. Katsumata M, Okudaira T, Samanta A et al. Prevention of breast tumour development in vivo by downregulation of the p185neu receptor. Nat Med 1995; 1(7):644-648.
22. Baselga J, Tripathy D, Mendelsohn J et al. Phase II study of weekly intravenous recombinant humanized anti-p185HER2 monoclonal antibody in patients with HER2/neu-overexpressing metastatic breast cancer [see comments]. J Clin Oncol 1996; 14(3):737-744.
23. Slamon D, Leyland-Jones B, Shak S et al. Addition of herceptin (humanized anti-HER2 antibody) to first line chemotherapy for HER2 overexpressing metastatic breast cancer (HER2+/MBC) markedly increases anticancer activity: a randomized, multinational controlled phase III trial. Proceddings of ASCO (Abstract) 1998; 17:98a.
24. Cobleigh MA, Vogel CL, Tripathy D et al. Efficacy and safety of herceptin (humanized anti-HER2 antibody) as a single agent in 222 women with HER2 overexpression who relapsed following chemotherapy for metastatic breast cancer. Proceedings of ASCO (Abstract) 1998; 17:97a.
25. Greenberg PD. Adoptive T cell therapy of tumors: mechanisms operative in the recognition and elimination of tumor cells. Adv Immunol 1991; 49:281-355.
26. Klarnet JP, Matis LA, Kern DE et al. Antigen-driven T cell clones can proliferate in vivo, eradicate disseminated leukemia, and provide specific immunologic memory. J Immunol 1987; 138(11):4012-4017.
27. Riddell S, Watanabe K, Goodrich J et al. Restoration of viral immunity in immunocompromised humans by the adoptive transfer of T cell clones. Science 1992; 257:238-241.
28. Walter EA, Greenberg PD, Gilbert MJ et al. Reconstitution of cellular immunity against cytomegalovirus in recipients of allogeneic bone marrow by transfer of T-cell clones from the donor [see comments]. N Engl J Med 1995; 333(16):1038-1044.
29. Rosenberg SA, Packard BS, Aebersold PM et al. Use of tumor-infiltrating lymphocytes and interleukin-2 in the immunotherapy of patients with metastatic melanoma. A preliminary report [see comments]. N Engl J Med 1988; 319(25):1676-1680.
30. van-der-Bruggen P, Traversari C, Chomez P et al. A gene encoding an antigen recognized by cytolytic T lymphocytes on a human melanoma. Science 1991; 254(5038):1643-1647.
31. Kawakami Y, Eliyahu S, Jennings C et al. Recognition of multiple epitopes in the human melanoma antigen gp100 by tumor-infiltrating T lymphocytes associated with in vivo tumor regression. J Immunol 1995; 154(8):3961-3968.
32. Bernards R, Destree A, McKenzie S et al. Effective tumor immunotherapy directed against an oncogene-encoded product using a vacinnia virus vector. Proc Natl Acad Sci USA 1987; 84:6854-6858.
33. Disis ML, Gralow JR, Bernhard H et al. Peptide-based, but not whole protein, vaccines elicit immunity to HER-2/neu, an oncogenic self-protein. J Immunol 1996; 156:3151-3158.
34. Sercarz EE, Lehmann PV, Ametani A et al. Dominance and crypticity of T cell antigenic determinants. Annu Rev Immunol 1993; 11:729-766.
35. Cibotti R, Kanellopoulos JM, Cabaniols JP et al. Tolerance to a self-protein involves its immunodominant but does not involve its subdominant determinants. Proc Natl Acad Sci USA 1992; 89:416-420.
36. Nagata Y, Furugen R, Hiasa A et al. Peptides derived from a wild-type murine proto-oncogene c-erbB-2/HER2/neu can induce CTL and tumor suppression in syngeneic hosts. J Immunol 1997; 159(3):1336-1343.
37. Bourdette DN, Whitham RH, Chou YK et al. Immunity to TCR peptides in multiple sclerosis. I. Successful immunization of patients with synthetic V beta 5.2 and V beta 6.1 CDR2 peptides [published erratum appears in J Immunol 1994 Jul 15; 153(2):910]. J Immunol 1994; 152(5):2510-2519.
38. Dranoff G, Jaffee E, Lazenby A et al. Vaccination with irradiated tumor cells engineered to secrete murine granulocyte-macrophage colony-stimulating factor stimulates potent, specific, and long-lasting anti-tumor immunity. Proc Natl Acad Sci USA 1993; 90(8):3539-3543.

39. Reid CD, Stackpoole A, Meager A et al. Interactions of tumor necrosis factor with granulocyte-macrophage colony-stimulating factor and other cytokines in the regulation of dendritic cell growth in vitro from early bipotent CD34+ progenitors in human bone marrow. J Immunol 1992; 149(8):2681-2688.
40. Kaplan G, Walsh G, Guido LS et al. Novel responses of human skin to intradermal recombinant granulocyte/macrophage-colony-stimulating factor: Langerhans cell recruitment, keratinocyte growth, and enhanced wound healing. J Exp Med 1992; 175(6):1717-28.
41. Disis ML, Bernhard H, Shiota FM et al. Granulocyte-macrophage colony-stimulating factor: An effective adjuvant for protein and peptide-based vaccines. Blood 1996; 88(1):202-210.
42. Tarr PE, Lin R, Mueller EA et al. Evaluation of tolerability and antibody response after recombinant human granulocyte-macrophage colony-stimulating factor (rhGM-CSF) and a single dose of recombinant hepatitis B vaccine. Vaccine 1996; 14(13):1199-1204.
43. Hess G, Kreiter F, Kosters W et al. The effect of granulocyte-macrophage colony-stimulating factor (GM-CSF) on hepatitis B vaccination in haemodialysis patients. J Viral Hepat 1996; 3(3):149-153.
44. Feller D, Cruz DL. Identifying antigenic T cell sites. Nature 1991; 349:720-721.
45. Roscoe DM, Jung SH, Benhar I et al. Primate antibody response to immunotoxin: serological and computer-aided analysis of epitopes on a truncated form of Pseudomonas exotoxin. Infect Immun 1994; 62(11):5055-5065.
46. Bisset LR, Fierz W. Using a neural network to identify potential HLA-DR1 binding sites within proteins. J Mol Recognit 1993; 6(1):41-48.
47. Panina-Bordignon P, Tan A, Termijtelen A et al. Universally immunogenic T cell epitopes: promiscuous binding to human MHC class II and promiscuous recognition by T cells. Eur J Immunol 1989; 19(12):2237-2242.
48. Valmori D, Sabbatini A, Lanzavecchia A et al. Functional analysis of two tetanus toxin universal T cell epitopes in their interaction with DR1101 and DR1104 alleles. J Immunol 1994; 152(6):2921-2929.
49. Tuttle TM, Anderson BW, Thompson WE et al. Proliferative and cytokine responses to class II HER-2/neu-associated peptides in breast cancer patients. Clin Cancer Res 1998; 4(8):2015-24.
50. Leitzel K, Teramoto Y, Sampson E et al. Elevated soluble c-erb-2 antigen levels in the serum and effusions of a proportion of breast cancer patients. J Clin Oncol 1992; 10:1436-1443.
51. Lin Y, Langman R, Cohn M. The priming of cytotoxic T-cell precursors is strictly helper T cell-dependent. Scand J Immunol 1992; 35:621-626.
52. Falk K, Rötzschke O, Stevanovic S et al. Allele-specific motifs revealed by sequencing of self-peptides eluted from MHC molecules. Nature 1991; 351:290-296.
53. Zaks T, Rosenberg S. Immunization with a peptide epitope (p369-377) from HER-2/neu leads to peptide-specific cytotoxic T lymphocytes that fail to recognize HER-2/neu+ tumors. Cancer Research 1998; 58:4902-4908.
54. Widmann C, Romero P, Maryanski JL et al. T helper epitopes enhance the cytotoxic response of mice immunized with MHC class I-restricted malaria peptides. J Immunol Methods 1992; 155(1):95-99.
55. Herrington DA, Clyde DF, Losonsky G et al. Safety and immunogenicity in man of a synthetic peptide malaria vaccine against Plasmodium falciparum sporozoites. Nature 1987; 328(6127):257-259.
56. Fayolle C, Deriaud E, LeClerc C. *In vivo* induction of cytotoxic T cell response by a free synthetic peptide requires CD4+ T cell help. J of Immunol 1991; 147(12):4069-4073.
57. Lee RS, Grusby MJ, Glimcher LH et al. Indirect recognition by helper cells can induce donor-specific cytotoxic T lymphocytes in vivo. J Exp Med 1994; 179(3):865-872.
58. Shirai M, Pendleton CD, Ahlers J et al. Helper-cytotoxic T lymphocyte (CTL) determinant linkage required for priming of anti-HIV CD8+ CTL in vivo with peptide vaccine constructs. J Immunol 1994; 152(2):549-556.
59. Disis M, Grabstein K, Sleath P et al. HER-2/neu peptide vaccines elicit T cell immunity to the HER-2/neu protein in patients with breast and ovarian cancer. Proceedings of ASCO (Abstract) 1998; 17:97a.
60. Lehmann PV, Forsthuber T, Miller A et al. Spreading of T-cell autoimmunity to cryptic determinants of an autoantigen. Nature 1992; 358(6382):155-157.

61. Disis ML, Schiffman K, Gooley T et al. Delayed type hypersensitivity response (DTH) is an accurate predictor of peripheral blood T cell immunity after HER-2/neu peptide immunization. 2000; 6(4):1347-1350.
62. McNeel D, Schiffman K, Disis M. Immunization with rhGM-CSF as a vaccine adjuvant elicits both a cellular and humoral response to rhGM-CSF. 1999; 93(8):2653-2659.

CHAPTER 10

Peptides in Prostate Cancer

Michael L. Salgaller

Introduction

The timely detection and effective treatment of prostatic cancer is one of the major health problems faced in the United States and, to a comparable extent, the rest of the world. It is predicted that there will be over 180,000 newly diagnosed cases of prostate cancer in the U.S. in 1998, making it the most common type of cancer detected in males or females. Some estimate that almost 20% of men will be diagnosed with prostate cancer during their lifetime. Moreover, the estimated 39,200 deaths from prostate cancer make it the one of the most deadly forms of cancer in men, second only to lung cancer.[1] A swift increase in incident rates occurred during the late 1980s and early 1990s, followed by a marked decrease since that time. It is thought that this fluctuation most likely results from the advent of routine, widespread screening for Prostate-Specific Antigen (PSA). Due to enhanced observation, more and earlier stage prostate cancer would be detected. As the increase in PSA tests level off, so would newly diagnosed cases of various stages of prostate cancer.[2]

The stages of cancer of the prostate are summarized in Table 10.1. Although most newly diagnosed patients present with localized tumor, a significant percentage will progress to a metastatic disease. Conventional therapy (surgery, chemotherapy, hormone therapy and radiotherapy) controls even advanced disease for a finite period of time. To a significant degree, the growth and division of malignant prostate cells are hormone-dependent. So the most common treatment is androgen deprivation or ablation through surgery, resulting in a clinical response in approximately three-fourths of patients.[3] But the term of disease-free survival is not durable, and in a majority of cases it recurs with 12 to 16 months in a disseminated metastatic form that is largely resistant to all standard approaches.[4] This hormone-refractory stage is the most aggressive and has the poorest prognosis, with a medium survival time off less than one year.[5] Recent studies suggest that recurrent tumors have a greater cell proliferative capacity and a lower apoptotic index as compared with primary tumors from the same patient.[6] Also, advanced tumors overcome a microenvironment low in androgen concentration through elevated receptor expression, cross-activation by non-steroidal mechanisms, and mutations in its cognate receptor allowing for an expanded activation range.[7]

Once the patient has progressed to a condition of hormone-refractory disease, few promising strategies exist. Unlike other neoplasms, such as breast and colon cancer, systemic adjuvant therapy for prostate cancer following primary treatment does not substantially increase survival rates.[8] Testicular androgen ablation therapy following orchiectomy is often continued because of minimal treatment risk despite its slight benefit.[4] Complicating even this modest approach are studies indicating a significant decline in PSA in almost one-fourth of men who

Table 10.1. Stages of cancer of the prostate

Stage	Characteristics
I (A)	The tumor is not palpable and is asymptomatic. Often found during surgery during other procedures, such as for benign prostatic hyperplasia. Also may be discovered following a needle biopsy performed after the detection of a rising PSA level. Cancer cells are confined to the prostate, although they may be detected in several regions of the organ.
II (B)	As with stage I, disease is confined to the prostate. However, tumors exceeding 1.5 cm in diameter may exist. Tumor does not extend beyond the covering (capsule), although more than one lobe may be involved. This stage may, in fact, have lymph node metastases that are detected only during surgery or subsequent histologic examination.
III (C)	There is extracapsular extension of tumor. There is involvement of the tissues surrounding the prostate. There may be involvement of the seminal vesicles, the glands that produce semen. This stage may, in fact, have lymph node metastases that are detected only during surgery or subsequent histologic examination.
IV (D)	The tumor has spread to pelvic and/or distant lymph nodes. In addition, common sites of metastasis are the liver, and lung. Bony or visceral metastases may be observed. At this stage, the patient often presents to the clinician complaining of pain and/or ureteral obstruction that disrupts normal urine flow.

had discontinued their antiandrogen therapy[9] Other treatments are marginally effective, including secondary hormonal therapy, chemotherapy, and adjuvant administration of radioisotopes to palliate the painful bone metastasis that frequently occur in end-stage disease.[4]

As far as peptide targets of prostatic carcinoma immunotherapy are concerned, the majority of the clinically applied investigations to date has centered on four proteins: 1) Prostatic Acid Phosphatase (PAP), 2) Prostate-specific Antigen (PSA), 3) Prostate-specific Membrane Antigen, and 4) mucin-1 (MUC-1). Therefore, these proteins will be the focus of this Chapter. This emphasis is meant only to represent the current state of prostate peptide-based immunotherapeutics, and is not intended to imply the lack of potential importance of putative peptide antigens from other proteins.

Prostatic Acid Phosphatase (PAP)

PAP, initially discovered in 1936, is a prostate-specific isoenzyme among a heterologous collection of acid phosphatases produced by prostatic cells.[10] More recently, Northern blot analysis indicates that PAP expression may be exclusively organ-restricted, although its mean abundance in prostatic carcinoma samples is only half that seen for benign prostatic hyperplasia (BPH) specimens.[11] Elevated serum PAP levels are detected in 75% of men with metastatic prostate cancer, sometimes reaching 20-40 times the upper normal range. In contrast, those with localized disease may exhibit normal or slightly elevated serum PAP levels. In many cases enzymatic activity returns to baseline soon after prostatectomy or hormone ablation therapy. A retrospective analysis of three different enzyme-based immunoassays concluded that PAP determinations add prognostic information to pretreatment Prostate-Specific Antigen (PSA) values and might serve as an independent predictor of recurrence.[12] However, complicating this association is the fact that transient elevations can occur following certain clinical procedures, such as a prostatic needle biopsy, cytoscopy, or catheter-induced infarction.[13]

PAP has been well characterized at the DNA level. PAP encodes a 354-residue protein with a molecular mass of approximately 41kD. A 5'-end region codes for a signal peptide of 32 amino acids.[14] It has been known for some time that PAP expression is controlled by androgens, and recent experiments utilizing nuclear run-on assays show that the mechanism involves regulation of transcription in the promoter region.[15] PAP and human placental alkaline phosphatase possess an alu-type repetitive sequence about 900 base pairs downstream from the coding region in the 3'-untranslated region, but within the coding region, PAP has no more than 50% sequence homology to other known proteins.[14] As a result, it has been investigated with the hope that immune responses elicited against PAP would not cross-react against normal, non-prostatic tissues or organs.

The promise of PAP peptide as a target antigen for an immunotherapeutic approach to cancer treatment has been put forth by the work of Peshwa and associates.[10,16] Several peptides were selected that were predictive to be HLA-A2 binders according to the HLA-A2 binding motif algorithm (i.e., a leucine or isoleucine at position 2 and a valine, leucine, or methionine at position 9).[17] The binding strength of several candidate peptides was estimated by measuring the shift in fluorescence intensity in a T2 cell binding assay. A nonamer designated PAP-5 (Ala-Leu-Asp-Val-Tyr-Asn-Gly-Leu-Leu) displayed the strongest HLA-A2 binding. Moreover, PAP-5 was chosen for additional study since it was located on the secreted molecule.[10]

PAP-5-specific CTLs were generated from the peripheral blood mononuclear cells (PBMC) of normal donors using stimulation in vitro (IVS) with autologous dendritic cells (DCs) pulsed with peptide. Details of the method for stimulation in vitro was not provided, and thus it was not possible to conclude how immunostimulatory PAP-5 was in this preclinical study. However, data that was including suggests that the peptide was weakly immunogenic in vitro. First, the authors were able to elicit PAP-5-specific CTLs in only four of 12 normal individuals, and results were shown for only two. Second, cytolytic activity was moderate at best: 1) at the highest effector:target ratio shown (80:1), percent lysis never exceeded 25% and was accompanied by as much as 30% background (unpulsed target); 2) in an endogenous presentation experiment, lysis of JY cells infected with vaccinia-PAP was much greater than 50% over background (JY cells infected with vaccinia-b-gal) in only one donor; 3) recognition of endogenously synthesized PAP was not strong, since lysis of the PAP^+ cell line LNCaP was >50% over background in only one donor.[10] This low level of immunogenicity likely reflects the biological reality of the molecule rather than any deficiency with the assay system employed.

The attempt to elicit an immune response to PAP-derived peptides or protein selectively expressed in the prostate is still an attractive proposition. Despite the moderate immune activity of the anti-PAP-5 effector cells, it should be noted that the aforementioned report was the first demonstration that PAP could serve as a target of cellular immunotherapy. The same group had earlier demonstrated that allogeneic dendritic cells exogenously presenting viral antigens could elicit antigen- (and not allo-HLA-)restricted CTL,[16] lending hope that even those with functionally defective DCs might benefit from such an approach.[18] Finally, there is the issue of overcoming tolerance to a shared antigen. In a syngeneic murine model, it has been demonstrated to the preexisting tolerance to PAP could be overcome yet not result in autoimmune prostatitis.[19] So it has been established that PAP-based immune responses can be generated, and such enhanced immunity is of potential benefit to prostate cancer patients in general.

Prostate-Specific Antigen (PSA)

PSA was initially identified and characterized during the 1970s. It is a single-chain glycoprotein of 240 amino acids that expresses a 34 kD protein product.[20] It is a member of the serine protease family, possessing both trypsinlike and chemotrypsinlike activity. PSA has 57% and 78% homology to human pancreatic/renal and glandular kallikrein, respectively.[19] PSA is secreted as a normal component of seminal fluid and uses proteolysis to liquefy the seminal coagulum. It was thought as exclusively produced in the prostate epithelium and therefore considered tissue-specific. Yet subsequent studies indicate that PSA expression at the mRNA can be detected in non-prostatic cancers,[21] and it has been found in the saliva, serum, and urine of normal women.[22] As a result, it is no longer regarded as prostate-specific.

The importance of PSA in the screening and early diagnosis of prostate cancer is well known. PSA determinations can disclose early-stage cancer, when it is confined to the prostate. Along with digital rectal examination, it is a crucial component to detect localized disease.[22] Its presence in serum can be quantified by using commercially available immunoassay kits that use monoclonal antibodies directed towards unique epitopes. One of the most established clinical applications of measuring serum PSA during prostate cancer treatment is in the detection of relapse following radical prostatectomy. Monitoring postoperative patients with highly sensitive PSA assays—some with a detection limit of 0.001 ug/l—increase the likelihood of early intervention.[23] Indeed, histologic examination of tissues following prostate removal has shown that as many as 71% of the cases of organ-confined carcinoma can be discovered using a comprehensive PSA screening program – compared with 33% of cases discovered by digital rectal examination.[24] Limitations exist despite the improvement in detection methods to high sensitivities and specificity. It is not prostate- or disease-specific. Elevated PSA titers can be detected in sera of normal men and women, as well as in 25-86% of men with BPH or prostatitis. But to date, PSA has proven to have greater sensitivity and specificity for screening and diagnosis than any other tumor marker.[22] And so throughout the years the importance of regular PSA testing has been a fundamental part of the clinician's advice to their male patients 50 years of age and over.

So as to develop a peptide-based immunotherapy approach, the amino acid sequence of PSA was studied for peptide motifs most likely to bind to HLA-A2.[25] A synthetic PSA peptide representing residues 146-154 (KLQCVDLHV) was manufactured and tested for binding in an HLA-stability assay. It appeared to be a strong binder, as it stabilized HLA-A2 more effectively than the highly immunogenic peptide from the human immunodeficiency virus (HIV RT 476-484).[26] Stimulation in vitro to generate PSA-specific CTL was similar to previously published methodologies.[27-29] Briefly, PBMC were separated by density gradient centrifugation and cultured in multiwell plates in the presence of exogenously added peptide. The initial stimulation did not include the addition of antigen-presenting cells (APCs). Weekly restimulation with peptide-pulsed, autologous irradiated PBMC was performed for eight cycles. Subsequently, effectors were restimulated in the presence of peptide-pulsed, irradiated autologous LCL (Epstein-Barr virus-transformed B-lymphoblastoid) cells. Interleukin-2 was added three days following the initial set-up, and then at the time of each restimulation. Chromium-release assays were used to assess specific lytic activity.[26]

Although $PSA_{146\text{-}154}$-specific, HLA-A2-restricted $CD8^+$ T-cells were produced, it was not clear how many cycles of in vitro stimulation were required to generate the lytic activity shown in the report. Also, data originated from only one donor. But the potential utility of PSA-derived peptides as targets of antigen-directed cancer vaccines was demonstrated, and was soon confirmed by a second study that involved the same epitope.[30] Correale and coworkers used a similar technique of repeated stimulations in vitro to generate CTL recognizing $PSA_{141\text{-}150}$ and $PSA_{154\text{-}163}$ in two disease-free subjects and one donor with prostate cancer. Three distinct CTL lines were produced that recognized $PSA_{154\text{-}163}$ as well as a fourth specific for $PSA_{141\text{-}150}$. These effectors could lyse the PSA^+ prostate cell line, LNCaP. They were able to demonstrate recognition of endogenous PSA using a PSA^- cell line infected with a vaccinia virus-PSA construct. Finally, MHC restriction was demonstrated by the significant lytic activity that could be blocked by HLA-A2 antibodies.[30]

Shortly after the first studies with PSA peptides were published, controversy arose concerning whether or not the monomeric form of PSA had been recognized by specific CTL.

Potential problems with the work of Xue[26] and Correale[30] included whether or not the formation of peptide dimers – resulting from the cysteine residue in the PSA epitope – accounted for the observed reactivity.[31] There were conflicting reports about whether or not reactivity could be generated from the monomeric form.[32] That the synthetic peptide used in the initial report was in its monomeric was confirmed by mass spectroscopy,[33] reestablishing the original claim of PSA nonamer immunogenicity. But persuasive evidence of its immunogenicity followed a report by Alexander et al[31] that expanded on previous investigations. Their results also confirmed that PSA-derived, HLA-A2-restricted peptides were immunogenic in vitro, albeit weakly. Although four peptides from various regions of the PSA molecule were studied ($PSA_{29\text{-}37}$, PSA $_{98\text{-}106}$, $PSA_{141\text{-}150}$, and $PSA_{146\text{-}154}$) using PBMC from seven patients with prostate adenocarcinoma, the investigators detected peptide-specific reactivity for only one peptide ($PSA_{141\text{-}150}$) in only one of seven subjects. Lysis of PSA-expressing targets was demonstrated, and was blocked by an antibody to human MHC class I but not to human MHC class II. Unlike earlier reports, the PSA-specific T-cells could not recognize and lyse an HLA-A2$^+$ EBV-B cell lines expressing PSA endogenously. It is possible that the source of responder lymphocytes came from donors rendered immunodeficient by a combination of their tumor burden and advanced age. Supporting this premise is their finding that $Flu_{58\text{-}66}$-specific cellular immunity could be elicited from only three of seven patients.[31] Therefore, in general, immune recognition of PSA-derived, HLA-A2-restricted epitopes in both normal men as well as those with prostate cancer is an uncommon occurrence.

In an effort to expand the usefulness of PSA-peptide-based immunotherapeutics, researchers went on to use multiple or oligopeptides that were present in more than the 50% of North American Caucasians and 34% of African-Americans who express HLA-A2.[34] One study centered on PSA-OP, a PSA-derived oligoepitope consisting of amino acid residues 141-170. PSA-OP was a 30-mer constructed to contain theoretical binding motifs for the HLA-2, -A3, -A11, and –B53 alleles of MHC class I. In total, the four alleles were present in 90% of men worldwide. Normal HLA-A2 and –A3 subjects donated PBMC for the series of experiments. At day 0 of culture, peptide was exogenously added to microwells without any APCs. One round of stimulation consisted of a 5-day incubation with peptide and 11 days with IL-2. The first restimulation took place on day 16, and every 16 days thereafter. As before, a concerted effort was required – six to seven rounds of IVS (over 100 days in culture) were necessary to produce four PSA peptide-specific T-cell lines. Using peptide-pulsed C1R-A2 cells as targets, the strongest lytic activity was observed against PSA-OP and, to a lesser extent, the HLA-A2-resticted peptides $PSA_{141\text{-}150}$ and $PSA_{154\text{-}163.}$ Using peptide-pulsed C1R-A3 cells as targets, the strongest lytic activity was the HLA-A3-restricted peptide $PSA_{162\text{-}170}$, although a vibrant response was also observed against PSA-OP. MHC class I specificity could be abolished by adding an HLA-A2- or A3-restricted antibodies to the corresponding haplotype-specific effectors. All T-cell lines were tested for cytolytic activity against C1R-A2 and C1R-A3 cells infected with either wild-type vaccinia virus or rV-PSA. Two of the four lines could lyse C1R-A2 cells infected with rV, and two could lyse C1R-A3 cells similarly infected. LNCaP cells were also lysed, although a PSA$^-$ cell line was not tested.[35]

The in vivo immunogenicity of PSA-OP was demonstrated by the ability of T-cells harvested from HLA-A2.1/K^b transgenic mice to lyse PSA-OP-pulsed Jurkat A2/K^b cells following immunization with PSA-OP. In order to analyze whether or not PSA-OP was processed into shorter, HLA-A2-restricted peptides, several protease inhibitors were added to ^{111}In-release assays. Three different inhibitors were able to block the CTL activity of PSA-OP- specific cells against PSA-OP$^+$ targets. The addition of such inhibitors did not block the CTL activity of PSA-OP-specific cells against HLA-A2$^+$ targets, indicating that the 30mer could be cleaved into peptides that would fit into the HLA-A2 binding groove.[35] The study did not address to what extent intracellular processing may have taken place. However, the report did examine whether or not the problem of HLA-restriction—a drawback of most single-peptide therapy approaches—could be overcome by the use of a larger synthetic peptide containing several HLA binding motifs.

At first glance it might be disconcerting that so much effort was required to generate what turned out to be a low frequency of in vitro immune reactivity. Yet there is no conclusive data that immunogenicity in vitro strongly correlates with clinical benefit in vivo. In fact, it is interesting that the melanoma antigen MART-1/MelanA is so potently immunogenic in vitro[27,36,37] and yet has produced disappointing clinical results.[38] It is possible that poorly immunogenic peptides result in a low level of tolerance within the individual. The result could be a lower threshold that peptide-based cancer vaccines would have to overcome in order to evoke a clinically beneficial immune response.

Prostate-Specific Membrane Antigen (PSMA)

PSMA is a 750-amino acid type II transmembrane glycoprotein possessing 10 potential N-linked glycosylation sites. It contains three structural domains: a 707-amino acid extracellular region, a 24-amino acid transmembrane region, and a 19-amino acid intracellular region. The presence of hydrophobic amino acids from residues 20-43 bounded by basic residues indicates that PSMA does not primarily exist in a secreted form, in contrast to PAP and PSA. At the DNA level, a span of 450 amino acids within the coding region contains more than 54% homology with the human transferrin receptor—also a type II integral membrane protein with similar structure.[39] Chromosomal analysis places the PSMA gene in the area of chromosome 11p11.2.[40]

PSMA was first detected as the protein target of a murine IgG_1 monoclonal antibody, 7E11.C5. The antibody was purified from stable clones of murine hybridomas following immunization with the isolated membrane fractions of LNCaP cells.[41] The LNCaP cell line itself was established from a metastatic lymph node harvested from an individual with hormone-refractory prostate carcinoma.[42] Epitope mapping using synthetic peptides revealed that 7E11.C5 recognized the primary polypeptide chain in intracellular domain of the N-terminus, instead of a glycoprotein region as was originally theorized.[43] Later, when a second anti-PSMA antibody, 3F5.4G6, was described it made possible the development of a sandwich immunoassay for screening purposes.[44] In the initial report on PSMA expression, 20/43 patients with prostate cancer displayed detectable serum titers of PSMA as determined by a competitive antibody binding assay. A strong association was established between a positive test and the presence of prostate cancer, with the percentage of positive patients directly correlating with advanced disease.[41] PSMA could also be detected in the seminal plasma of men with prostate cancer—at a higher level than in serum.[45] Later, when a western blot assay was developed,[45] significant amount of serum PSMA was detected in men with prostate cancer as compared with those with benign prostatic hyperplasia (BPH) or normal subjects. Indications as to its great utility as a marker for disease progression resulted from an extensive analysis of 235 patients from eight different cancer centers.[46] The analysis was double-blinded and involved multiple time-points. High initial PSMA levels were predictive of lack of response to treatment, usually post-prostatectomy. Chronically elevated PSMA reflected a state of clinical progression, even three years following surgery and in the face of undetectable PSA levels. In contrast, PSA was reconfirmed as an important marker when limited to the diagnosis and staging of prostate cancer.[46]

Throughout the years since it was first detected, the nearly prostate-specific expression of PSMA at the protein level has become more established. Initial reports using the 7E11.C5 antibody and immunohistochemistry demonstrated strong PSMA staining in both normal and malignant human prostatic epithelium.[41,47] Strong staining of benign and neoplastic epithelial cells was detected in all sections from 184 radical prostatectomies obtained from patients with advanced adenocarcinoma of the prostate. Basal cells, prostate stroma, and urothelium generally did not stain. The percent of immunoreactive cells directly correlated with disease progress, as PSMA expression increased from benign tissue to prostatic intraepithelial neoplasia (PIN) to adenocarcinoma specimens.[47] In a retrospective survey of frozen sections, none of the 26 different non-prostatic tumors or 120/122 samples displayed positive immunoreactivity. Two of 14 normal kidneys were reactive;[41] a finding corroborated years later when extraprostatic expression was noted in a subset of proximal renal tubules. One laboratory reported that staining could also be observed in the endothelial cells of seven of 13 transitional cell cancers, eight of 17 renal cell carcinomas, and three of 19 colon carcinomas.[48] Therefore, although not exclusively organ-restricted, extraprostatic PSMA expression is highly limited.

The biological function of PSMA is still being elucidated. Carter and associates discovered a 1.4 Kb partial cDNA from a rat brain expression library that shared 86% homology with the PSMA cDNA. By immunoscreening the library, the partial cDNA was identified as NAALADase (N-acetylated a-linked acidic dipeptidase). Following this, a cDNA containing the entire open reading frame was isolated from LNCaP cells. Transient transfection of this cDNA into two NAALADase-negative cell lines conferred NAAG-hydrolyzing activity that was abolished by NAALADase inhibitors. Therefore, it was concluded that PSMA and NAALADase shared hydrolytic activity.[49] At the same time, Pinto and coworkers determined that only the $PSMA^+$ cell line, LNCaP, and not three other $PSMA^-$ prostate cancer cell lines, demonstrated folate hydrolase activity. This enzymatic activity was classified as an exopeptidase because of where it cleaved methotrexate moieties. The detailed analysis of different forms of methotrexate not only determined the cleavage site(s) on polyglutamated methotrexate; it also suggested a role for PSMA in cellular resistance to the chemotherapy agent.[50]

Most of the work with PSMA-derived peptides has utilized dendritic cells as APCs. At first, in order to develop a peptide-based approach to cellular immunotherapy, it was important to demonstrate that: 1) APCs could be generated from the precursor cells of prostate cancer patients, and 2) immunogenic prostate-associated peptides existed. Dendritic cells possessed unique capabilities for antigen uptake, processing, and recognition[51-55]—and thus were thought to be superior to PBMC or EBV-B cells as a cancer vaccine component. The PSA experiments of Correale et al had shown that autologous PBMC or EBV-B cells were effective APCs to elicit peptide-specific T-cells in vitro,[30,35] but their efficacy in vitro was not explored. In 1995, Tjoa et al published their findings that helped establish PSMA-derived peptides delivered by DCs as a clinically effective, technically feasible cancer vaccine.[56] In that report, DCs were cultured from adherent precursors in the peripheral blood of prostate cancer subjects using a combination of GM-CSF and IL-4.[56,57] Seven of ten donors had highly progressive disease, being classified as stage D_1 or D_2, and eight had hormone-refractory tumors. Despite their age (62-80 years; mean of 69) and advanced disease, DC yields from PBMC ranged from 2-14%. Cell surface phenotypic analysis showed the majority of cells to be $CD14^-$, $CD11c^+$, $CD1a^+$, $CD86^+$, and $HLA\text{-}DR^+$. One seven-day culture of DCs averaged $2\text{-}7 \times 10^6$ from only a 50 ml blood draw. Autologous T-cells from prostate cancer patients' PBMC were activated against tetanus toxoid presented by day seven cultured DCs, as measured by a proliferation assay measuring ^{3}HTdR incorporation. Also, a significant increase in ^{3}HTdR incorporation was seen in two cases when DCs were used to present prostate cancer proteins, in the form of LNCaP cell lysates to, autologous T-cells.[56] Since proliferation often measures largely CD4-mediated immunity, follow-up studies looked at CD8-mediated cytotoxicity. In that subsequent study, DCs which were osmotically loaded with autologous cell lysate derived from primary prostate tumors were preferentially lysed by autologous T-cells.[58] Therefore, it was shown that DCs derived from the peripheral blood of prostate cancer patients even with end-stage disease could provide immunocompetent APCs. This functionality was important to establish, since DCs harvested from other neoplasms had displayed aberrantly low levels of antigen presentation.[18] The initial work of Tjoa and associates also revealed that it was technically feasible to generate large numbers of DCs from a blood draw or leukapheresis.[56]

Subsequently, a phase I clinical trial was begun to examine whether the administration of autologous dendritic cells exogenously pulsed with HLA-A2.01-specific peptides derived from PSMA could result in therapeutic benefit.[59] DCs were generated from adherent monocytic cells following six days in culture with GM-CSF and IL-4. Twelve putative HLA-A2 nonamers were identified within the open reading frame of PSMA. Two PSMA-derived peptides were chosen for analysis, one from the amino- and one from the carboxyl-terminus. Each peptide, designated PSM-P1 (LLHETDSAV) and -P2 (ALFDIESKV) contained favorable HLA-A2.01 binding residues at positions 2 and 9.[25] Competitive binding analysis revealed PSM-P1 and -P2 as strong MHC class I binders.[60] All the subjects enrolled had hormone-refractory prostate cancer. A majority had previously undergone one or more primary treatments, such as hormone therapy, chemotherapy, radiotherapy, or surgery (prostatectomy or orchiectomy). Fewer than one-quarter of patients was regarded as strongly immunocompetent, as determined by an overall lack of response to several highly immunogenic recall antigens given via skin test (delayed-type hypersensitivity; DTH).[61]

Participants were divided into five groups receiving four-six infusions of peptides alone (PSM P1 or -P2; groups 1 and 2, respectively), autologous DCs alone (group 3), or DCs presenting PSM-P1 or -P2 (groups 4 and 5, respectively). Patients were vaccinated every six weeks. In groups 1 and 2, participants were further subdivided for purposes of dose-escalation. Those men received either 0.2, 2.0, or 20 ug peptide in saline. In groups 3-5, participants received from 0.1-2.0 x 10^7 DCs. Members of groups 4 and 5 had their DCs exogenously pulsed for two hours with 1 ug/ml peptide prior to infusion. The vaccine was administered into the lateral entry port of a fast-flowing intravenous infusion of 0.9% saline. Clinical monitoring for toxicity, hematological studies, and prostate markers occurred regularly throughout the 370 days of the study.[59] Fifty- one patients completed a full treatment regimen and were available for evaluation.

Acute toxicity monitored during and for four hours following treatment included the development of allergic and anaphylactic reactions—none of which were observed. The most frequent side-effect was quite mild: a moderate, transient hypotension whose incidence was highest during the time of the first infusion and significantly declined thereafter. Only eight of 51 subjects experienced hypotension on more than one occasion, and only one patient experienced hypotension during the third infusion. All blood pressure elevations returned to normal within four hours of vaccination. Six percent of participants complained of post-treatment fatigue lasting as long as eight days. No significant increase in serum TNFα or INFγ was seen.[54] After four cycle of infusions, the highest dose (2 x 10^7 DCs and 20 μg peptide) had been given, and yet, due to the lack of toxicity, a true maximum tolerated dose had not been attained.[61] In general, quality of life during treatment was quite favorable, especially compared to many other approaches for adjuvant therapy of cancer.[62]

Clinical response or lack thereof was determined according to PSA values and criteria of the National Prostate Cancer Project (NPCP); summarized in Table 10.2.[63] These criteria take in account essential individual parameters, including the extent of measurable and osseous disease, together with biochemical markers that: 1) are predominantly tumor-specific, 2) reflect tumor burden, and 3) mirror the secondary changes of the tumor on the patient.[64] The results of the phase I clinical trial using autologous DCs and/or PSMA-derived peptides is shown in Table 10.3.[59,61] In total, seven cases of partial regression were observed, along with 11 patients with stabilized disease, and 33 with progressive disease. The minimum length of response was 100 days, observed in one patient. Two patients had a length of response of 200 days, while the remaining four subjects responded to treatment for over a year and were subsequently enrolled in the phase II study of DCs with PSMA-derived peptides. It should be mentioned

Table 10.2. NPCP response criteria for evaluating patient response to prostate cancer treatment

Complete Response

All of the following:

1. Tumor masses, if present, totally disappeared and no new lesions appeared.
2. Elevated prostate specific antigen (PSA), if present, returned to normal.
3. Osteolytic lesions, if present, recalcified.
4. Osteoblastic lesions, if present, disappeared with a negative bone scan.
5. If hepatomegaly is a significant factor, there must be a complete return in size of the liver to normal (as measured by distention below both costal margins at mid-clavicular lines and from the tip of the xiphoid process during quiet respiration of liver function, including bilirubin (mg per dl) and SGOT.
6. No significant cancer-related deterioration in weight (greater than 10 per cent), symptoms, or performance status.
7. Repeat ProstaScint® scan showing complete reduction in lymph node disease.

Partial Regression

Any of the following:

1. Recalcification of one or more of any osteolytic lesions.
2. A reduction by 50 per cent or more in the number of increased uptake areas on the bone scan.
3. Decrease of 50 per cent or more in cross-sectional area of any measurable lesion.
4. If hepatomegaly is a significant indicator, there must be at least a 30 per cent reduction in liver size as indicated by a change in the measurements, function, including bilirubin (mg per dl) and SGOT.
5. Repeat ProstaScint® scan showing significant reduction in lymph node disease.

All of the following:

6. No new sites of disease.
7. PSA returned to normal or was reduced by greater than 50 per cent (on several occasions for at least 30 days)
8. No significant cancer-related deterioration in weight (greater than 10 per cent), symptoms, or performance status.

Objectively Stable

All of the following:

1. No new lesions occurred and no measurable lesions increased more than 25 per cent in cross-sectional area.
2. Elevated PSA, if present, decreased, though need not have returned to normal or decreased by less than 50 per cent.
3. Osteolytic lesions, if present, did not appear to worsen.
4. Osteoblastic lesions, if present, remained normal on bone scan.
5. Hepatomegaly, if present, did not appear to worsen by more than a 30 per cent.
6. No significant cancer-related deterioration in weight (greater than 10 per cent), symptoms, or performance status.
7. No significant change in repeat ProstaScint® scan.

Objectively Stable

Any of the following:

1. Significant cancer-related deterioration in weight (greater than 10 per cent), symptoms, or performance status.
2. Appearance of new areas of malignant disease by bone scan or x-ray or soft tissue by other appropriate techniques.
3. Increase in any previously measurable lesion by greater than 25 per cent in cross-sectional area.
4. Development of recurring anemia secondary to prostatic cancer (not related to treatment; protocols for patients with metastatic disease who have not failed hormone therapy.)
5. Development of ureteral obstruction (protocols for patients as in No. 4 above).
6. PSA increase of greater than 50 per cent.
7. ProstaScint® scan showing new lymph node disease.

Note: An increase in acid or alkaline phosphatase alone is not to be considered an indication of progression. These should be used in conjunction with other criteria.

Table 10.3. Results of phase 1 clinical trial

Group	Progressive Disease	Stable	Partial Responder	Total	HLA-A2 +/-	Length of Response (days)
1	7	3	1	11	7/4	200
2	5	3	1	9	5/4	>200
3	10	2	0	12	11/12	
4	3	2	4	9	3/6	100-260+
5	8	1	1	10	6/3	>230

that establishing the cause of the one partial response in group 2 is complicated by the single dose of hormone therapy (1 mgm per qd DES) given at the time of his third infusion.[54] Three of the partial responders, all from group 4, were actually HLA-A2 negative. It is thought that they might have responded to treatment despite their HLA status because: 1) it could not be ruled out that PSM-P1 and -P2 might bind to other HLA molecules present on the patient's cells, and 2) evidence exists that untreated DCs by themselves or in the presence of antigen-pulsed DCs can be of therapeutic benefit.[65-67] Finally, in this initial study, T-cell tolerization following peptide vaccination—observed in one comprehensive murine study—was not observed.[68]

As a whole, PSA values rose slightly in group 1-3 as well as group 4 patients with a starting PSA value >10. PSA values were constant overall in group 4 patients with an initial PSA value of 0-10, as well as in group 5 patients. The levels of alkaline phosphatase, hematocrit, and absolute lymphocytes stayed constant overall.[59] However, differences in certain factors were detected when subjects were grouped according to clinical response or non-response. Mean alkaline phosphatase levels decreased following the initial infusion within the responder group, while it rose slightly among the non-responders. Mean PSA values for those with either low (0-19 ng/ml) or high (>19 ng/ml) pretreatment levels increased steadily during treatment. An average decrease in PSA from 60 to 24 ng/ml was observed among responders. Responders had a higher, although relatively unchanging, average PSMA level compared with non- responders throughout the duration of the study.[61]

Most of the PBMC collected for the study was dedicated to the generation of DCs for infusion. However, sufficient cells remained for proliferation assays to be performed. They revealed a slight enhancement of cellular immunity in groups 4 and 5, reflected by an average stimulation ratio of 1.9 and 2.8 after the second vaccination with DCs plus PSM-P1 or -P2, respectively.[59]

Clinical studies with DCs and PSMA-derived peptides continued with a phase II investigation of 107 subjects. Sixty-six subjects had hormone refractory prostate cancer at the time of enrollment, while 41 had locally recurrent disease. Half of the enrolled patients were randomly selected to receive systemic adjuvant therapy with GM-CSF. Those subjects who were on hormonal therapy and wished to continue during immunotherapy were allowed to continue treatment; those whose anti-androgen therapy was discontinued waited three months before their first vaccination. Infusions were given six times every six weeks. The treatment regimen was similar to the phase I protocol, with the exception that all patients received DCs, and all patients received both PSMA-derived peptides. An average of 17×10^6 DCs were administered, with a range of $5\text{-}24 \times 10^6$, dependent on the amount of blood able to be collected along with the number of cells obtained from each culture. Day 7 DCs were pulsed with 10 ug/ml of PSM-P1 and P2 for two hours immediately prior to harvest.[69]

Of the 33 patients with hormone-refractory metastatic prostate cancer, 25 patients completed at least one infusion cycle and were evaluated for response. Their clinical status was recently reported.[69] As before, NPCP criteria were used to determine patient response,[63] in addition to PSA measurements and ProstaScint® scans to evaluate the extent of nodal disease. Six patients (24%) were identified as partial responders, two (8%) as complete responders. One complete responder showed resolution of his nodal disease, while the second complete responder showed resolution of his bony metastases. Among the eight total responders, three participants demonstrated >50% decline in their serum PSA while the other five subjects returned to and maintained baseline PSA serum levels throughout the study. Among the non-responders, 16 patients (64%) demonstrated disease progression and one (4%) exhibited stable disease. Among those subjects classified as having progressive disease, only seven (44%) could complete all six infusions. Seven subjects (44%) died during their course of treatment, while another two (13%) withdrew from the study.[69]

In a follow-up report from the same study, one complete response and 10 partial responders were obtained from 37 prostate patients with locally recurrent disease after primary treatment.[70] Therefore, almost 30% of patients (28/95) were identified as clinical responders during an average total evaluation period of almost one year. The patient with a complete response demonstrated a decrease in PSA as well as a complete resolution of his nodal disease.

DTH testing performed before, during, and after therapy revealed a moderate association between general cellular immune reactivity and clinical response. Only twenty-five percent (2/8) of responders exhibited a diminishing level of skin reactivity, while sixty-six percent (4/6) of non-responders exhibited decreased responses.[69] None of the patients in the initial phase II report were leukapheresed; therefore, the collected PBMC had to be dedicated to the generation of autologous DCs and could not be used for in vitro immune monitoring. The patients with locally recurrent disease were leukapheresed, and so immune monitoring is ongoing. Already cellular immune enhancement during the course of treatment has been detected, as assessed by PSM-P1- and -P2-specific reactivity in ELISA and ELISPOT assays.[60] It is essential to determine whether or not any observed immune improvement correlates with clinical outcome. This is being done for all subjects undergoing immune monitoring. And, ultimately, treatment benefit will have to be evaluated in large, multi-center clinical trial with overall survival as the seminal endpoint.[64]

Mucin Gene-1 (MUC-1)

Mucins are large integral membrane glycoproteins that contain an extensive polypeptide nucleus that is abundantly glycosylated by O-linked carbohydrates. Most of its molecular weight results from the carbohydrate side-chains, which are linked to threonine and serine amino acids of the polypeptide nucleus. Mucins originating in different sources display distinct patterns of post-translational modification. They are present on the luminal surface of ductal epithelial cells and on the tumor cells originating from them.[71] Several characteristics of mucins such as MUC-1 make it a particularly attractive target for immunotherapy: 1) it is truly tumor-specific in that it is aberrantly (hypo)glycosylated as compared with its normal counterpart,[72] 2) the overproduced protein is extensively deposited along the cell surface,[73] and 3) the epitope recognized by T-cells is found on the 5' end of a 20mer peptide which can recur up to 200 times per molecule.[74] In fact, epithelial tumors without this tandem repeat cannot be lysed by CTL.[71] Although studies have demonstrated the overexpression of several members of the MUC gene family (primarily MUC-2 and -3),[75,76] clinical trials to date have been conducted exclusively with MUC-1.

Goydos and associates[75] were the first to conduct a phase I clinical trial with MUC-1-derived peptides in an attempt to exploit the hypoglycosylated nature of tumor mucins. They were able to detect T-cell responses in a majority of patients with adenocarcinoma of the pancreas, breast, or colon using a 105 amino acid synthetic peptide containing five repetitions of conserved, tandem repeats.[77] Following up this initial success, a truncated form of the MUC-1 peptide was used in a dose-escalating study of 20 patients with prostate cancer.[78] Subjects received an amino acid consisting of 32 residues containing the 20mer tandem repeat unit of the MUC-1 peptide. In order to enhance immunogenicity, the peptide was conjugated with keyhole limpet hemocyanin (KLH)[79] as a carrier molecule. This complex was administered with a powerful adjuvant, QS-21[80], to further augment T- and B-cell stimulation. A majority of participants had minimal residual disease with escalating PSA levels following radiotherapy or prostatectomy. The study endpoints of this dose-escalation trial included safety as well as B- and T-cell responses.[78]

All 20 subjects given the MUC-1 peptide vaccine had detectable IgM and IgG titers following seven weeks of treatment involving three immunizations. Peak serum antibodies were detected by week 13, after which they declined by week 19. Their fifth and last immunization, given around week 19, caused another rise in antibody titers. Importantly, high IgM and IgG titers (1:640 and 1:1280, respectively) were maintained as long as one year in a majority of subjects.[78] Assessment of clinical benefit remained to be conducted. Such findings offer promise of another clinically relevant target of prostate peptide immunotherapy, despite the concern about the potential of MUC1 as a vaccine target due to its ability to downregulate human T-cell responses.[81]

Future Directions

Interestingly, the future direction of many antigen-directed prostate cancer vaccines involves the use of proteins rather than peptides. Proteins as targets of immunotherapy have several advantages over peptide-based therapeutics including, but not limited to: 1) lack of severe HLA haplotype restrictions (thus making treatment immediately relevant to a larger patient population), 2) potential helper and adjuvant epitope effects, and 3) lack of need to define the target antigen(s). However, the disadvantages of abandoning peptide-based immunotherapeutics cannot be discounted. The possibility of generating immune responses to dominant, but clinically irrelevant, heterologous epitopes is one of the major drawbacks. Moreover, the likelihood of cross-reactivity with normal components or non-prostate tissue is greater. The cost of developing quality controls and producing clinical grade material for human trials is much higher when using proteins. Since investigators hope to eventually advance their preliminary work to a multicenter phase III clinical study involving hundreds of subjects, these are not insignificant considerations. In any case, the lack of emphasis on developing peptide-based vaccines for prostate cancer is a relative one, and comes to the fore when compared with the numerous peptide-based experimental human studies underway or planned for melanoma, renal cell carcinoma, breast, and colon cancer.

It is much easier to detect protein expression than peptide abundance on the surface of a cell. The presence and strength of protein expression can be assessed via techniques such as immunohistochemistry that are relatively straightforward. Once an overexpressed protein is detected, and its frequency across and within tumor types is established, it appears as though most laboratories are choosing protein-based cancer vaccines as their initial clinical approach. This is most likely because of the difficulty in defining the precise epitopes responsible for immunogenicity. The technologies required to define such epitopes, among them tandem mass spectroscopy,[82] are much more difficult and time-consuming than histologically based techniques. As a result, several candidate prostate-associated proteins are being proposed for clinical development, including the pancarcinoma glycoprotein, KSA[83], sialyted Tn[84], GM2 ganglioside[76], and human chorionic gonadotropin.[77]

As methodologies for the molecular profiling of tumor genes become more efficient, e.g., using differential display[85] and microarray analysis[86], putative markers for prostate cancer have been put forth. It is crucial that the potential immunogenicity of such genes been established as early as possible. It is counterproductive to initiate a comprehensive study of a gene whose overexpression may be irrelevant to cancer. Using such criteria, two of the more promising human prostate genes are prostate carcinoma antigen 1 (PCTA-1)[87] and thymosin β 15.[88] PCTA-1, selectively expressed in prostate carcinoma versus normal or benign tissue, was identified using expression cloning of cDNAs and antibodies reacting with its encoded protein. Thymosin β 15, an actin-binding moiety, helps regulate tumor cell motility in metastatic prostate cancer.[88] Both of these gene products, by demonstrating their relevance to tumorgenicity, have been validated as potentially important for scientists to direct anti-tumor immune reactivity. Finally, research groups are also looking into recombinant viral vectors that express high levels of prostate-associated antigens as a method of using cancer vaccines to generate clinically beneficial anti-tumor immune responses.[89]

Several other laboratories are using the interesting approach of trying to induce autoimmune prostatitis as a way to generate an anti-prostate tumor response for patients following prostatectomy. One candidate autoantigen, prostatic steroid binding protein (PSBP) is a major target of autoimmunity following injection of human tissue extracts such as the male sex accessory glands (consisting of all lobes of the prostate, in addition to the seminal vesicles) into rats.[90] PSBP is capable of eliciting both humoral and cell-mediated immune responses that are, importantly, restricted to prostatic tissues.[89,90] Severe prostatitis was detected after a single vaccination with purified PSBP, and the adoptive transfer of splenic T-cells from vaccinated rats efficiently transferred the condition into naïve recipients.[92] It is hoped that the vibrant autoimmune response directed against surface epitopes on normal prostate cells might concurrently destroy malignant cells. The human homologue to rat PSBP has not been found, and it is unclear whether autoimmune reactivity would cross-react with cancer cells.[73] Still, this is just one of many attractive approaches towards generating a clinically beneficial anti-tumor immune response by targeting organ- rather than tumor-specific antigens.

Some special problems exist which complicate the development of peptide-based immunotherapy against prostatic carcinoma. One-third of all males over the age of 50 is diagnosed when their cancer is asymptomatic. Its insidious nature is further underscored by the uncertainty by which more aggressive disease stages are activated.[93] Prostate cancer is usually slow growing, and its poor immunogenicity is reflected by the lack of documented spontaneous remission.[94] Also, it is not established whether or not a durable anti-tumor response could be generated against bone lesions, its preferred sites of metastasis. But evidence exists that an immune response may be crucial in prostate cancer. Early studies, such as those demonstrating that adjuvant treatment with Bacillus calmette-guerin (BCG; a non-specific immunostimulatory agent) could enhance patient survival[95] suggested that an immunogenic approach was feasible. Tumor-reactive immune cells such as tumor infiltrating lymphocytes (TIL) have been noted in tumor or nodal biopsies, and their relevance to tumor progression was demonstrated in a large clinical study.[96] However, the development of antigen-specific therapies is hampered by the paucity of information concerning prostate cancer immunology, including the identification, abundance, and immunodominance of tumor-associated or -specific prostate cancer epitopes. Some progress has been made: the loss of HLA expression[97] and aberrant antigen processing[98] by prostate tumor cells has been observed. Yet, especially when compared to other carcinomas, much work needs to be done in the realm of understanding the relationship between the immune system and prostate cancer. With more insight, it will also be of great interest to study if peptide-based vaccines for prostate cancer will be effective as a stand-alone treatment for advanced disease or be limited to an adjuvant setting for the treatment of minimal tumor burden.[8]

References

1. Landis SH, Murray T, Bolden S et al. Cancer statistics, 1998 [published erratum appears in CA Cancer J Clin 1998 May-Jun; 48(3):192]. CA Cancer J Clin 1998; 48:6-29.
2. Wingo PA, Landis S, Ries LA. An adjustment to the 1997 estimate for new prostate cancer cases. CA Cancer J Clin 1997; 47:239-242.
3. Gittes RF. Carcinoma of the prostate [see comments]. N Engl J Med 1991; 324:236-245.
4. Oh WK, Kantoff PW. Management of hormone refractory prostate cancer: current standards and future prospects [In Process Citation]. J Urol 1998; 160:1220-1229.
5. Stearns ME, McGarvey T. Prostate cancer: Therapeutic, diagnostic, and basic studies. Lab Invest 1992; 67:540-552.
6. Koivisto P, Visakorpi T, Rantala I et al. Increased cell proliferation activity and decreased cell death are associated with the emergence of hormone-refractory recurrent prostate cancer. J Pathol 1997; 183:51-56.
7. Culig Z, Hobisch A, Hittmair A et al. Expression, structure, and function of androgen receptor in advanced prostatic carcinoma. Prostate 1998; 35:63-70.
8. Nelson WG, Simons JW. New approaches to adjuvant therapy for patients with adverse histopathologic findings following radical prostatectomy. Urol Clin North Am 1996; 23:685-696.
9. Small EJ, Vogelzong NJ. Second-line hormonal therapy for advanced prostate cancer: A shifting paradigm. J Clin Oncol 1997; 15:382-388
10. Peshwa MV, Shi JD, Ruegg C et al. Induction of prostate tumor-specific CD8+ cytotoxic lymphocytes in vitro using antigen-presenting cells pulsed with prostatic acid phosphatase peptide. Prostate 1998; 36:129-138.
11. Solin T, Kontturi M, Pohlmann R et al. Gene expression and prostate specificity of human prostatic acid phosphatase (PAP): evaluation by RNA blot analyses. Biochim Biophys Acta 1990; 1048:72-77.
12. Moul JW, Connelly RR, Perahia B et al. The contemporary value of pretreatment prostatic acid phosphatase to predict pathological stage and recurrence in radical prostatectomy cases. J Urol 1998; 159:935-940.
13. Moss DW, Henderson AR, Kachmar JF. Enzymes. In: Tietz NW, ed. Textbook of clinical chemistry. Ninth ed. Philadelphia: W.B. Saunders Company, 1986:619-762.
14. Vihko P, Virkkunen P, Henttu P et al. Molecular cloning and sequence analysis of cDNA encoding human prostatic acid phosphatase. FEBS Lett 1988; 236:275-281.
15. Zelivianski S, Comeau D, Lin MF. Cloning and analysis of the promoter activity of the human prostatic acid phosphatase gene. Biochem Biophys Res Commun 1998; 245:108-112.
16. Peshwa MV, Benike C, Dupuis M et al. Generation of primary peptide-specific CD8+ cytotoxic T-lymphocytes in vitro using allogeneic dendritic cells. Cell Transplant 1998; 7:1-9.
17. Rammensee HG, Falk K, Rotzschke O. Peptides naturally presented by MHC class I molecules. Annu Rev Immunol 1993; 11:213-244.
18. Gabrilovich DI, Corak J, Ciernik IF et al. Decreased antigen presentation by dendritic cells in patients with breast cancer. Clin Cancer Res 1997; 3:483-490.
19. Fong L, Ruegg CL, Brockstedt D et al. Induction of tissue-specific autoimmune prostatitis with prostatic acid phosphatase immunization—Implications for immunotherapy of prostate cancer. J Immunol 1997; 159:3113-3117.
20. Wang MC, Kuriyama M, Papsidero LD et al. Prostate antigen of human cancer patients. Methods Cancer Res 1982; 19:179-197.
21. Smith MR, Biggar S, Hussain M. Prostate-specific antigen messenger RNA is expressed in non-prostate cells: implications for detection of micrometastases. Canc Res 1995; 55:2640-2644.
22. Gao X, Porter AT, Grignon DJ et al. Diagnostic and prognostic markers for human prostate cancer. Prostate 1997; 31:264-281.
23. Yu H, Diamandis EP, Wong PY et al. Detection of prostate cancer relapse with prostate specific antigen monitoring at levels of 0.001 to 0.1 microg./l [see comments]. J Urol 1997; 157:913-918.
24. Amling CL, Blute ML, Lerner SE et al. Influence of prostate-specific antigen testing on the spectrum of patients with prostate cancer undergoing radical prostatectomy at a large referral practice [see comments]. Mayo Clin Proc 1998; 73:401-406.

25. Falk K, Rotzschke O, Stevanovic S et al. Allele-specific motifs revealed by sequencing of self-peptides eluted from MHC molecules. Nature 1991; 351:290-296.
26. Xue BH, Zhang Y, Sosman JA et al. Induction of human cytotoxic T lymphocytes specific for prostate-specific antigen. Prostate 1997; 30:73-78.
27. Rivoltini L, Kawakami Y, Sakaguchi K et al. Induction of tumor reactive CTL from peripheral blood and tumor infiltrating lymphocytes of melanoma patients by in vitro stimulation with the human melanoma antigen MART-1. J Immunol 1995; 154:2257-2265.
28. Salgaller ML, Afshar A, Marincola FM et al. Recognition of multiple epitopes in the human melanoma antigen gp100 by peripheral blood lymphocytes stimluated in vitro with synthetic peptides. Canc Res 1995; 55:4972-4979.
29. Salgaller ML, Marincola FM, Cormier JN et al. Immunization against epitopes in the human melanoma antigen gp100 following patient immunization with synthetic peptides. Canc Res 1996; 56:4749-4757.
30. Correale P, Walmsley K, Nieroda C et al. In vitro generation of human cytotoxic T lymphocytes specific for peptides derived from prostate-specific antigen [see comments]. J Natl Cancer Inst 1997; 89:293-300.
31. Alexander RB, Brady F, Leffell MS et al. Specific T cell recognition of peptides derived from prostate- specific antigen in patients with prostate cancer. Urology 1998; 51:150-157.
32. Alexander RB. Letter to the editor. The Prostate 1998; 32:73
33. Peace DJ. Response to letter to the editor. The Prostate 1997; 32:74
34. Lee TD. Distribution of HLA antigens in North American Caucasians, North American Blacks and Orientals. In: Lee J, ed. Distribution of HLA antigens. New York: Springer-Verlag, 1990:141-178.
35. Correale P, Walmsley K, Zaremba S et al. Generation of human cytolytic T lymphocyte lines directed against prostate-specific antigen (PSA) employing a PSA oligoepitope peptide. J Immunol 1998; 161:3186-3194.
36. Rivoltini L, Loftus DJ, Barracchini K et al. Binding and presentation of peptides derived from melanoma antigens MART-1 and Glycoprotein-100 by HLA-A2 subtypes. J Immunol 1996; 156:3882-3891.
37. Cormier JN, Salgaller ML, Prevette T et al. Enhancement of cellular immunity in melanoma patients immunized with a peptide from MART-1/Melan A. Cancer J Sci Amer 1997; 3:37-44.
38. Marincola F. Dendritic cells and melanoma. Proceedings of the Fifth international symposium on dendritic cells in fundamental and clinical immunology. 1998; S18.
39. Israeli RS, Powell CT, Fair WR et al. Molecular cloning of a complementary DNA encoding a prostate-specific membrane antigen. Canc Res 1993; 53:227-230.
40. Gregorakis AK, Holmes EH, Murphy GP. Prostate-specific membrane antigen: Current and future utility. Semin Urol Oncol 1998; 16:2-12.
41. Horoszewicz JS, Kawinski E, Murphy GP. Monoclonal Antibodies to a New Antigenic Marker in Epithelial Prostatic Cells and Serum of Prostatic Cancer Patients. Anticancer Researchers 1987; 7:927-936.
42. Fair WR, Israeli RS, Heston WD. Prostate-specific membrane antigen. Prostate 1997; 32:140-148.
43. Ruppert J, Sidney J, Celis E et al. Prominent role of secondary anchor residues in peptide binding to HLA-A2.1 molecules. Cell 1993; 74:929-937.
44. Murphy GP, Tino WT, Holmes EH et al. Measurement of prostate-specific membrane antigen in the serum with a new antibody. The Prostate 1996; 28:266-271.
45. Rochon YP, Horoszewicz JS, Boynton AL et al. Western blot assay for prostate-specific membrane antigen in serum of prostate cancer patients. The Prostate 1994; 25:219-223.
46. Murphy G, Ragde H, Kenny G et al. Comparison of prostate specific membrane antigen, and prostate specific antigen levels in prostatic cancer patients. Anticancer Researchers 1995; 15:1473-1479.
47. Bostwick DG, Iczkowski KA, Amin MB et al. Malignant lymphoma involving the prostate: report of 62 cases. Cancer 1998; 83:732-738.
48. Silver DA, Pellicer I, Fair WR et al. Prostate-specific membrane antigen expression in normal and malignant human tissues. Clin Cancer Res 1997; 3:81-85.

49. Carter RE, Feldman AR, Coyle JT. Prostate-specific membrane antigen is a hydrolase with substrate and pharmacologic characteristics of a neuropeptidase. Proc Natl Acad Sci USA 1996; 93:749-753.
50. Pinto JT, Suffoletto BP, Berzin TM et al. Prostate-specific membrane antigen: A novel folate hydrolase in human prostatic carcinoma cells. Clin Cancer Res 1996; 2:1445-1451.
51. Steinman RM. The dendritic cell system and its role in immunogenicity. Ann Rev Immunol 1991; 9:271-296.
52. Young JW, Inaba K. Dendritic cells as adjuvants for class I major histocompatibility complex-restricted antitumor immunity. J Exp Med 1996; 183:7-11.
53. Marland G, Bakker ABH, Adema GJ et al. Dendritic cells in immune response induction. Stem Cells 1996; 14:501-507.
54. Salgaller ML, Tjoa BA, Lodge PA et al. Dendritic cell-based immunotherapy of prostate cancer. Crit Rev Immunol 1998; 18:109-119.
55. Banchereau J, Steinman RM. Dendritic cells and the control of immunity. Nature 1998; 392:245-252.
56. Tjoa B, Erickson S, Barren R3 et al. In vitro propagated dendritic cells from prostate cancer patients as a component of prostate cancer immunotherapy. Prostate 1995; 27:6369.
57. Romani N, Reider D, Heuer M et al. Generation of mature dendritic cells from human blood. An improved method with special regard to clinical applicability. J Immunol Meth 1996; 196:137-151.
58. Tjoa B, Boynton A, Kenny G et al. Presentation of prostate tumor antigens by dendritic cells stimulates T-cell proliferation and cytotoxicity. Prostate 1996; 28:65-69.
59. Murphy G, Tjoa B, Ragde H et al. Phase I clinical trial: T-cell therapy for prostate cancer using autologous dendritic cells pulsed with HLA-A0201-specific peptides from prostate-specific membrane antigen. Prostate 1996; 29:371-380.
60. Salgaller ML, Lodge PA, McLean JG et al. Report of immune monitoring of prostate cancer patients undergoing T- cell therapy using dendritic cells pulsed with HLA-A2-specific peptides from prostate-specific membrane antigen (PSMA). Prostate 1998; 35:144-151.
61. Tjoa BA, Erickson SJ, Bowes VA et al. Follow-up evaluation of prostate cancer patients infused with autologous dendritic cells pulsed with PSMA peptides. The Prostate 1997; 32:272-278.
62. Salgaller ML, Lodge PA. Use of cellular and cytokine adjuvants in the immunotherapy of cancer. J Surg Oncol 1998; 68:122-138.
63. Murphy GP, Slack NH. Response criteria for the prostate of the USA National Prostatic Cancer Project. Prostate 1980; 1:375-382.
64. Scher HI, Mazumdar M, Kelly WK. Clinical trials in relapsed prostate cancer: Defining the target. J Natl Cancer Inst 1996; 88:1623-1634.
65. Knight SC. Dendritic cells as initiators of tumor immunity. Immunol Today 1995; 16:547
66. Macatonia SE, Patterson S, Knight SC. Primary proliferative and cytotoxic T-cell responses to HIV induced in vitro by human dendritic cells. Immunol 1991; 74:399-406.
67. Knight SC, Iqball S, Roberts MS et al. Transfer of antigen between dendritic cells in the stimulation of primary T cell proliferation. Eur J Immunol 1998; 28:1636-1644.
68. Toes RE, van d, V, Schoenberger SP et al. Enhancement of tumor outgrowth through CTL tolerization after peptide vaccination is avoided by peptide presentation on dendritic cells. J Immunol 1998; 160:4449-4456.
69. Murphy GP, Tjoa BA, Simmons SJ et al. Infusion of dendritic cells pulsed with HLA-A2-specific prostate-specific membrane antigen peptides: A phase II prostate cancer vaccine trial involving patients with hormone-refractory metastatic disease. Prostate 1999; 38:73-78.
70. Murphy GP, Tjoa BA, Simmons SJ et al. Phase II prostate cancer vaccine trial: Report of a study involving 37 patients with disease recurrence following primary treatment. Prostate 1999; 39:54-59.
71. Finn OJ, Jerome KR, Henderson RA et al. MUC-1 epithelial tumor mucin-based immunity and cancer vaccines. Immunol Rev 1995; 145:61-89.
72. Magarian-Blander J, Hughey RP, Kinlough C et al. Differential expression of MUC1on transfected cell lines influences its recognition by MUC1 specific T cells. Glycoconjugate J 1996; 13:749-756.
73. Naitoh J, Witte O, Belldegrun A. The University of California, Los Angeles/Jennifer Jones Simon Foundation symposium on prostate cancer and epithelial cell biology: Bringing together basic scientists and clinicians in the fight against advanced prostate cancer. Cancer Res 1998; 58:2895-2900.

74. Jerome KR, Barnd DL, Bendt KM et al. Cytotoxic T-lymphocytes derived from patients with breast adenocarcinoma recognize an epitope present on the protein core of a mucin molecule preferentially expressed by malignant cells. Cancer Res 1991; 51:2908-2916.
75. Ho SB, Niehans GA, Lyftogt C et al. Heterogeneity of mucin gene expression in normal and neoplastic tissues. Cancer Res 1993; 53:641-651.
76. Zhang S, Zhang HS, Reuter VE et al. Expression of potential target antigens for immunotherapy on primary and metastatic prostate cancers. Clin Cancer Res 1998; 4:295-302.
77. Goydos JS, Elder E, Whiteside TL et al. A phase I trial of a synthetic mucin peptide vaccine. Induction of specific immune reactivity in patients with adenocarcinoma. J Surg Res1996; 63:298-304.
78. Slovin SF, Kelly WK, Scher HI. Immunological approaches for the treatment of prostate cancer. Semin Urol Oncol 1998; 16:53-59.
79. Livingston PO. Augmenting the immunogenicity of carbohydrate tumor antigens. Semin Cancer Biol 1995; 6:357-366.
80. Kensil CR, Patel U, Lennick M et al. Separation and characterization of saponins with adjuvant activity from Quillaja saponaria Molina cortex. J Immunol 1991; 146:431-437.
81. Agrawal B, Krantz MJ, Reddish MA et al. Cancer-associated MUC1 mucin inhibits human T-cell proliferation, which is reversible by IL-2. Nature Med 1998; 4:43-49.
82. Wang W, Man S, Gulden PH et al. Class I-restricted alloreactive cytotoxic T lymphocytes recognize a complex array of specific MHC-associated peptides. J Immunol 1998; 160:1091-1097.
83. Litvinov SV, Velders MP, Bakker HA, et al. Ep-CAM: a human epithelial antigen is a homophilic cell-cell adhesion molecule. J Cell Biol 1994; 125:437-446.
84. Miles DW, Towlson KE, Graham R et al. A randomised phase II study of sialyl-Tn and DETOX B adjuvant with or without cyclophosphamide pretreatment for the active specific immunotherapy of breast cancer. Br J Cancer 1996; 74:1292-1296.
85. Caetano-Anolles G. Scanning of nucleic acids by in vitro amplification: New developments and applications. Nat Biotechnol 1996; 14:1668-1674.
86. Kononen J, Bubendorf L, Kallioniemi A et al. Tissue microarrays for high-throughput molecular profiling of tumor specimens [In Process Citation]. Nat Med 1998; 4:844-847.
87. Su ZZ, Lin J, Shen R et al. Surface-epitope masking and expression cloning identifies the human prostate carcinoma tumor antigen gene PCTA-1 a member of the galectin gene family. Proc Natl Acad Sci U S A 1996; 93:7252-7257.
88. Bao LR, Loda M, Janmey PA et al. Thymosin 15: A novel regulator of tumor cell motility upregulated in metastatic prostate cancer. Nature Med 1996; 2:1322-1328.
89. Hodge JW, Schlom J, Donohue SJ et al. A recombinant vaccinia virus expressing human prostate-specific antigen (PSA) safety and immunogenicity in a non-human primate. Int J Cancer 1995; 63:231-237.
90. Maccioni M, Rivero VE, Riera CM. Prostatein (or rat prostatic steroid binding protein) is a major autoantigen in experimental autoimmune prostatitis. Clin Exp Immunol 1998; 112:159-165.
91. Casas-Ingaramo A, Depiante-Depaoli M, Pacheco-Rupil B. Activation of cytotoxic cells by syngeneic prostate antigens in experimental autoimmune vesiculo-prostatitis. Autoimmunity 1991; 9:151-157.
92. Liu KJ, Chatta GS, Twardzik DR et al. Identification of rat prostatic steroid-binding protein as a target antigen of experimental autoimmune prostatitis - Implications for prostate cancer therapy. J Immunol 1997; 159:472-480.
93. Chiarodo A. National cancer institute roundtable on prostate cancer: Future research directions. Cancer Res 1991; 51:2498-2505.
94. Hrouda D, Muir GH, Dalgleish AG. The role of immunotherapy for urological tumors. Br J Urol 1997; 79:307-316.
95. Guinan P, Toronchi E, Shaw M et al. Bacillus calmette-guerin (BCG) adjuvant therapy in stage D prostate cancer. Urology 1982; 20:401-403.
96. Vesalainen S, Lipponen P, Talja M et al. Histological grade, perineural infiltration, tumor-infiltrating lymphocytes and apoptosis as determinants of long-term prognosis in prostatic adenocarcinoma. Eur J Cancer 1994; 30A:1797-1803.
97. Blades RA, Keating PJ, McWilliam LJ et al. Loss of HLA class I expression in prostate cancer: Implications for immunotherapy. Urology 1995; 46:681-6; discussion 686-7.
98. Sanda MG, Restifo NP, Walsh JC et al. Molecular characterization of defective antigen processing in human prostate cancer. J Natl Cancer Inst 1995; 87:280-285.

Chapter 11

Peptides in Cervical Cancer

Maaike E. Ressing, Remco M.P. Brandt, Joan H. de Jong, Rienk Offringa, Cornelis J.M. Melief and W. Martin Kast

Introduction

Observations that susceptibility to several cancer types is increased in immunocompromised individuals have led to the assumption that immune responses are able to interfere with tumor development.[1] Early attempts focussed on the general activation of the patient's immune system as a means of cancer therapy, for instance by administration of bacterial extracts, Bacillus Calmette-Guérin, or high dose interleukin-2.[2-4]

In the last decades, our knowledge on the specificity and regulation of the immune system has expanded tremendously. The major effectors involved in the eradication of tumor cells are likely to be T lymphocytes, which can recognize subtle intracellular changes showing up as antigenic peptides presented by major histocompatibility complex (MHC) class I and class II molecules. MHC class II molecules are mainly found on cells of the immune system with a specialized antigen presenting function, including dendritic cells (DC). Upon triggering by class II/peptide complexes, $CD4^+$ T helper (Th) cells can orchestrate the action of numerous immune cells. MHC class I molecules are expressed on the surface of virtually all nucleated cells and, through these, $CD8^+$ cytotoxic T lymphocytes (CTL) can screen almost all cells of the body for antigenic peptides that may be presented as a consequence of viral infection or malignant transformation. Thus, CTL represent a major effector subset of specific T cells potentially able to eradicate virus-infected or malignantly transformed cells.

Besides this, understanding of molecular processes leading to tumor formation has contributed to identifying specific tumor antigens. Particularly attractive immunotherapeutic targets are viral proteins essential for transformation because evasion of immune recognition through loss of (onco)protein expression would also entail loss of the transformed state.

The combined insights provide new opportunities to develop intervention schemes to trigger or amplify tumor-specific immune responses.

Cervical Cancer

Carcinoma of the cervix uteri is the second most common cause of cancer-related death in women worldwide.[5] In early stages of cervical disease, primary treatment consisting of surgery and/or radiotherapy is curative in most cases. However, the five-year survival of treated patients progressively deteriorates from 90-95% for FIGO stage I to less than 20% for stage IV,[6] and current treatment results in recurrent disease are generally very poor.[7] Furthermore, side effects still constitute a severe complication of conventional cancer treatment including radiotherapy and chemotherapy. For these reasons, new or additional treatment modalities are actively pursued to combine effective anti-tumor activity with a reduction of the side effects.

Peptide-Based Cancer Vaccines, edited by W. Martin Kast. ©2000 Eurekah.com.

Human Papillomaviruses

Epidemiological studies have identified infection with high-risk or oncogenic types of human papillomaviruses (HPV) as a major risk factor for the development of cervical carcinoma.[8] HPV constitute a heterogeneous group of small circular DNA viruses infecting epithelia.[9, 10] In vitro analyses have demonstrated that oncogenic HPV E6 and E7 expression is required for immortalization of primary cells as well as for maintenance of the transformed state,[11, 12] through a number of molecular interactions, among which interference by E6 and E7 with the cellular tumor suppressor proteins, p53 and retinoblastoma protein (pRB), respectively. DNA of the oncogenic HPV types is detected in more than 95% of the cervical carcinomas worldwide, with HPV16 accounting for at least 50%.[13] (Genital) HPV infection occurs most frequently at young age and disappears seemingly spontaneously in most cases.[14] HPV-infected cells can undergo hyperproliferation, leading to cervical intraepithelial neoplasia (CIN), but only a limited fraction of lesions in time acquires additional cellular changes and progresses to malignant cervical disease. These findings suggest that natural immunity against HPV is capable of clearing HPV or HPV-infected cells. The increased incidence of HPV infections and related proliferative lesions of the cervix observed in immunocompromised patients[15] further supports the view that cellular immune recognition of HPV is important in the elimination of HPV-infected cells.

Natural Immunity

Evidence for natural immunity in patients with HPV DNA-positive lesions is now accumulating with the detection of virus-directed immune responses. In patients with cervical lesions (CIN and early stage cervical carcinoma), natural CTL responses against HPV16 and HPV18 have been demonstrated that were absent in healthy control donors.[16-18] The frequency of CTL responding to a recombinant vaccinia virus expressing the HPV16 and HPV18 E6/E7 gene products was found to be higher in tumors and lymph nodes compared with that of peripheral blood.[17] These results establish the existence of natural HPV E7-specific CTL immunity in humans. In addition, HPV16 E6 and E7-directed Th cells have been identified in patients with abnormal cervical cytology and/or with serological responses against HPV16 proteins.[19-22] These HPV16 Th responses were not detectable in individuals without indications of viral infection, and therefore represent virus-specific memory Th responses primed in vivo. In a nonintervention follow-up study, HPV16 E7-specific Th responses were more frequent in patients with progressing CIN and persistent HPV infection, implying that Th immunity largely develops as a consequence of increased viral antigen availability and may be induced too late to be effective.[19, 22]

Taken together, HPV-specific immunity can be raised in response to the appearance of viral proteins, which can result in protection against (the development of) tumors. However, in patients with cervical lesions, the HPV-directed immunity has obviously failed in clearing the virus(-infected cells) resulting in persistence of infection with an increased risk of additional oncogenic changes, eventually leading to the development of cancer.

Induction of (Tumor) Antigen-Specific CTL

To exploit the full potential of the specific immune system to reject (virus-induced) tumors (and to understand why natural responses fail to do so), knowledge of T cell induction is crucial.

Initiation of antigen-specific T lymphocyte responses in vivo usually requires priming of naive T cells by two signals: antigen-specific triggering of the TCR by peptide-MHC complexes and simultaneous costimulation by, for instance, B7-CD28 interactions.[23-25] A general mechanism implicated in induction of immunity appears to be represented by a form of exogenous antigen presentation referred to as "cross-priming." [26-29] The favorite APC responsible for cross-priming are DC, as these cells are specialized to prime helper and killer T cells in vivo.[30, 31] Cross-priming can be envisaged as follows. Langerhans cells, the immature DC in the skin, can capture antigen locally and upon migration to the draining lymph nodes will present these antigens to naive T cells. This should only happen when potentially noxious agents are present. Such "danger" to the host may occur in the form of tissue destruction, for instance during infection.[32] A resulting local inflammatory response initiates the maturation of DC and their migration to the lymph nodes where they can induce adaptive immunity mediated by antigen-specific T cells. Moreover, generation of effective CTL responses often requires "help" from $CD4^+$ Th lymphocytes. Recently, a major pathway of T help for CTL priming was shown to be mediated through CD40-CD40 ligand (CD154) interactions [33-35]: (antigen-specific) recognition of a Th epitope on the professional APC allows interaction of CD40 ligand on the activated Th cell with CD40 on the DC thereby activating the DC to express costimulatory molecules and/or cytokines. A different way of activating DC consists of direct infection by viruses.[33] In turn, the activated DC is enabled to prime naive CTL. From this it follows that immune intervention schemes aiming at the induction of tumor-specific CTL immunity should ensure full activation of DC.

Productive HPV infection exploits the normal differentiation of keratinocytes, resulting in release of virus particles from the shedded upper layers of the epithelium with little or no local tissue damage.[36] Furthermore, infected keratinocytes do not have all the features of professional APC, whereas the number of Langerhans cells is reduced in HPV-induced cervical lesions.[37, 38] Both the absence of a lytic infection and the lack of virally infected professional APC may explain why HPV infection is initially ignored by the immune system. In addition, HPV may have evolved strategies to escape from immune recognition. Upon large overexpression of viral proteins after integration of HPV DNA or upon tissue damage (e.g., by biopsy; thus providing a "danger signal"), the Langerhans cells in the squamous epithelium are alerted to capture viral antigens for transport to the draining lymph nodes and subsequent induction of HPV-directed T cells. However, due to this late arousal of immune effector cells (if at all) and potential immunosuppression accompanying tumor outgrowth, the HPV-infected or already - transformed cells are likely to outpace the immune system.

In contrast to the natural immunity which may come too late, active immune intervention may provide adequate and effective (T cell-mediated) immunity in time to prevent or treat early HPV-associated neoplasia.

Immunotherapeutic Approaches to Cervical Carcinoma

Because of their close and causative relation with cervical disease, HPV-encoded proteins are the major targets to be addressed by immune intervention against cervical carcinoma and its precursor lesions. Furthermore, viral antigens allow specific discrimination between tumor cells and their (non-infected) healthy counterparts. For the early stages of (productive) viral infection, neutralizing antibodies against the viral capsid proteins induced with virus-like particles (VLP)) may be of use, as well as T cells specific for early proteins abundantly expressed in replicating keratinocytes (e.g., E1, E2, E5). For the treatment of established cervical carcinoma, it is necessary to direct the immune system towards the viral E6 and E7 proteins in view of their high levels of expression and, particularly, their essential role in sustaining cellular transformation. In all cases, it will be important to maintain activated immune responses against the viral proteins until full clearance of either infection or cancer cells is achieved, to prevent escape of virus-infected cells upon inadequate eradication.

Peptide-Based Vaccination Strategies against HPV16⁺ Tumors in Animal Models

Immune responses to species-specific papillomaviruses have been studied in a number of animals as described elsewhere (cattle,[39,40] rabbits,[41,42] dogs[43]). HPV16 also transforms rodent cells, in addition to human keratinocytes, which has permitted the generation of HPV-induced tumor models in mice[44-46] and in rats.[47] In murine models, targeting of the immune system towards these antigens is feasible by vaccination, among others with peptides.

Subcutaneous vaccination with a synthetic peptide that represents an HPV16-encoded CTL epitope ($E7_{49-57}$) emulsified in incomplete Freunds adjuvant (IFA) protected mice from a subsequent challenge with a lethal dose of HPV16-transformed syngeneic tumor cells[46] CTL induced by this immunization protocol were capable of lysing peptide-pulsed target cells as well as tumor cells. Indeed, the naturally processed peptide has been eluted from the presenting $H\text{-}2D^b$ molecules.[48] These studies show that selective stimulation of a T cell specificity of choice by vaccination with a peptide derived from an oncogenic protein is effective in inducing protective T cell immunity without the introduction of the entire oncogene.

Successful prophylactic induction of anti-viral and anti-tumor immunity by vaccination with peptides has been achieved in several murine models,[49] but examples of eradication of pre-existing tumors are scarce. Particularly, therapeutic immune intervention appears to require optimal presentation of the tumor antigen to the immune system and in this respect the administration of professional antigen presenting cells as carriers appears effective.[50] For instance, intravenous injection of bone marrow-derived dendritic cells (BMDC) prepulsed with the HPV16 $E7_{49-57}$ epitope into mice bearing established $HPV16^+$ tumors resulted in tumor regression and long-term immunity.[51] Treatment was also achieved through vaccination with BMDC loaded with unfractionated acid- eluted peptides from an HPV16-containing tumor.[52] The latter finding indicates that, when autologous tumor cells are the only known source of tumor antigens, this strategy of tumor treatment with naturally presented peptides can be used. Besides with peptide vaccines emulsified in numerous adjuvants, protective immunity against several HPV16-induced tumors has been achieved by vaccination with a recombinant adenovirus encoding a minimal CTL epitope, HPV16 E7 protein in adjuvant or fed to DC, a vaccinia virus containing the E7 gene with or without the lysosomal targeting protein LAMP1, and (E7-chimeric) VLP.[53-56]

In addition to the effects of anti-tumor immunity against a challenge with HPV16-transfected cells, interference with cells endogenously expressing the viral oncoproteins has been studied using HPV16 transgenic mice. Various strains of HPV16 transgenic mice resemble human HPV-induced carcinogenesis to some extent, because the prolonged expression of the HPV16 (E6 and/or) E7 protein in epithelial cells causes spontaneous lens, skin, or mucosal hyperproliferation and tumor development.[57-63] Immunological outcome appears to depend on the level, site, and time of transgene expression. In line 19 mice, expressing HPV16 E6 and E7 ORF in lens and skin driven from the aA- crystallin promoter, parenteral E7 immunization induces B cell immunity as well as proliferative and cytotoxic T cell responses, comparable to nontransgenic littermates.[64] These (CTL) responses result in tumor rejection without affecting the skin.[65] The combined results indicate that the lymphocytes of line 19 animals are not tolerant of E7. In contrast, K14 mice, with thymic and skin expression of the HPV16 E7 protein under control of a keratinocyte-specific promoter, display a reduced E7-specific cytotoxic T cell function together with enhancement of E7-specific Th repertoire, whereas B cell responses are unaffected when compared to nontransgenic littermates.[66, 67] In these animals, split tolerance appears the result to transgene expression. Thus, both line 19 and K14 HPV16 transgenic animals provide useful tools to study conditions of induction of T cell (un)responsiveness to (E6 and) E7, but may not directly provide an in vivo model to evaluate immunotherapeutic strategies against cervical cancer.

In summary, the animal studies described show that induction of anti-tumor immunity is feasible with peptide-based vaccines, albeit more difficult in the tumor-bearing state.

Towards Human Application

In the past year, the first evidence for successful prevention of cancer as the result of immune intervention has been obtained: the incidence of hepatocellular carcinoma in children has declined since Taiwan implemented a universal HBV vaccination programme.[68] The effectiveness of peptides as vaccines to induce specific immunity in human beings was first demonstrated for HBV in healthy volunteers. Injection of a covalently linked construct containing an HLA-A2-restricted HBV core CTL epitope appeared to be both safe and capable of inducing virus-specific CTL responses of a strength comparable to CTL immunity detected upon natural HBV infection.[69,70] Although the in vivo relevance of peptide-induced CTL awaits further analysis, these studies raise hope for peptide-based vaccines against (cervical) cancer.

Preclinical Studies with HPV16-Encoded Targets for Human T Cells

Human Response Inductions

The identification of human HPV16-encoded CTL epitopes has not been productive until recently, an important reason being the difficulty in obtaining (large quantities of) infectious virus, HPV16-infected or -transformed cells, and HPV16-derived proteins.

We have employed peptide-MHC binding and immunogenicity studies[71, 72] to identify human CTL epitopes within the HPV16 E6 and E7 proteins. HPV16 E7-directed T cell cultures were obtained in vitro from PBMC of healthy donors by stimulation with selected peptides capable of high affinity binding to HLA-A2. Thus, a number of HPV16-encoded human CTL epitopes were defined ($E7_{11\text{-}20}$, $E7_{82\text{-}90}$, $E7_{86\text{-}93}$).[73] Human $CD8^+$ CTL clones against the three E7 peptides lysed an HLA-A*0201$^+$, HPV16$^+$ cervical carcinoma cell line [73] and $E7_{11\text{-}20}$-specific CTL were also capable of lysing an E7- transfected keratinocyte cell line.[74] Among the lymphocytes from patients with proven HPV16 infections, HLA-A2-restricted recall CTL against two HPV16 E7-encoded peptides ($E7_{11\text{-}20}$ and $E7_{86\text{-}93}$) were occasionally detected, which were absent in healthy donors.[16,17] The combined results suggest these peptides to represent naturally processed human CTL epitopes of HPV16 E7. In addition, an HPV16-encoded MHC class I-presented peptide sequence $_{(18\text{-}26)}$has been determined through an alternative approach, i.e., by analysis of peptides eluted from HLA-A*0201$^+$ EBV- transformed B cells infected with a vaccinia-E6 construct [(75)]. Because no T cells specific for this E6 peptide were generated so far, it is not clear whether this HLA class I ligand represents a CTL epitope.

The establishment of human HPV16-specific T cells in vitro in order to identify T cell epitopes should be improved. First, the use of stimulator cells expressing the HPV antigens endogenously are preferred. Accordingly, virus-specific CTL were generated by coculture with APC that expressed the HPV16 and/or HPV18 E6/E7 gene products upon infection with either recombinant vaccinia virus [17] or adenovirus constructs;[18] epitope specificity of these HPV-directed CTL awaits further analysis. Autologous cervical cancer cells could be employed for generating tumor-specific T cells in vitro, similar to the successful stategies leading to identification of melanoma antigens. Second, the efficiency of T cell culture can be improved by employing responder cells enriched for expression of the relevant TCR. In addition to using (cancer) patient PBMC or TIL, CTL could be selected on the basis of TCR interaction with specific antigenic complexes in the form of tetramers.[76] Recently, a combination of the tetramer-based technology to detect TCR-mediated antigen- specific binding and the single-cell ELISPOT assay staining antigen-stimulated cytokine release allowed isolation of specific and functional CTL. Moreover, during acute virus infections, the anti-viral CTL were shown to undergo tremendous proliferation and to represent a very significant proportion of T cells.[77] This was much larger than originally assumed on the basis of methods that depend on (short term) in vitro expansion induced by APC, such as in limiting dilution assays. Thus, MHC tetrameric complexes are highly sensitive for specific TCR and their potential application for enrichment of CTL of choice deserves to be explored.

Studies in HLA-Transgenic Mice

Vaccination of, for instance, HLA-A2 transgenic mice yields results that appear translatable to the human situation.[78-84;85-88] Indeed, three HLA-A2-binding peptides encoded by HPV16 E7 that were immunogenic to human PBMC also elicited CTL responses in vivo in HLA-A2K^b transgenic mice.[73] Furthermore, to directly assess the effects of various HPV16-based vaccination strategies on the outgrowth of HLA-A2$^+$ HPV16-induced tumors, in vivo tumor models in HLA-A2 transgenic animals are being developed. First, K14 HPV16 E7 transgenic mice have been crossed with HLA-A2 transgenic mice. Immunization of the HLA-A2 x HPV16 transgenic mice with two of the HLA-A*0201-binding E7-derived peptides ($E7_{82-90}$ and $E7_{86-93}$) - comparable to an H-2D^b-binding peptide ($E7_{49-57}$) - did not result in detectable T cell responses, whereas specific CTL were induced in non- HPV16 transgenic littermates.[66] The data imply that these peptides are naturally presented in the HPV16 transgenic mice and tolerize the specific CTL, as has been described for H-2-restricted CTL responses. Secondly, challenge experiments with HLA-A2$^+$, HPV16$^+$ tumor cells can provide information on both protective and therapeutic vaccination regimens in HLA-A2 transgenic animals. As a third approach, SCID mice - deficient in both B cell and T cell immunity - may prove useful for adoptive transfer experiments to determine the in vivo effectiveness of human T cells against human tumor cells.

In conclusion, preclinical studies have resulted in the characterization of a number of HPV16 E7-encoded human CTL epitopes in the context of HLA-A*0201.[16, 73,17, 66, 74, 89, 90] The next step is to use these peptides in appropriately formulated vaccines for clinical use aiming at the induction or augmentation of protective or therapeutic T cell-mediated immunity against HPV16-induced cervical lesions.

Trials on Vaccination against HPV in Patients with Cervical Carcinoma

Four phase I/II clinical studies have been performed that aim at inducing or enhancing T cell-mediated immunity against HPV-associated cervical carcinoma.[91, 92]

In Australia, five patients with stage III or IV cervical carcinoma were immunized at monthly intervals with an HPV16 E7-GST fusion protein emulsified in Algammulin (an adjuvant based on alum and inulin). Protein doses of 5 to 100 µg were safe as well as immunogenic, inducing both antibody and proliferative T cell responses against E7 in the majority of evaluable patients (I.H. Frazer et al, abstract 15th International Papillomavirus Workshop). The second trial with cervical carcinoma patients (in the UK) employed a recombinant vaccinia virus containing modified HPV16 and 18 E6 and E7 genes (designated TA-HPV) as a means to induce HPV-directed immunity.[93] Out of 58 screened patients, only eight were included that showed sufficient immunocompetence (as measured by CD4 counts) to clear vaccinia virus infection. A single vaccination with TA-HPV resulted in a tumor-free status of two patients up to 21 months after vaccination and HPV18 E6/E7-specific CTL were detected in one of these patients.[93] However, it is difficult to envisage that these T cells were responsible for tumor regression since the tumor biopsy sample of this patient did not contain HPV18 (but HPV16) DNA and the E6 and E7 proteins of HPV16 and 18 display little homology. Two other phase I/II clinical studies were based on vaccination with synthetic peptides representing HPV16 E7-encoded CTL epitopes. In the Netherlands, 15 patients with advanced, HPV16⁺ cervical carcinoma received repeated injections of two HLA-A*0201-restricted, HPV16 E7- encoded CTL epitopes administered in combination with a universal, non-HPV-related, T helper peptide - PADRE - emulsified in Montanide ISA 51 adjuvant. Vaccination caused no side effects attributable to the adjuvant or the synthetic peptides at any dose (100, 300, 1000 μg). Two patients experienced stable disease for over one year after vaccination, but all other patients maintained progressive disease. Local infiltrations of T cells at the site of vaccination were observed.[65] No HPV16-specific cytolytic activity was demonstrable in peripheral blood samples of any patient following short term in vitro peptide restimulation. A large proportion of screened patients displayed relative lymphopenia. When we analyzed the immune status of the fifteen vaccinated patients by screening a combination of parameters, we observed a reduction in influenza virus-specific CTL, proliferative responses against conventional antigens, and signalling z chain expression in the patient group, supportive of a more general state of immunosuppression in patients with residual or recurrent cervical carcinoma failing conventional treatment. Despite this, lymphoproliferation against the PADRE helper peptide was detected in PBMC of four patients taken after, but not before, vaccination. This demonstrated that peptide vaccination can induce specific T cell immunity, even in immunocompromised patients.[94]

The second peptide vaccination study (in the USA) tested the effectiveness of multiple vaccinations with a lipidated HPV16 E7 peptide epitope linked to PADRE.[95] Twelve patients expressing the HLA-A2 restiction element (two of these expressed subtypes other than HLA-A*0201) were included with recurrent or refractory cervical or vaginal cancer that in most cases contained HPV16 DNA (although this was not an inclusion criterium). No obvious clinical responses or treatment toxicities were observed. The HLA-A*0201⁺ patients were able to mount a cellular immune response to the control influenza virus matrix epitope as measured by peptide-specific IFN-γ release. Since the strength of these patient responses was not related to those of healthy control donors employing a more sensitive cytokine release assay (compared to cytotoxocity) in combination with in vitro restimulation with peptide-pulsed autologous DC (rather than autologous PBMC), these observations are not necessarily in contradiction with our results. Weak responses to HPV16 E7-specific peptide stimulation were measured, some of which were supposed to be induced by vaccination. However, the responses were not always consistent and no statistically significant differences existed between pre- and post-vaccination groups. Finally, the CTL epitope within the lipopeptide used for vaccination contained a modified amino acid, which rendered cross reactive lysis of tumor cells less likely.

The combined results of clinical studies in advanced cervical carcinoma patients so far indicate that this patient group is generally immunosuppressed. Only few vaccine-specific immune responses were observed after immunization, perhaps due to the difficulties in inducing and/or detecting HPV-directed T cell immunity in these immunocompromised patients. Because none of the immunotherapeutic schemes caused toxicity attributable to the vaccines, clinical trials are now feasible in earlier stages of the disease, where they are more likely to be effective.

Prospects for Immune Intervention against HPV16-Associated Cervical Disease

Together with results obtained for other malignancies (described elsewhere), the first immunotherapy trials against cervical carcinoma are both encouraging and instructive for further anti-tumor vaccine development. Based on these, the following considerations may aid to design of future (peptide-based) vaccination against HPV-induced lesions.

Selection of Patients to be Vaccinated

For molecularly defined vaccines, the patient group to be treated can be selected on the basis of antigen expression by their tumor cells using PCR technology. In addition, if a certain restriction element is obliged, HLA type can be assessed serologically.

The extreme polymorphism of MHC poses a significant obstacle to the development of broadly efficacious peptide-based immunotherapeutics. HLA class I (and class II) molecules show large inter- and intra-ethnic differences in their distribution,[96,97] also among populations where HPV infection and cervical carcinoma are more prevalent.[5,13] To achieve wide applicability, a subunit vaccine should comprise a "cocktail" of multiple epitopes restricted by a number of HLA alleles. Together, five common HLA-A alleles (A1, A2, A3, A11, and A24) cover a majority of the human population and HPV16 E6- and E7-encoded peptide ligands were selected on the basis of their capacity to bind to these five MHC class I molecules.[98,99] In addition, T cell epitopes capable of degeneration binding to different MHC alleles could be broadly employable. A large proportion of HLA class I alleles has been grouped into four supertypes with overlapping peptide specificities (the A2, A3, B7, and B44 supertypes; [100-103]). These supertypes and resulting "supermotifs" for binding would provide a framework for identification of degenerate CTL epitopes.[104,105] We[106] and others[107,108] found large differences in peptide binding specificity between a number of HLA-A2 variants that were reported to belong to the A2 supertype. These data imply that promiscuous binding cannot be generalized among HLA class I alleles belonging to the A2 supertype, but should be tested separately for each peptide/HLA-A2 allele combination. In a vaccine for HPV16-associated cancer, the binding specificities of two HPV16 E7 peptides indicate that they may be useful in the context of HLA-A*0201, A*0202, A*0203, A*0204, and A*0209 (but not A*0205, A*0206, A*0207, or A*0208).[106] For these reasons, genomic subtyping of serologically HLA-A2$^+$ subjects is required.

Based on an occurrence of 60% HPV16 DNA in cervical carcinoma cells[13] and of 40% HLA- A*0201 in the West-European and Northern American populations,[96,97] it was estimated that 25% of cervical carcinoma patients referred could be included in the peptide vaccination trials. Indeed, these numbers were largely reached.[94,95] However, a considerable number of patients meeting these criteria was not vaccinated due to factors such as poor clinical condition or disease progression, showing that the actual numbers of patients included will be more limited than those selected solely on the basis of HLA type and tumor antigen expression.

Immune Evasion of HPV16-Associated Cervical Lesions

A topic of further investigation concerns tumor escape from (natural) immunity, since this may also hamper the effectivity of T cells induced by immunotherapy. Detailed knowledge of the mechanisms responsible for virus and tumor evasion from host defense can guide the design of effective intervention strategies. Loss or downregulation of HLA class I molecules has been described for a wide variety of tumor types[109] including cervical carcinoma.[110-113] Defects in HLA class I expression can be genetic or regulatory in nature and may vary with respect to the number of alleles affected (from one to all). T cell-mediated anti-tumor immunity is prohibited by complete HLA loss, but can be compensated for by NK-mediated lysis to which class I-negative cells are more susceptible.[114,115] For cervical carcinoma, (allele-specific) downregulation or loss of HLA class I expression occurs more frequently in metastases compared to primary tumors and is associated with a worse prognosis.[116-119] Mechanisms responsible for loss of MHC class I expression have not been studied to a sufficient extent, but include at least genetic defects[120] as well as down-regulation of the transporter associated with antigen presentation (TAP).[111] (Persistent) viruses have evolved sophisticated strategies to interfere with antigen presentation in order to hide from immune recognition.[121,122] In the case of cervical carcinoma, the prolonged presence of HPV - starting before tumor development—may have interfered with host immunity. So far, class I downregulation has not been found to correlate with papillomavirus persistence,[123] whereas a possible role for viral interference with class II presentation has been suggested for the E5 and E6 proteins.[124, 125] Furthermore, viral mutations can change a CTL epitope into an antagonist that prevents T cell activation and function, thereby avoiding immune detection[126, 127] and this was suggested for a mutated HPV16 E6-derived HLA-B7 ligand.[128] However, since HPV in general mutate extremely slowly, mutations in the E6 protein are rather indicative of intratype variants which are detected more frequently in cervical carcinomas than in their precursor lesions.[129] This suggests that specific variants have an advantage over the prototype due to subtle differences in transforming potential or to immune evasive capacity (if the prototype sequence is recognized by CTL). Providing multiple T cell epitopes (for both antigen-specific CTL and Th cells) within a vaccine ensures effectivity in the case of epitope loss due to antigenic mutation or single HLA allele downregulation. Loss of class I expression due to regulatory defects may be overcome by appropriate stimuli, for example (co)administration of IFN–γ. To minimize the risk for immune escape, anti-cancer immunotherapy should preferably be started during early stages of (cervical) disease, because progression is accompanied by genetic instablility and concomitant tumor heterogeneity.

Furthermore, the immune status of patients is likely to be better in less advanced cancer stages. Tumor development may have been feasible due to a state of reduced immunocompetence in the host. Patients with cervical carcinoma display a reduction of delayed type hypersensitivity reactions towards recall antigens.[130] Indeed, during patient accrual for vaccination trials, a large proportion of screened patients with cervical carcinoma had a poor immune status. The persistence of HPV, the cancer-bearing state, as well as previous treatment regimens including radio- and/ or chemotherapy are likely to have contributed to this immunosuppression. Immune function appears to deteriorate with progressive cervical disease. Healthy donors and patients with CIN showed higher ζ chain expression as well as higher levels of TNF-α and IFN-γ production in response to polyclonal stimulation than patients with (advanced) cervical carcinoma. In patients with early stage cervical lesions, cytolytic responses and lymphoproliferation to control antigens were only marginally decreased when compared to healthy control donors [16,131] (M.E.R. et al unpublished results). Others have reported that the severity of HPV16-induced lesions could be correlated with a shift in cytokine release from Th1-like to Th2-like.[132] and that reduced CD4/CD8 ratios were found in cervical lesions which reverted to normal upon treatment.[133] Taken together, an inadequate immune status in patients permits enhanced tumor growth and renders successful immune intervention more difficult. Therefore, active immunotherapy should preferably be performed during earlier stages of disease.

In any case, cancers that do not naturally result in immune response induction can still be susceptible to T cell-mediated eradication when activated CTL are generated by vaccination.[134-136]

Formulation of Anticancer Vaccines

The mode of antigen delivery through vaccines to induce T cell immunity deserves sufficient attention, especially in the tumor-bearing state. Recent findings imply that activation of CTL requires antigen to be presented on fully activated professional APC.[35] In vivo, APC can be converted into such an activated state—in a potentially noxious milieu—through signals such as T cell help (predominantly via CD40-CD40 ligand interactions) [34,35] or viral infection.[33]

As a source of T cell help, the universal PADRE helper peptide was contained in both HPV16 peptide-based vaccines employed in phase I/II trials and PADRE-specific proliferation was indeed induced in some immunized patients with cervical cancer in our study.[94] The peptide vaccines, however, may be improved by providing HPV16-directed T help. For peptide-induced anti-tumor immunity, antigen-specific T help was shown superior over non-specific T help, which may be explained by a dual effectiveness during the CTL priming as well as effector phases.[137] $CD4^+$ T cell- mediated proliferation in response to HPV-derived recombinant proteins or synthetic peptides has been observed,[19-22] but no undisputed class II-restricted Th epitope was identified that could be used as a vaccine component. Recently, we have generated an HPV16 E7 peptide-specific, HLA-DR-restricted CD4+ T cell line that also responds to naturally processed recombinant E7 protein (M.E.R. et al unpublished results). The (minimal) epitope recognized by such T helper cells could be included in future vaccines to induce HPV16-specific help. Th cells can also be activated upon vaccination with protein (fragments), which obviates the need to match both MHC class I and class II molecules for presentation of viral antigen-specific CTL and Th epitopes. Indeed, animal experiments demonstrated that effective anti-tumor immunity was induced by vaccination with protein either in IFA or pulsed onto DC.[54]

Alternatively, direct activation of DC (in vivo) could be pursued with activating anti-CD40 antibodies[34,35] or by viral infection.[33] It will be very interesting to know whether infection with VLP formed by HPV16 L1 (and L2) proteins is able to activate DC. Chimeric VLP immunization generated HPV16-specific CTL in a CD4-independent way,[138] thus suggesting that VLP can bypass T cell help through CD40 and can directly activate the (infected) DC. If recombinant viruses such as vaccinia, canary pox, or fowl pox are capable of physiologically relevant infection with concomitant activation of professional APC, they are promising vehicles to selectively deliver tumor antigens to the immune system. By employing different virus constructs to prime or boost immune responses, possible negative effects of neutralizing antibodies can be reduced to a minimum. Lastly, autologous DC, that are activated and antigen-pulsed ex vivo, can be reinfused as a means to induce effective T cell responses.[139]

Evaluation of Effects Induced by Vaccination

Assessment of effectiveness of the immunotherapy in patients constitutes a major issue. Several methods are used to evaluate in vivo induction of immune responses in vaccinated subjects. In vivo parameters consist of monitoring tumor regression and delayed type hypersensitivity responses to immunizing agents. In vitro, T cell reactivity can be measured by antigen-specific proliferation, cytotoxicity, or secretion of cytokines. These approaches all differ in specificity and sensitivity. Most studies so far have used PBMC, while it is conceivable that a majority of effective tumor-specific CTL reside in the tumor lesions or draining lymph nodes.[17] The use of short term peptide PBMC cultures was feasible for the detection of virus-specific memory CTL responses[16,17] However, this method may not be sufficiently sensitive to detect low numbers of CTL present in the blood of advanced cervical carcinoma patients whether pre-existing or induced by vaccination[94] A method employing TCR interactions with antigen-MHC class I tetramers [76] may allow detection of low numbers of (viral) peptide-specific CTL. Finally, determination of vaccination effects in populations at risk or with early cervical lesions will be particularly important, yet even more difficult to accomplish due to the relatively small proportion of HPV16-infected people having lesions that progress to malignancy and the long lag period before transformation and tumor progression occur. Nonetheless, this area requires detailed attention, since early vaccination or combinational therapy are most promising with respect to clinical effectiveness/ achieving (detectable) anticancer effects.

Immunotherapy, at which Stage?

Given the absence of serious toxicity in the clinical vaccination trials described above in advanced cervical carcinoma patients, these anti-HPV vaccination strategies can now be applied to patients at earlier stages of cervical disease, who have a limited tumor load, little or no immune escape, a better immune status, and received less potentially immunosuppressive treatment.

Prevention of HPV-induced lesions could be attempted by vaccinating healthy individuals, e.g., before onset of sexual activity. Although prophylactic vaccines were successfull against HBV infection in human beings[68] and against virus infection as well as tumor challenge in animals,[49] it remains to be established whether vaccination-induced memory CTL would protect against a tumor arising without local inflammation, thus rendering activation of local professional APC unlikely. In this context, the success of vaccines for healthy individuals to prevent HPV-induced cervical cancer hinges on (re)activation of virus-specific memory T cells when encountering a natural HPV-infection which does not cause much tissue damage. The presence of activated T cells at the time of infection could be enforced by periodic booster immunizations.

For therapeutic purposes, vaccination should preferably be conducted in the phases of productive viral infection with oncogenic HPV types or during early stages of cellular transformation. Patients with CIN lesions or non-disseminated cervical carcinoma are less immunocompromised than patients with residual or recurrent cervical carcinoma[16, 131, 132] (M.E.R. et al unpublished results). Adjuvant vaccination could be performed, for instance, immediately following surgery for CIN or early stage invasive cervical carcinoma. Besides affecting the cervix, HPV16 is also detectable in vulvar intraepithelial neoplasms (VIN) [140-143], which proceed to malignancy if left untreated[(144)]. Evaluation of HPV-based immunotherapeutics in VIN lesions is an attractive alternative because of the high levels of MHC class I expression on the affected cells[145] as well as the ease of monitoring vulvar dysplasia. If future immune interventions will be conducted in patients with advanced stage cervical carcinoma, they should be combined with decreasing tumor load through surgery, radio-, or chemotherapy. Such studies could substantiate the alleged synergistic effect of combinational therapy which was suggested by the tumor regressions observed in three cases of cervical carcinoma treated with both HPV-based vaccination and chemotherapy[94] (Dr. L.K. Borysiewicz, personal communication). Finally, in patients with advanced carcinoma for whom active vaccination is unlikely to result in induction of tumor-specific T cells, adoptive transfer of ex vivo activated CTL is a possibility to provide potent anticancer immunity.

Conclusion

Taken together, the first steps towards the development of HPV-peptide-based vaccines have been discussed. A number of HPV16 E7-encoded CTL epitopes were identified and two of these were employed for vaccination in phase I/II clinical studies on peptide vaccination. Improvement of vaccine formulations and of the methods to evaluate anti-tumor effects will provide a framework for successful trial design. Moreover, both the immune status and tumor burden of the patients at start of vaccination are important factors for the outcome. Future trials in which the given immunotherapy is compared to relevant controls will allow definite conclusions on the effectiveness of T cell-mediated prevention and therapy of malignancies.

Acknowledgments

The studies described in this Chapter were supported in part by the Dutch Cancer Society (grant RUL 95-1089 to R.O. and C.J.M.M.), Cytel Corporation, the IDPH, and the NIH of the USA (grants RO1-CA57933, RO1-CA74397, PO1CA74182, RO1CA/AI78399 to W.M.K.).

References

1. Klein, G. Immunovirology of transforming viruses. Curr Opin Immunol 1991; 3:665-673.
2. Coley, W.B. A report of recent cases of inoperable sarcoma successfully treated with mixed toxins of erysipelas and Bacillus prodigiosus. Surg Gynecol Obstet 1911; 13:174-190.
3. Rosenberg, S.A. Immunotherapy of cancer using interleukin 2: current status and future prospects. Immunol Toda 1988; 9:58-62.
4. Pardoll, D.M. New strategies for enhancing the immunogenicity of tumors. Curr Opin Immunol 1993; 5:719-725.

5. Parkin, D.M., Pisani, P., and Ferlay, J. Estimates of the worldwide incidence of eighteen major cancers in 1985. Int J Cancer 1993; 54:594-606.
6. Wright, T.C., Ferenzy, A., and Kurman, R.J. Carcinoma and other tumors of the cervix. In: R.J. Kurman (eds.) Blaustein's pathology of the female genital tract., pp. 279-326. New York: Springer-Verlag, 1994.
7. Larson, D.M., Copeland, L.J., Stringer, C.A., Gerhenson, D.M., Malone, J.M., and Edwards, C.L. Recurrent cervical carcinoma after radical hysterectomy. Gynecol Oncol 1988;30: 381-387.
8. IARC. Human papillomaviruses. Lyon: International Agency for Research on Cancer., 1995 Monographs on the evaluation of the carcinogenic risk of chemicals to humans.; vol 64.
9. Zur Hausen, H. Human papillomaviruses in the pathogenesis of anogenital cancer. Virology 1991;184: 9-13.
10. Zur Hausen, H. Papillomavirus infections—a major cause of human cancers. Biochem Biophys Acta 1996;1288: F55-78.
11. Seedorf, K., Oltersdorf, T., Krämmer, G., and Röwekamp, W. Identification of early proteins of the human papillomaviruses type 16 (HPV16) and type 18 (HPV18) in cervical carcinoma cells. EMBO J 1987;6: 139-144.
12. Von Knebel Doeberitz M, Bauknecht T, Bartsch D et al. Influence of chromosomal integration on glucocorticoid-regulated transcription of growth-stimulating papillomavirus genes E6 and E7 in cervical carcinoma cells. Proc Natl Acad Sci USA 1991; 88:1411-1415.
13. Bosch FX, Manos MM, Munoz N et al. Prevalence of human papillomavirus in cervical cancer: a worldwide perspective. J Natl Cancer Inst 1995;87:796-802.
14. Melkert PW, Hopman E, Van den Brule, AJ et al. Prevalence of HPV in cytomorphologically normal cervical smears as determined by the polymerase chain reaction, is age dependent. Int J Cancer 1993; 53:919-923.
15. Benton C, Shahidullah H, and Hunter JAA. Human papillomavirus in the immunosuppressed. Papillomavirus Rep 1992; 3:23-26.
16. Ressing ME, Van Driel WJ, Celis E et al. Occasional memory cytotoxic T-cell responses of patients with human papillomavirus type 16-positive cervical lesions against a human leukocyte antigen-A *0201-restricted E7-encoded epitope. Cancer Res 1996;56: 582-588.
17. Evans EM, Man S, Evans AS et al. Infiltration of cervical cancer tissue with human papillomavirus-specific cytotoxic T-lymphocytes. Cancer Res 1997; 57:2943- 2950.
18. Nimako M, Fiander AN, Wilkinson GWG et al. Human papillomavirus-specific cytotoxic T lymphocytes in patients with cervical intraepithelial neoplasia grade III. Cancer Res 1997; 57:4855-4861.
19. De Gruijl TD, Bontkes HJ, Stukart MJ et al. T cell proliferative responses against human papillomavirus 16 E7 oncoprotein are most prominent in cervical intraepithelial neoplasia patients with a persistent viral infection. J Gen Virol 1996; 77:2183-2191.
20. Luxton, JC, Rowe AJ, Cridland JC et al. Proliferative T cell responses to the human papillomavirus type 16 E7 protein in women with cervical dysplasia and cervical cacinoma and in healthy individuals. J Gen Virol 1996; 77:1585-1593.
21. Kadish, A.S., Ho, G.Y.F., Burk, R.D et al. Lymphoproliferative responses to human papillomavirus (HPV) type 16 proteins E6 and E7: Outcome of HPV infection and associated neoplasia. J Natl Cancer Inst 1997; 89:1285-1293.
22. De Gruijl TD, Bontkes HJ, Walboomers JMM et al. Differential T helper cell responses to human papillomavirus type 16 E7 related to viral clearance or persistence in patients with cervical neoplasia: a longitudinal study. Cancer Res 1998; 58:1700-1706.
23. Schwartz, R.H. Costimulation of T lymphocytes: The role of CD28, CTLA-4, and B7/BB1 in interleukin-2 production and immunotherapy. Cell 1992; 71:1065-1068.
24. Jenkins MK, Johnson JG. Molecules involved in T-cell costimulation. Curr Opin Immunol 1993; 3:361-367.
25. Allison JP. CD28-B7 interactions in T-cell activation. Curr Opin Immunol 1994; 6:414-419.
26. Bevan MJ. Cross-priming for a secondary cytotoxic response to minor H antigens with H-2 congenic cells which do not cross-react in the cytotoxic assay. J Exp Med 1976; 143:1283-1288.
27. Bennett SRM, Carbone FR, Karamalis F et al. Induction of a $CD8^+$ cytotoxic T lymphocyte repsonse by cross-priming requires cognate $CD4^+$ T cell help. J Exp Med 1997; 186:65-70.

28. Huang AYC, Golumbek P, Ahmadzadeh M et al. Role of bone marrow-derived cells in presenting MHC class I-restricted tumor antigens. Science 1994; 264:961-965.
29. Toes REM, Blom RJJ, Van der Voort EIH, Offringa R, Melief CJM, Kast WM. Protective antitumor immunity induced by immunization with completely allogeneic tumor cells. Cancer Res 1996; 56:3782-3787.
30. Steinman, R.M. The dendritic cell system and its role in immunogenicity. Annu. Rev. Immunol 1991; 9:271-296.
31. Cella M, Sallusto F, Lanzavecchia A. Origin, maturation and antigen presenting function of dendritic cells. Curr Opin Immunol 1997; 9:10-16.
32. Matzinger P. Tolerance, danger, and the extended family. Annu Rev Immunol 1994; 12:991-1045.
33. Ridge JP, Di Rosa F, Matzinger P. A conditioned dendritic cell can be a temporal bridge between a $CD4^+$ T-helper and a T-killer cell. Nature 1998; 393:474-478.
34. Bennett SRM, Carbone FR, Karamalis F et al. Help for cytotoxic-T-cell responses is mediated by CD40 signalling. Nature 1998; 393:478-480.
35. Schoenberger, S.P., Toes, R.E.M., Van der Voort, E.I.H., Offringa, R., and Melief, C.J.M. T- cell help for cytotoxic T lymphocytes is mediated by CD40-CD40L interactions. Nature 1998;393: 480-483.
36. Tindle, R.W. Human papillomavirus vaccines for cervical cancer. Curr Opin Immunol 1996;8: 643-650.
37. McArdle, J.P., and Muller, K. Quantative assessment of Langerhans' cells in human cervical intraepithelial neoplasia and wart virus infection. Am J Obstet Gynecol 1986; 154:509-515.
38. Viac J, Guerin-Reverchon Y, Chardonnet Y et al. Langerhans cells and epithelial cell modifications in cervical intraepithelial neoplasia: correlation with human papillomavirus infection. Immunobiol 1990; 180:328-338.
39. Jarrett WFH, Smith KT, O'Neil BW et al. Studies on vaccination against papillomaviruses: prophylactic and therapeutic vaccination with recombinant structural proteins. Virology 1991; 184:33-42.
40. Campo MS, Grindlay GJ, O'Neil BW et al. Prophylactic and therapeutic vaccination against a mucosal papillomavirus. J Gen Virol 1993; 74:945-953.
41. Lin YL, Borenstein LA, Selvakumar R et al. Effective vaccination against papilloma development by immunization with L1 or L2 structural protein of cottontail rabbit papillomavirus. Virology 1992; 187:612-619.
42. Selvakumar R, Borenstein LA, Lin YL et al. Immunization with non-structural proteins E1 and E2 of cottontail rabbit papillomavirus stimulates regression of virus-induced papillomas. J Virol 1995; 69:602-605.
43. Suzich JA, Ghim SJ, Palmer-Hill FJ et al. Systemic immunization with papillomavirus L1 protein completely prevents the development of viral mucosal papillomas. Proc Natl Acad Sci USA 1995; 92:11553-11557.
44. Chen LP, Thomas EK, Hu SL et al. Human papillomavirus type 16 nucleoprotein E7 is a tumor rejection antigen. Proc Natl Acad Sci USA 1991; 88:110-114.
45. McLean CS, Sterling JS, Mowat J et al. Delayed-type hypersensitivity response to the human papillomavirus type 16 E7 protein in a mouse model. J Gen Virol 1993; 74:239-245.
46. Feltkamp MCW, Smits HL, Vierboom MPM et al. Vaccination with a cytotoxic T lymphocyte epitope-containing peptide protects against a tumor induced by human papillomavirus type 16-transformed cells. Eur J Immunol 1993; 23:2242-2249.
47. Meneguzzi G, Cerni C, Kieny MP et al. Immunization against human papillomavirus type 16 tumor cells with recombinant vaccinia viruses expressing E6 and E7. Virology 1991; 181:62-69.
48. Sadovnikova E, Zhu X, Collins S et al. Limitations of predictive motifs revealed by cytotoxic T lymphocyte epitope mapping of the human papillomavirus E7 protein. Int Immunol 1994; 6:289-296.
49. Melief CJM, Offringa R, Toes REM et al. Peptide-based cancer vaccines. Curr Opin Immunol 1996; 8:651-657.
50. Mayordomo JI, Zorina T, Storkus WJ et al. Bone marrow-derived dendritic cells serve as potent adjuvants for peptide-based anti-tumor vaccines. Stem Cells 1997; 15:94-103.
51. Mayordomo JI, Zorina T, Storkus WJ et al. Bone marrow-derived dendritic cells pulsed with synthetic tumour peptides elicit protective and therapeutic antitumour immunity. Nat Med 1995; 1:1297-1302.

52. Zitvogel L, Mayordomo JI, Tjandrawan T et al. Therapy of murine tumors with tumor peptide-pulsed dendritic cells: dependence of T cells, B7 costimulation, and T helper cell 1-associated cytokines. J Exp Med 1996; 183:87-97.
53. Toes REM, Hoeben R, Van der Voort EIH et al. Protective anti-tumor immunity induced by vaccination with recombinant adenoviruses encoding multiple tumor-associated cytotoxic T lymphocyte epitopes in a string-of-beads fashion. Proc Natl Acad Sci USA 1997; 94:14660-14665.
54. De Bruijn MLH, Schuurhuis DS, Vierboom MPM et al. Immunization with human papillomavirus type 16 (HPV16) oncoprotein-loaded dendritic cells as well as protein in adjuvant induces MHC class I-restricted protection to HPV16-induced tumor cells. Cancer Res 1998; 58:724-731.
55. Lin KY, Guarnieri FG, Stavely-O'Carroll KF et al. Treatment of established tumors with a novel vaccine that enhances major histocompatibility class II presentation of tumor antigen. Cancer Res 1996; 56:21-26.
56. De Bruijn MLH, Greenstone HL, Vermeulen H, Melief CJM, Lowy DR, Schiller JT, Kast WM. L1-specific protection from tumor challenge elicited by HPV16 virus-like particles. Virology 1998; 250:371-376.
57. Griep AE, Herber R, Jeon S et al. Tumorigenicity by human papillomavirus type 16 E6 and E7 in transgenic mice correlates with alterations in epithelial cell growth and differentiation. J Virol 1993; 67:1373-1384.
58. Lambert PF, Pan H, Pitot HC et al. Epidermal cancer associated with expression of human papillomavirus type 16 E6 and E7 oncogenes in the skin of transgenic mice. Proc Natl Acad Sci USA 1993; 90:5583-5587.
59. Auewarakul P, Gissmann L, Cid-Arregui A. Targeted expression of the E6 and E7 oncogenes of human papillomavirus type 16 in the epidermis of transgenic mice elicits generalized epidermal hyperplasia involving autocrine factors. Mol Cell Biol 1994; 14:8250-8258.
60. Arbeit JM, Münger K, Howley PM et al. Progressive squamous epithelial neoplasia in K14-human papillomavirus type 16 transgenic mice. J Virol 1994; 68:4358-4368.
61. Arbeit JM, Howley PM, Hanahan D. Chronic estrogen-induced cervical and vaginal squamous carcinogenesis in human papillomavirus type 16 transgenic mice. Proc Natl Acad Sci USA 1996; 93:2930-2935.
62. Coussens LM, Hanahan D, Arbeit JM. Genetic predisposition and parameters of malignant progression in K14-HPV16 transgenic mice. Am J Pathol 1996; 149:1899-1917.
63. Herber R, Liem A, Pitot H et al. Squamous epithelial hyperplasia and carcinoma in mice transgenic for the human papillomavirus type 16 E7 oncogene. J Virol 1996; 70:1873-1881.
64. Herd K, Fernando GJ, Dunn LA et al. E7 oncoprotein of human papillomavirus type 16 expressed constitutively in the epidermis has no effect on E7-specific B- or Th-repertoires or on the immune response induced or sustained after immunization with E7 protein. Virology 1997; 231:155-165.
65. Dunn L, Evander M, Tindle R et al. Presentation of the HPV16E7 protein by skin grafts is insufficient to allow graft rejection in an E7-primed animal. Virology 1997; 235:94-103.
66. Doan T, Chambers M, Street M et al. Mice expressing the E7 oncogene of HPV16 in epithelium show central tolerance, and evidence of peripheral anergising tolerance, to E7-encoded cytotoxic T-lymphocyte epitopes. Virology 1998; 244:352-364.
67. Frazer I, Fernando G, Fowler N et al. Split tolerance to a viral antigen expressed in thymic epithelium and keratinocytes. Eur Immunol 1998; 28:2791-2800.
68. Chang MH, Chen CJ, Lai MS et al. Universal Hepatitis B vaccination in Taiwan and the incidence of hepatocellular carcinoma in children. N Engl J Med 1997; 336:1855-1859.
69. Vitiello A, Ishioka G, Grey HM et al. Development of a lipopeptide-based therapeutic vaccine to treat chronic HBV infection. I. Induction of a primary cytotoxic T lymphocyte response in humans. J Clin Invest 1995; 95:341-349.
70. Livingston BD, Crimi C, Grey H et al. The Hepatitis B virus-specific CTL responses induced in humans by lipopeptide vaccination are comparable to those elicited by acute viral infection. J Immunol 1997; 159: 1383-1392.
71. Melief CJM, Kast WM. Prospects for T cell immunotherapy of tumors by vaccination with immunodominant and subdominant peptides. In: Chadwick DJ, Marsh J, eds. Vaccines against virally induced cancers, 97-112. Chichester: Wiley, 1994.

72. Celis E, Tsai V, Crimi C et al. Induction of anti-tumor cytotoxic T lymphocytes in normal humans using primary cultures and synthethic peptide epitopes. Proc Natl Acad Sci USA 1994; 91: 2105-2109.
73. Ressing ME, Sette A, Brandt RMP et al. Human CTL epitopes encoded by human papillomavirus type 16 E6 and E7 identified through in vivo and in vitro immunogenicity studies of HLA-A*0201-binding peptides. J Immunol 1995; 154:5934-5943.
74. Konya, J., Eklund, C., af Geijersstam, V., Yuan, F., Stuber, G., and Dillner, J. Identification of a cytotoxic T-lymphocyte epitope in the human papillomavirus type 16 E2 protein. J Gen Virol 1997; 78:2615-2620.
75. Bartholomew, J.S., Stacey, S.N., Coles, B., Burt, D.J., Arrand, J.R., and Stern, P.L. Identification of a naturally processed HLA-A*0201-restricted viral peptide from cells expressing human papillomavirus type 16 E6 oncoprotein. Eur J Immunol 1994; 24:3175-3179.
76. Altman, J.D., Moss, P.A.H., Goulder, P.J.R., Barouch, D.H., McHeyzer-Williams, M.G., Bell, J.I., McMichael, A.J., and Davis, M.M. Phenotypic analysis of antigen-specific T lymphocytes. Science 1996; 274:94-96.
77. McMichael AJ, O'Callaghan CA. A new look at T cells. J Exp Med 1998; 187:1367-1371.
78. Le AXT, Bernhard EJ, Holterman MJ et al. Cytotoxic T cell responses in HLA A2.1 transgenic mice. Recognition of HLA alloantigens and utilization of HLA-A2.1 as a restriction element. J Immunol 1989; 142:1366-1371.
79. Vitiello A, Marchesini D, Furze J et al. Analysis of the HLA-restricted influenza-specific cytotoxic T lymphocyte response in transgenic mice carrying a chimeric human-mouse class I major histocompatibility complex. J Exp Med 1991; 173:1007-1015.
80. Engelhard VH, Lacy E, Ridge JP. Influenza A-specific, HLA-A2.1-restricted cytotoxic T lymphocytes from HLA-A2.1 transgenic mice recognize fragments of the M1 protein. J Immunol 1991; 146:1226-1232.
81. Newberg MH, Ridge JP, Vining DR, Salter RD, and Engelhard, V.H. Species specificity in the interaction of CD8 with the alpha 3 domain of MHC class I molecules. J Immunol 1992;149: 136-142.
82. Man, S., Ridge, J.P., and Engelhard, V.H. Diversity and dominance among TCR recognizing HLA-A2.1+ influenza matrix peptide in human MHC class I transgenic mice. J Immunol 1994; 153: 4458-4467.
83. Lehner PJ, Wang ECY, Moss PAH et al. Human HLA-A*0201-restricted cytotoxic T lymphocyte recognition of influenza A is dominated by T cells bearing the V beta 17 gene segment. J Exp Med 1995; 181:79-91.
84. Shirai M., Arichi T, Nishioka M et al. CTL responses of HLA-A2.1-transgenic mice specific for hepatitis C viral peptides predict epitopes for CTL of humans carrying HLA-A2.1. J Immunol 1995; 154:2733-2742.
85. Sette A., Vitiello A., Reherman B et al. The relationship between class I binding affinity and immunogenicity of potential cytotoxic T cell epitopes. J Immunol 1994;153: 5586-5592.
86. Theobald M, Biggs J, Dittmer D et al. Targeting p53 as a general tumor antigen. Proc Natl Acad Sci USA 1996; 92:11993-11997.
87. Wentworth PA, Vitiello A, Sidney J et al. Differences and similarities in the A2.1-restricted cytotoxic T cell repertoire in humans and human leukocyte antigen-transgenic mice. Eur J Immunol 1996; 26:97-101.
88. Vitiello A, Sette A, Yuan L et al. Comparison of cytotoxic T lymphocyte responses induced by peptide or DNA immunization: implications on immunogenicity and immunodominance. Eur J Immunol 1997; 27:671-678.
89. Alexander M, Salgaller ML, Celis E et al. Generation of tumor-specific cytolytic T lymphocytes from peripheral blood of cervical cancer patients by in vitro stimulation with a synthetic human papillomavirus type 16 E7 epitope. Am J Obstet Gynecol 1996; 175:1586-1593.
90. Kaufmann AM, Gissmann L, Schreckenberger C et al. Cervical carcinoma cells transfected with the CD80 gene elicit a primary cytotoxic T lymphocyte response speciric for HPV16 E7 antigens. Cancer Gene Ther 1997; 4:377-382.
91. McNeil, C. HPV vaccine treatment trials proliferate, diversify. J Natl Cancer Inst 1997; 89:280-281.
92. McNeil, C. HPV vaccines for cervical cancer move toward clinic, encounter social issues. J Natl Cancer Inst 1997; 89:1664-1666.

93. Borysiewicz LK, Fiander A, Nimako M et al. A recombinant vaccinia virus encoding human papillomavirus types 16 and 18, E6 and E7 proteins as immunotherapy for cervical cancer. Lancet 1996; 347:1523-1527.
94. Ressing ME, Van Driel, WJ, Brandt RMP et al. Detection of T helper responses, but not of HPV-specific CTL responses, following peptide vaccination of cervical carcinoma patients. J Immunother 2000; 23:255-266.
95. Steller, M., Gurski, K., Murakami, M et al. Cell-mediated immunological responses in cervical and vaginal cancer patients immunized with a lipidated epitope of human papillomavirus type 16 E7. Clin Cancer Res 1998; 4:2103-2109.
96. Imanishi I, Akaza T, Kimura A et al. Allele and haplotype frequencies for HLA and complement loci in various ethnic groups. In: K. Tsuji, M. Aizawa, and T. Sasazuki, eds. HLA 1991. Proceedings of the eleventh international histocompatibility workshop and conference, 1065-1220. Oxford: Oxford University Press, 1992.
97. Browning M, Krausa P. Genetic diversity of HLA-A2: Evolutionary and functional significance. Immunol. Today 1996; 17:165-170.
98. Kast WM, Brandt RMP, Drijfhout JW et al. Human leukocyte antigen- A2.1 restricted candidate cytotoxic T lymphocyte epitopes of human papillomavirus type 16 E6 and E7 proteins identified by using the processing-defective human cell line T2. J Immunother 1993; 14:115-120.
99. Kast WM, Brandt RMP, Sidney J et al. Role of HLA-A motifs in identification of potential CTL epitopes in human papillomavirus type 16 E6 and E7 proteins. J Immunol 1994; 152:3904-3912.
100. Del Guercio MF, Sidney J, Hermanson G et al. Binding of a peptide antigen to multiple HLA alleles allows definition of an A2-like supertype. J Immunol 1995; 154:685-693.
101. Sidney J, Grey HM, Kubo RT et al. Practical, biochemical and evolutionary implications of the discovery of HLA class I supermotifs. Immunol Today 1996; 17:261-266.
102. Sidney J, Del Guercio MF, Southwood S et al. The HLA-A*0207 peptide binding repertoire is limited to a subset of the A*0201 repertoire. Hum Immunol 1997; 58:12-20.
103. Sette A, Sidney J. HLA supertypes and supermotifs: a functional perspective on HLA polymorphism. Curr Opin Immunol 1998; 10:478-482.
104. Bertoni R, Sidney J, Fowler P et al. Human histocompatibility leukocyte antigen-binding supermotifs predict broadly cross-reactive cytotoxic T lymphocyte responses in patients with acute hepatitis. J Clin Invest 1997; 100:503-513.
105. Doolan DL, Hoffman SL, Southwood S et al. Degenerate cytotoxic T cell epitopes from P. falciparum restricted by multiple HLA-A and HLA-B supertype alleles. Immunity 1997; 7:97-112.
106. Ressing ME, De Jong JH, Brandt REM et al. Differential binding of viral peptides to HLA-A2 alleles. Implications for HPV16 E7 peptide-based vaccination against cervical carcinoma. Eur J Immunol 1999; 29:1292-1303.
107. Barouch D, Friede T, Stevanovic S et al. HLA-A2 subtypes are functionally distinct in peptide binding and presentation. J Exp Med 1995; 182:1847-1856.
108. Rivoltini L, Loftus DJ, Barracchini K et al. Binding and presentation of peptides derived from melanoma antigens MART-1 and glycoprotein-100 by HLA-A2 subtypes. Implications for peptide-based immunotherapy. J Immunol 1996; 156:3882-3891.
109. Browning M, Dunnion D. HLA and cancer: Implications for cancer immunotherapy and vaccination. Eur J Immunogenet 1997; 24:293-312.
110. Connor ME, Stern PL. Loss of MHC class I expression in cervical carcinomas. Int J Cancer 1990; 46:1029-1034.
111. Cromme FV, Airey J, Heemels MT et al. Loss of transporter protein, encoded by the TAP-1 gene, is highly correlated with loss of HLA expression in cervical carcinomas. J Exp Med 1994; 179:335-340.
112. Duggan-Keen M, Keating PJ, Cromme FV et al. Alterations in major histocompatibility complex expression in cervical cancer: Possible consequences for immunotherapy. In: C. Lacey, eds. Papillomavirus reviews: current research on papillomaviruses., pp. 141-150. Leeds: Leeds University Press, 1996.
113. Honma S, Tsukada S, Honda S et al. Biological-clinical significance of selective loss of HLA class I allelic product expression in squamous cell carcinoma of the uterine cervix. Int J Cancer 1994; 57:650-655.

114. Kärre K, Ljunggren HG, Piontek G et al. Selective rejection of H-2 deficient lymphoma variants suggests alternative immune defense strategy. Nature 1986; 319:675-678.
115. Garrido F, Ruiz-Cabello F, Cabrera T et al. Implications for immunosurveillance of altered HLA class I phenotypes in human tumors. Immunol Today 1997; 18:89-95.
116. Cromme FV, Van Bommel PF, Walboomers JMM et al. Differences in MHC and TAP expression in cervical cancer lymph node metastases as compared with the primary tumors. Br J Cancer 1994; 69:1176-1181.
117. Hilders CGJM, Morgado Munoz I, Nooyen Y et al. Altered HLA expression by metastatic cervical carcinoma cells as a factor in impaired immune surveillance. Gynecol Oncol 1995; 57:366-375.
118. Van Driel WJ, Tjiong MY, Hilders CGJM et al. Association of allele-specific HLA-expression and histopathologic progression of cervical carcinoma. Gynecol Oncol 1996; 62:33-41.
119. Duggan-Keen MF, Keating PJ, Stevens FRA et al. Immunogenetic factors in HPV-associated cervical cancer: Influence on disease progression. Eur J Immunogenet 1996; 23:275-284.
120. Koopman LA, Mulder A., Corver WE et al. HLA class I phenotype and genotype alterations in cervical carcinomas and derivative cell lines. Tissue Antigens 1998; 51:623-636.
121. Hengel H, Kozinowski UH. Interference with antigen processing by viruses. Curr Opin Immunol 1997; 9:470-476.
122. Ploegh HL. Viral strategies of immune evasion. Science 1998; 280:248-253.
123. Cromme FV, Meijer CJLM, Snijders PJF et al. Analysis of MHC class I and II expression in relation to HPV genotypes in premalignant and malignant cervical lesions. Br J Cancer 1993; 67:1372-1380.
124. Andresson T, Sparkowski J, Goldstein DJ et al. Vacuolar H+-ATPase mutants transform cells and define a binding site for the papillomavirus E5 oncoprotein. J Biol Chem 1995; 270:6830-6837.
125. Tong X, Boll W, Kirchhausen T et al. Interaction of the bovine papillomavirus E6 protein with the clathrin adaptor complex AP-1. J Virol 1998; 72:476-482.
126. Klenerman P, Rowland-Jones S, McAdam S et al. Cytotoxic T-cell activity antagonized by naturally occurring HIV-1 gag variants. Nature 1994; 369:403-407.
127. Bertoletti A, Sette A, Chisari FV et al. Natural variants of cytotoxic epitopes are T cell receptor antagonists for antiviral cytotoxic T cells. Nature 1994; 369:407-410.
128. Ellis JRM, Keating PJ, Baird J et al. The association of an HPV16 oncogene variant with HLA-B7 has implications for vaccine design in cervical cancer. Nat Med 1995; 1:464-470.
129. Zehbe I, Wilander E, DeliusH et al. Human papillomavirus 16 E6 variants are more prevalent in invasive cervical carcinoma than the prototype. Cancer Res 1998; 58:829-833.
130. Wagner, G. Hautteste als Immunparameter beim Zervixkarzinom. Wiener Klinische Wochenschrift 1984; 96:467-473.
131. Nieland JD, Loviscek K, Kono K et al. PBLs of early breast carcinoma patients with a high nuclear grade tumor unlike PBLs of cervical carcinoma patients do not show a decreased T expression but are functionally impaired. J Immunother 1998; 21:317-322.
132. Clerici, M., Merola, M., Ferrario, E., Trabattoni, D., Villa, M.L., Stefanon, B., Venzon, D.J., Shearer, G.M., De Palo, G., and Clerici, E. Cytokine production patterns in cervical intraepithelial neoplasia: association with human papillomavirus infection. J Natl Cancer Inst 1997;89: 245-250.
133. Soutter, W.P., and Kesic, V. Treatment of cervical intraepithelial neoplasia reverses CD4/CD8 lymphocyte abnormalities in peripheral venous blood. J Gynecol Oncol 1994;4: 279-282.
134. De Bruijn, M.L.H., Schumacher, T.N.M., Nieland, J.D., Kast, W.M., and Melief, C.J.M. Peptide loading of empty major histocompatibility complex molecules on RMA-S cells allows the induction of primary cytotoxic T lymphocyte responses. Eur J Immunol 1991;21: 2963-2970.
135. Koeppen, H., Acena, M., Drolet, A., Rowley, D.A., and Schreiber, H. Tumors with reduced expression of a cytotoxic T cell recognized antigen lack immunogenicity but retain sensitivity to lysis by cytotoxic T cells. Eur J Immunol 1993;23: 2770-2776.
136. Speiser, D.E., Miranda, R., Zakarin, A., Bachmann, M.F., McKall-Faienza, K., Odermatt, B., Hanahan, D., Zinkernagel, R.M., and Ohashi, P.S. Self antigen expressed by solid tumors do not efficiently stimulate naive or activated T cells. Implications for immunotherapy. J Exp Med 1997;186: 645-653.

137. Ossendorp F, Mengedé E, Camps M et al. Specific T helper cell requirement for optimal induction of cytotoxic T lymphocytes against major histocompatibility complex class II negative tumors. J Exp Med 1998; 187:693-702.

138. Greenstone HL, Nieland JD, De Visser KE et al. Chimeric papillomavirus virus-like particles elicit antitumor immunity against the E7 oncoprotein in an HPV16 tumor model. Proc Natl Acad Sci USA 1998; 95:1800-1805.

139. Hsu FJ, Benike C, Fagnoni F et al. Vaccination of patients with B-cell lymphoma using autologous antigen-pulsed dendritic cells. Nat Med 1996; 2:52-58.

140. Hording, U., Daugaard, S., and Iversen, A.K.N. Human papillomavirus type 16 in vulvar carcinoma, vulvar intraepithelial neoplasia, and associated cervical neoplasia. Gynecol Oncol 1991; 42: 22-26.

141. Hording U, Kringsholm B, Andreasson B et al. Human papillomavirus in vulvar squamous-cell carcinoma and in normal vulvar tissues: A search for a possible impact of HPV on vulvar cancer prognosis. Int J Cancer 1993; 55:394-396.

142. Van Beurden M, Ten Kate FWJ, Smits HL et al. Multifocal vulvar intraepithelial neoplasia grade III and multicentric lower genital tract neoplasia is associated with transcriptionally active human papillomavirus. Cancer 1995; 75:2879-2884.

143. Kagie MJ, Kenter GG, Zomerdijk-Nooijen Y et al. Human papillomavirus infection in squamous cell carcinoma of the vulva, in various synchronous epithelial changes and in normal vulvar skin. Gynecol Oncol 1997; 67:178-183.

144. Jones RW, Rowan DM. Vulvar intraepithelial neoplasia III: a clinical study of the outcome in 113 cases with relation to the later development of invasive vulvar carcinoma. Obstet Gynecol 1994; 84:741-745.

145. Jochmus I, Dürst M, Reid R. et al. Major histocompatibility complex and human papillomavirus type 16 E7 expression in high grade vulvar lesions. Hum Pathol 1993; 24:519-524.

Chapter 12

Peptide Vaccines for the Treatment of Melanoma

Willem W. Overwijk and Nicholas P. Restifo

Introduction

The development of cancer vaccines has been greatly advanced by the recent identification of many tumor-associated antigens (TAA) recognized by T cells.[1,2] A majority of these antigens have been cloned from melanoma, including the MAGE family, which are also expressed in normal testes, as well as melanocyte differentiation antigens (MDA), which are expressed by normal melanocytes. (Table 12.1, and refs. 3,4) An advantage of these antigens is that they are shared among tumors from a majority of patients. Since they are also expressed in normal tissues, they are considered "self". In attempts to target MDA with cancer vaccines, one of the pitfalls may be the different immunologic nature of "self" antigens when compared to unique antigens such as mutated or viral gene products. "Self"-reactive T cells may have been physically or functionally deleted in the thymus or in the periphery, leaving behind only limited numbers of functionally impaired T cells. Therefore, the major focus of cancer vaccine development has shifted from the identification of target proteins to the induction of immune responses that can eliminate established metastatic tumors in patients. One approach toward this goal is the use of synthetic peptide vaccines.

Tolerance to "Self" Antigens

The identification of human tumor antigens recognized by T cells has enabled the development of recombinant and synthetic cancer vaccines. Attempts to induce therapeutic immune responses to these antigens initially focused on strategies that had been highly successful in inducing powerful immune responses in vivo, such as recombinant viruses and 'naked' DNA.[5,6] Recombinant poxviruses, recombinant adenovirus (rAd), and influenza virus, could induce strong $CD8^+$ T cell responses against encoded model antigens, capable of curing established tumors.[5] Concurrent administration of cytokines such as IL-2 and IL-12, or the co-encoding into the viral genome of IL-2, IL-12, and/or the costimulatory molecule CD80/B7-1, further enhanced therapeutic efficacy.[7,8]

One drawback of these early studies was the use of model tumor antigens such as *Escherichia coli*—galactosidase, chicken ovalbumin, or influenza nucleoprotein. These proteins are "foreign" to the host, while MDA, constitutively expressed on normal melanocytes, are "self". Indeed, when immunizing experimental animals against normal autoantigens, the induction of a detectable immune response appears to be an exception rather than the rule.[9-12] This observation was mirrored in cancer patients, where it appeared difficult if not impossible to induce strong immune responses to several candidate tumor antigens in a variety of studies,

Peptide-Based Cancer Vaccines, edited by W. Martin Kast. ©2000 Eurekah.com.

Table 12.1. Selected melanoma/melanocyte differentiation antigens recognized by melanoma-reactive T cells

Gene	Restriction Element	Peptide Epitope	Reference
gp100/ pmel17	HLA-A*0201	KTWGQYWQV	65,66
		AMLGTHTMEV	67
		MLGTHTMEV	67
		ITDQVPFSV	65
		YLEPGPVTA	65,67
		LLDGTATLRL	65
		VLYRYGSFSV	65
		SLADTNSLAV	67
		RLMKQDFSV	68
		RLPRIFCSC	68
	HLA-A3	ALLAVGATK	69
		LIYRRRLMK	68
	HLA-A*2402	VYFFLPDHL	70
	H-2D^b	KVPRNQDWL EGSRNQDWL	9
Tyrosinase	HLA-A*0201	MLLAVCYLL	71
		YMDGTMSQV	71
	HLA-A1	DAEKCDKTDEY	72
		SSDYVIPIGTY	68
	HLA-A*2402	AFLPWHRLF	73
	HLA–B44	SEIWRDIDF	74
	HLA-DR4	QNILLSNAPLGP QFP SYLQDSPDSFQD	75
MART-1/ Melan-A	HLA-A*0201	AAGIGILTV	76
	HLA-B*4501	AEEAAGIGILT	77
TRP-1/gp75	HLA-31	MSLQRQFLR	61
TRP-2	HLA-A*0201	SVYDFFVWL	78
	HLA-A31	LLPGGRPYR	79
	HLA-A33	LLPGGRPYR	79
	H2-K^b	SVYDFFVWL	80

with only an occasional patient demonstrating induction of specific T cells that recognize tumors.[13-16] Clearly, strategies that are sufficient to immunize against foreign antigens fall short when the target antigen is "self."

The absence of an immune response to a defined "self" antigen can be due to negative selection of self-antigen-specific T cells during maturation in the thymus, termed "central" tolerance.[17] However, a low level of autoreactivity is required for positive selection in the thymus,[17-19] and T cells with limited reactivity to autoantigens thus persist. Mature T lymphocytes with reactivity to these "self" antigens may remain in a functionally tolerant state, termed "ignorance", if they do not traffic to antigen bearing cells, or if the target antigen is not processed and presented to a level that can trigger the specific T cell receptor (TCR). Mature self-reactive T cells that encounter antigen on normal tissues in the absence of an activating co-stimulatory micro environment can be physically or functionally eliminated by anergization or by deletion, thus effecting extra-thymic or "peripheral" tolerance.[17,20,21]

Much of our current understanding of tolerance to tissue antigens has been obtained using transgenic models in which mice are genetically engineered to express a single TCR, sometimes together with their cognate antigen. Most often, these models utilize defined "foreign" antigens made "self" by expressing them in normal tissues under the control of constitutive or tissue-specific promoters.[22,23] While responsible for greatly increased understanding of the mechanisms behind the immune response to tissue differentiation antigens, it is difficult to know to what extent these models reflect "real world" immune responses to "self" antigens such as MDA. For example, questions that remained unanswered by the early models include how the expression levels of an antigen driven by a viral or altered mammalian promoter compare to the varying levels of a naturally expressed antigen. Further, it is not clear whether very low levels of endogenous, MDA-specific T cell precursors with heterogeneous TCRs will react similarly to peripheral antigen as large numbers of genetically identical, transgenic T cells. New mouse models that more closely paralleled the immunologic situation in patients with cancer were needed.

Lessons from gp100 as a Murine Melanoma Tumor Rejection Antigen

We attempted to induce MDA-specific cytotoxic T cells using recombinant vaccinia virus (rVV), a method proven highly successful in inducing therapeutic $CD8^+$ T cells against "foreign" antigens such as -galactosidase. However, we were unable to detect any $CD8^+$ T cells upon in vitro stimulation with B16 of splenocytes from mice immunized with rVV encoding murine gp100/pmel-17 (mgp100), the mouse homologue of human gp100 (hgp100).[24] Likewise, immunization with recombinant fowlpoxvirus or pDNA encoding mgp100 did not induce detectable levels of mgp100-specific $CD8^+$ T cells. Since the work of Mamula and others had previously shown that xenoimmunization could sometimes break the barriers of tolerance to "self" proteins, we used plasmid DNA encoding hgp100 to immunize mice, then stimulated splenocytes with murine B16 melanoma or DC infected with rVVhgp100.[10,11,25]

Using this approach, we elicited cytotoxic $CD8^+$ T cells that recognized both hgp100 and mgp100-transfected cells, as well as the B16 melanoma.[9] Using a computer-generated epitope-forecasting algorithm,[26,27] we identified a H-$2D^b$-restricted peptide, EGSRNQDWL in mgp100 that was recognized by gp100-reactive T cells. Interestingly, the corresponding human peptide, KVPRNQDWL, differed only in its 3 N-terminal amino acids. Computer algorithms predicted the human peptide to bind significantly better to H-$2D^b$. This enhanced binding was confirmed by in vitro MHC Class I stabilization assays[9] This enhanced binding resulted in a half-maximal recognition of the $hgp100_{25-33}$ peptide at a concentration 2-3 orders of magnitude lower than required for the parental $mgp100_{25-33}$. Indeed, the resulting 100-fold increase in MHC Class I stabilization appeared to be responsible for the high immunogenicity of the hgp100$_{25-33}$ peptide in mice. Splenocytes from mice immunized with rVVhgp100 and stimulated with the mgp100$_{25-33}$ peptide displayed strong gp100-specific cytotoxicity, and upon adoptive transfer could destroy established pulmonary melanoma deposits. Thus, gp100 functions as a tumor rejection antigen in mice.[9] Furthermore, this mouse model of melanoma vaccine therapy may aid in the identification of new strategies for the immunotherapy of human melanoma (Table 12.2).

Clinical Applications: Results of Virus- and Peptide-Based Cancer Vaccine Trials

Based on these results, clinical trials were started with rVV and rAd encoding hgp100 or hMART-1. Although administration of these viruses was well tolerated, the first results from a trial using recombinant adenovirus were disappointing, with little objective tumor regression and little or no proof of enhanced gp100 or MART-1-specific T cell responses in immunized patients.[28] One reason for the stark difference in efficacy between viral immunization in mice and man may be the high levels of neutralizing adenovirus-specific antibodies that were found in the large majority of patients. Indeed, preliminary results indicate the existence of high levels of neutralizing vaccinia-specific antibodies in sera from patients, some of whom have not been exposed to the virus in several decades. These neutralizing antibodies may severely limit the use of recombinant vaccinia viral vectors in patients. Efforts are underway to develop viral vectors to which no pre-existing immunity exists, such as avian poxviruses.

In order to circumvent issues of vector-specific pre-existing immunity, we attempted to develop a vaccine based on the minimal peptide epitopes recognized by $CD8^+$ T cells. In the mouse, we had identified the mgp100$_{25-33}$ peptide epitope and its more immunogenic homologue, hgp100$_{25-33}$. Results of immunization with gp100 peptide emulsified in incomplete Freund's adjuvant (IFA) mirrored those obtained with full-length gp100 encoded in rVV: hgp100$_{25-33}$ peptide induced mgp100-specific, tumoricidal $CD8^+$ T cells whereas mgp100$_{25-33}$ did not (our unpublished results).

Table 12.2. Characteristics of human melanoma patients and gp100 mouse melanoma model

Human	Mouse	Reference
Poor immunization with parental peptide unpublished	hgp100$_{209-217}$	mgp100$_{25-33}$
Improved immunization with modified peptide unpublished	hgp100$_{209-217}$-2M	hgp100$_{25-33}$
CTL precursors alone do not reduce tumor burden unpublished [309]	hgp100$_{209-217}$-2M	hgp100$_{25-33}$
CTL transfer reduces tumor burden	TIL1200	Clone 9[659]
Peptide immunization induces tumor regression unpublished [30]	hgp100$_{209-217}$-2M	ESmgp100$_{25-33}$

While the discovery of hgp100$_{25-33}$ as a peptide homologue with higher affinity to MHC Class I was a fortuitous one, the principle can be applied in the rational design of similar peptide homologues of other potential target peptides. Indeed, when the hgp100 epitope hgp100$_{209-217}$, was modified to enhance binding to the MHC Class I molecule HLA-A201, its ability to induce tumor-specific, CD8$^+$ T cells was dramatically increased.[29] Paralleling the results in mice, the modified hgp100$_{209-217}$-2M peptide, which contained a methionine instead of a threonine in the second position, was highly successful (91%) in immunizing melanoma patients as evidenced by increased CTL precursor levels in the blood.[30] When IL-2 was administered following gp100 peptide vaccination, 13 of 31 patients experienced an objective tumor regression (Table 12.3).[30] This was a phase I study with a small number of patients, and results will have to be confirmed in a larger, randomized study.

An interesting observation was made in patients receiving IL-2 or intravenous IL-12 upon peptide vaccination: when either cytokine was added to the peptide vaccination regimen, we were no longer able to induce measurable T cells in peripheral blood lymphocytes from immunized patients.[30] It is unclear whether these cytokines cause the differential distribution of

Table 12.3. Characteristics of the 13 patients with metastatic malignant melanoma who had objective responses to modified gp100 peptide (hgp100$_{209-217}$-2M) plus IL-2

Patient	Age (yr)/Sex	Site of Tumor	Type	Response Duration (mo)
a	48/M	Lung	CR	6
b	51/M	Lung, Subcutaneous	PR	6
c	44/M	Lymph node, Lung, Subcutaneous, Cutaneous	PR	2+
d	45/F	Lymph node, Bone, Subcutaneous	PR	4
e	42/F	Cutaneous, Subcutaneous, Liver	PR	7+
f	41/F	Lung	PR	5+
g	39/M	Lung	PR	6
h	22/F	Lung hilum	PR	5
i	48/M	Cutaneous	PR	6
j	43/M	Subcutaneous	PR	5+
k	42/F	Lung, Lymph node, Liver, Brain	PR	2
l	47/F	Lung, Lymph node	PR	5+
m	59/M	Subcutaneous	PR	3+

specific T cells upon vaccination, or whether T cells are being deleted in the periphery, perhaps as a result of activation induced cell death. We have recently shown that addition of IL-2 to viral immunization in mice leads to the induction of a suppressive population of antigen presenting cells that is capable of inducing apoptosis in activated T cells.[31] Furthermore, addition of high- dose IL-12 to viral immunization abrogates the ability to detect CTL in secondary cultures from immunized mice (W.W.O., unpublished results). Though their importance is unclear at the moment, these observations may aid in our understanding of the interplay between T cells, APC, and cytokines during and following immunization.

The Addition of Leader Sequences to Peptides

The promising results of peptide-based vaccines in melanoma patients warrant their continuing study and improvement. In mice, immunization with hgp$100_{25\text{-}33}$ in IFA induced mgp100- specific, $CD8^+$ T cells, yet at levels lower than those induced with rVVhgp100. Furthermore, mice bearing pulmonary or subcutaneous B16 melanoma did not experience any therapeutic effect from vaccination, even upon addition of IL-2 or IL-12 (our unpublished results). To enhance induction of T cell responses, we employed a strategy developed previously using the "foreign" antigen, OVA. Addition of an 18 amino acid signal sequence from the adenoviral protein E3/19K (ES signal sequence) to the amino-terminus of an immunodominant 8 amino acid OVA peptide, had been shown to dramatically increase the induction of cytotoxic $CD8^+$ T cells upon immunization with the fusion construct in IFA.[32]

In direct accordance with these results, addition of the ES signal sequence to the hgp$100_{25\text{-}33}$ peptide significantly enhances its ability to induce T cell responses in mice compared to immunization with the hgp$100_{25\text{-}33}$ nonamer alone. More unexpected, addition of the ES signal sequence to the mgp$100_{25\text{-}33}$ peptide resulted in an equally strong $CD8^+$ T cell response to a previously non-immunogenic "self" peptide. Furthermore, immunization with the ES-hgp$100_{25\text{-}33}$ peptide, as well as with the ES-mgp$100_{25\text{-}33}$ peptide, resulted in significant inhibition of subcutaneous tumor growth in mice bearing established subcutaneous B16 melanoma (Overwijk et al, manuscript in preparation). Currently, the first patients are being treated with ES-hgp$100_{209\text{-}217}$-2M. Preliminary in vitro assays reveal successful induction of $CD8^+$ T cell responses in these patients, though the small number of patients does not yet permit an estimation of the efficacy of the ES-peptide construct (our unpublished observations).

It remains unclear exactly how the addition of an ES signal sequence to a MHC Class I restricted peptide enhances the induction of specific T cells. One trivial explanation could be that the ES signal sequence functions as a "helper" epitope for $CD4^+$ T cells. However, previous work showed that the improved induction of $CD8^+$ T cells using an ES-OVA peptide vaccine was retained in MHC Class II knockout mice, as well as in mice depleted of $CD4^+$ T cells by injection with anti-CD4 antibody.[32]

The natural function of the ES signal sequence is to efficiently direct the transport of the adenoviral E3/19K protein into the ER at the moment it emerges from the ribosome during translation. The ES signal sequence binds to the signal recognition particle (SRP), and translation temporarily halts as the complex of unfinished polypeptide is directed towards an oligomeric protein pore complex that mediates the actual translocation across the ER membrane. Translation then recommences and the growing polypeptide is threaded into the ER, where a signal peptidase removes the N-terminal signal sequence. A second, less well defined ER-targeting pathway exists where fully synthesized (poly)peptides can bind directly to the pore complex in a signal sequence-dependent manner. The peptide is then directed through the pore complex into the ER and the signal sequence cleaved off by signal sequence peptidase. In case of short peptides, these then may become available for direct loading onto MHC Class I. Experiments are underway to determine whether other N-terminal extensions can enhance the generation of $CD8^+$ T cells to peptide determinants. Furthermore, preliminary data suggest that addition of an ES signal sequence to peptides decreases their susceptibility to proteolytic cleavage, possibly by retarding their degradation by N-terminal aminopeptidases ubiquitously present in sera and inside cells (our unpublished results).

Induction of CD4⁺ T Cell "Help"

It is not known why anti-tumor reactivities are often directed at "self" antigens in melanoma patients. It has been noted that many of the molecular targets for this reactivity are generally resident in the melanosome.[2] These intracellular organelles are the sites where pigment biosynthesis occurs. Importantly, they are biochemically related to the acidic, proteolytic, "compartment for peptide loading", where class II molecules are complexed with peptides.[3] Immunohistochemical analysis of the borders of progressively spreading skin lesions in patients with vitiligo reveals a marked local upregulation of MHC class II expression and IL-2 production, as well as infiltration of T-lymphocytes and phagocytes.[33-35] Upon treatment of melanoma patients with interleukin-2 (IL-2), approximately 20% of responding melanoma patients, but none of responding renal cancer patients, developed vitiligo.[36] The relationship, if any, between vitiligo and tumor regression has not been elucidated, but it is interesting that IL-2, one of the prime cytokines secreted by $CD4^+$ T lymphocytes, along with antigen on the tumor, can induce autoimmune vitiligo while mediating tumor regression. It is conceivable that the inclusion of $CD4^+$ T cell "helper" epitopes in peptide vaccines may improve their efficacy.

$CD4^+$ T cells can dramatically enhance $CD8^+$ T cell responses to antigens in animal models and humans, and a hallmark of many types of autoimmune disease in mouse and man is the critical role played by $CD4^+$ T lymphocytes.[37-40] Autoimmune diseases such as experimental allergic encephalomyelitis (EAE), systemic lupus erythematosus (SLE) and diabetes can often be transferred to naive mice with purified, "self" reactive $CD4^+$ splenocytes or specific $CD4^+$ T lymphocyte clones. Conversely, active disease can sometimes be suppressed with $CD4^+$ T cell populations.[41-45] In general, a majority of experimental autoimmune diseases is characterized by a pronounced dependency on $CD4^+$ T lymphocytes. Several recent reports indicate critical roles for $CD4^+$ T cells in the anti-tumor response to specific antigens as well, extending classic studies where $CD4^+$ T cells were shown to be of major importance to the anti-tumor immune response in less defined systems.[46-48] In addition, we and others have demonstrated the essential role of MHC Class II and $CD4^+$ T cells in immunity to a "self" protein, the MDA TRP-1.[12,49] Vaccination with rVV encoding murine TRP-1 results in autoimmune depigmentation (vitiligo), and protection against B16 melanoma in a $CD4^+$ T cell-dependent manner.[12] Together, these observations stress the importance of $CD4^+$ T cells in the anti-tumor response, and current efforts are aimed at identifying MHC Class II restricted T cell epitopes in murine and human MDA, as well as strategies to induce effective $CD4^+$ T cell responses in vivo.

Molecularly Defined Adjuvants

Mouse models remain the best pre-clinical means of identifying new adjuvants capable of augmenting peptide vaccination. For example, IL-2, IL-12 and GM-CSF have all been shown to increase the magnitude of virus- and peptide-induced, antigen-specific T cell responses in mice.[7,50-52] Other candidate molecules with adjuvant activity include costimulatory molecules, such as B7-1 (CD80) and B7-2 (CD86), both of which can be induced on APC. These molecules are capable of activating or inhibiting T cell function, depending on their interaction with CD28 or CTLA4, respectively.[17] Indeed, blockade of CTLA4 has been reported to potentiate immune responses to tumor cells.[53]

Other molecules currently under investigation are CD40L, a molecule found on activated T cells that is capable of activating APC, inducing them to increase the expression of costimulatory molecules and secrete IL-12. We have found that the addition of CD40L trimer can enhance antitumor efficacy of DNA vaccination.[54] A different strategy to increase presentation of peptide vaccines includes the administration of FLT3 ligand, which induces proliferation and differentiation of dendritic cells, and has been shown to enhance anti-tumor immunity in a tumor model with undefined antigen specificity. Our preliminary experiments suggest that administration of FLT3 also enhances the effects of antigen-specific peptide vaccination, presumably through the increased number and/or activity of dendritic cells in treated animals.

Observations using experimental animals do not necessarily translate to efficacy in the clinic. For example, when patients were immunized with the hgp100$_{209\text{-}217}$-2M peptide, addition of IL-2 enhanced the clinical anti-tumor response. In contrast, the addition of IL-12 and GM-CSF in the clinical setting did not measurably increase the priming of CD8$^+$ T cell responses, nor did it increase the rate of clinical responses, even though each of these cytokines had shown beneficial effects in mice. Clearly, more predictive animal models would greatly aid the discovery and development of new adjuvants and vaccine strategies.

Tumor Escape

Despite the recent advances in treating melanoma patients with peptide vaccines, a significant portion of patients do not respond favorably to treatment. In those patients that do respond in a less than complete way, responses can be short-lived. One reason for this observation may be the phenomenon of tumor escape. In patients treated with chemotherapy, tumor cells often adapt or mutate to become resistant to previously effective treatment, for example by inducing the expression of the multidrug resistance gene. Likewise, tumor cells could develop mechanisms to escape a potentially therapeutic immune response. Recent reports suggest that immunologic pressure may exist that can lead to systematic selection for tumor cells that somehow evade recognition by the immune system.[55,56] One mechanism by which tumor cells could escape immune destruction is by inducing the death of lymphocytes that infiltrate the tumor mass.

Expression of FasL/CD95L by tumor cells has been proposed as a means by which tumor cells destroy activated T cells that bear FAS/CD95 on their surface.[57] However, recent results from our laboratory indicate that none of 19 human melanoma lines tested killed the FAS$^+$ targets in a sensitive functional assay, and none of 26 melanoma lines expressed FasL mRNA as evaluated by reverse-transcription polymerase chain reaction.[58] Still, tumor cells clearly can diminish the effects of an anti-tumor immune response, for example through decreased expression of the molecules that mediate their recognition by T cells, such as β_2-microglobulin and MHC Class I, TAP-1, as well as the ability to process antigens.[59,60] More recently, biopsies of remaining tumor nodules after partially successful treatment with gp100 peptide vaccine revealed a selectively reduced expression of gp100, but not of other MDA, suggesting that immunologic selection for antigen-loss variants does take place in vivo (A.I. Riker, F.M. Marincola, Unpublished data).[59,61,62]

Tumor escape poses a significant problem for virtually all non-surgical therapies of cancer, including immunotherapy. However, the variety of immune mechanisms that can potentially target tumor cells provides hope for the treatment of tumors that have become resistant to one form of immune intervention. For example, tumor cells that have lost MHC Class I expression may become more susceptible to lysis by Natural Killer (NK) cells, through decreased engagement of killer inhibitory receptors on NK cells by MHC Class I on tumor cells.[63,64] Furthermore, tumor cells that have lost the expression of one antigen, for example gp100, often still express other MDA which may be targeted with peptide vaccine therapy.[62] It remains therefore important, despite the increasing number of identified human tumor antigens, to continue the search for new tumor antigens, especially those frequently expressed in a variety of tumors.

Conclusions

Great progress has been made in the identification of antigens recognized by T cells not only in melanoma but in a variety of other cancers as well. Knowledge of the peptide sequence together with a clearer understanding of the interaction between a peptide antigen and MHC class I or class II molecules, makes it possible to manipulate peptides in order to improve binding affinity and stability. Studies in experimental animals are continuing to increase our understanding of the immune response to "self" tumor antigens, and allow the direct testing of a variety of experimental vaccines that are currently under development. Future efforts in the development of peptide vaccine development will likely include the identification of additional murine "self" tumor antigens, the use of molecularly defined adjuvants to enhance peptide vaccine therapy, and an increased understanding of the fate of "self"-reactive T cells in the tumor bearing host.

References

1. Pardoll DM. Cancer vaccines. Nat Med 1998; 4:525-531.
2. Restifo NP. Cancer vaccines '98: A reductionistic approach. Mol Med Today 1998; 4:327
3. Boon T, Old LJ. Cancer Tumor antigens. Curr Opin Immunol 1997; 9:681-683.
4. Rosenberg SA. Cancer vaccines based on the identification of genes encoding cancer regression antigens. Immunol Today 1997; 18:175-182.
5. Restifo NP. The new vaccines: Building viruses that elicit antitumor immunity. Curr Opin Immunol. 1996; 8:658-663.
6. Irvine KR, Rao JB, Rosenberg SA et al. Cytokine enhancement of DNA immunization leads to effective treatment of established pulmonary metastases. J Immunol 1996; 156:238-245.
7. Bronte V, Tsung K, Rao JB et al. IL-2 enhances the function of recombinant poxvirus- based vaccines in the treatment of established pulmonary metastases. J Immunol 1995; 154:5282-5292.
8. Carroll MW, Overwijk WW, Surman DR et al. Construction and characterization of a triple-recombinant vaccinia virus encoding B7-1, interleukin 12, and a model tumor antigen. J Natl Cancer Inst 1998; 90:1881-1887.
9. Overwijk WW, Tsung A, Irvine KR et al. gp100/pmel 17 is a murine tumor rejection antigen: Induction of "self"- reactive, tumoricidal T cells using high-affinity, altered peptide ligand. J Exp Med 1998; 188:277-286.
10. Mamula MJ. Lupus autoimmunity: from peptides to particles. Immunol Rev 1995; 144:301-314.
11. Fong L, Ruegg CL, Brockstedt D, et al: Induction of tissue-specific autoimmune prostatitis with prostatic acid-phosphatase immunization—implications for immunotherapy of prostate-cancer. J Immunol 1997; 159:3113-3117.
12. Kim CJ, Prevette T, Cormier J, et al. Dendritic cells infected with poxviruses encoding MART-1/ Melan A sensitize T lymphocytes in vitro. J Immunother 1997; 20:276-286.
13. Tsang KY, Zaremba S, Nieroda CA et al. Generation of human cytotoxic T cells specific for human carcinoembryonic antigen epitopes from patients immunized with recombinant vaccinia-CEA vaccine. J Natl Cancer Inst 1995; 87:982-990.
14. Cormier JN, Salgaller ML, Prevette T et al. Enhancement of cellular immunity in melanoma patients immunized with a peptide from MART-1/Melan A. Cancer J Sci Am 1997; 3:37-44.
15. Zaks TZ, Rosenberg SA. Immunization with a peptide epitope (p369-377) from HER-2/neu leads to peptide-specific cytotoxic T lymphocytes that fail to recognize HER- 2/neu+ tumors. Cancer Res 1998; 58:4902-4908.
16. Goydos JS, Elder E, Whiteside TL et al. A phase I trial of a synthetic mucin peptide vaccine. Induction of specific immune reactivity in patients with adenocarcinoma. J Surg Res 1996; 63:298-304.
17. Van Parijs L, Abbas AK. Homeostasis and self-tolerance in the immune system: Turning lymphocytes off. Science 1998; 280:243-248.
18. Sant'Angelo DB, Waterbury PG, Cohen BE, et al. The imprint of intrathymic self-peptides on the mature T cell receptor repertoire. Immunity 1997; 7:517-524.
19. Hu Q, Bazemore WC, Girao C et al. Specific recognition of thymic self-peptides induces the positive selection of cytotoxic T lymphocytes. Immunity. 1997; 7:221-231.

20. Wallace PM, Rodgers JN, Leytze GM et al. Induction and reversal of long-lived specific unresponsiveness to a T-dependent antigen following CTLA4Ig treatment. J Immunol 1995; 154:5885-5895.
21. Schwartz RH. T cell clonal anergy. Curr Opin Immunol 1997; 9:351-357.
22. Kurts C, Heath WR, Carbone FR et al. Constitutive class I-restricted exogenous presentation of self antigens in vivo. J Exp Med 1996; 184:923-930.
23. Carbone FR, Kurts C, Bennett SR et al. Cross-presentation: a general mechanism for CTL immunity and tolerance. Immunol Today 1998; 19:368-373.
24. Zhai Y, Yang JC, Spiess P et al. Cloning and characterization of the genes encoding the murine homologues of the human melanoma antigens MART1 and gp100. J Immunother 1997; 20:15-25.
25. Mamula MJ, Lin RH, Janeway CAJ, et al: Breaking T cell tolerance with foreign and self co-immunogens. A study of autoimmune B and T cell epitopes of cytochrome c. J Immunol 1992; 149:789-795.
26. Parker KC, Bednarek MA, Coligan JE. Scheme for ranking potential HLA-A2 binding peptides based on independent binding of individual peptide side-chains. J Immunol 1994; 152:163-175.
27. Parker, K. C. HLA peptide binding predictions. http://bimas.dcrt.nih.gov/molbio/hla_bind .1996.
28. Rosenberg SA, Zhai Y, Yang JC, et al: Immunizing patients with metastatic melanoma using recombinant adenoviruses encoding MART-1 or gp100 melanoma antigens. J Natl Cancer Inst 1998; 90:1894-1900.
29. Parkhurst MR, Salgaller ML, Southwood S et al. Improved induction of melanoma- reactive CTL with peptides from the melanoma antigen gp100 modified at HLA-A*0201-binding residues. J Immunol 1996; 157:2539-2548.
30. Rosenberg SA, Yang JC, Schwartzentruber DJ et al. Immunologic and therapeutic evaluation of a synthetic peptide vaccine for the treatment of patients with metastatic melanoma. Nat Med 1998; 4:321-327.
31. Bronte V, Wang M, Overwijk WW et al. Apoptotic death of CD8+ T lymphocytes after immunization: induction of a suppressive population of Mac-1+/Gr-1+ cells. J Immunol 1998; 161:5313-5320.
32. Minev BR, McFarland BJ, Spiess PJ et al. Insertion signal sequence fused to minimal peptides elicits specific $CD8^+$ T-cell responses and prolongs survival of thymoma-bearing mice. Cancer Res 1994; 54:4155-4161.
33. Le Poole IC, van den Wijngaard RM, Westerhof W et al. Presence of T cells and macrophages in inflammatory vitiligo skin parallels melanocyte disappearance. Am J Pathol 1996; 148:1219-1228.
34. Badri AM, Todd PM, Garioch JJ et al. An immunohistological study of cutaneous lymphocytes in vitiligo. J Pathol 1993; 170:149-155.
35. Hann SK, Park YK, Lee KG et al. Epidermal changes in active vitiligo. J Dermatol 1992; 19:217-222.

Chapter 13

Gp100 and G250: Towards Specific Immunotherapy Employing Dendritic Cells in Melanoma and Renal Cell Carcinoma

Joost L.M.Vissers, I. Jolanda M. de Vries, Egbert Oosterwijk, Carl G. Figdor and Gosse J. Adema

Summary

A long history of studies demonstrate the capacity of the immune system to develop specific reactivity against antigens foreign to the host, like viral and bacterial antigens. During the last decade it is becoming more and more apparent that an immune response can also be mounted against tumor-associated antigens. The molecular identification of several tumor antigens has boosted research to address means by which long lasting T cell-mediated immunity against tumors can be induced. In this Chapter, we focus on the identification of the tumor-associated antigens gp100 and G250 and recent developments in the application of dendritic cell-based vaccines.

Immunogenicity Of Human Tumors

During the last decades evidence has accumulated indicating that the immune system can play a significant role in the defense against tumors in man.[1,2] In a small group of cancer patients 'spontaneous' partial or complete remissions have been observed due to the action of the immune system.[3-6] Melanoma and renal cell carcinoma (RCC) belong to this group, and nonspecific- or specific immunotherapy may increase the reactivity of the immune system against these tumors. Melanoma is an aggressive, rapidly metastasizing tumor that arises from melanocytes, the pigment producing cells in the skin and eye. Primary treatment for melanoma is surgery, but additional therapy is needed since 80% of the patients develop metastatic disease. RCC accounts for about 2% of all adult cancers and is the most frequently occurring malignant tumor involving the kidney.[7] In the past years, the incidence of RCC has increased dramatically (54% from 1975 to 1990). RCC originates from the kidney parenchyma and the majority has a typical clear-cell histology.[8] As in melanoma, surgery is the standard treatment for primary disease. In time, however, 30% of the RCC patients develop metastases in lungs, bone and brain.[9] Recent studies implicate that tumors of other origins can be immunogenic as well e.g., prostate cancer, head and neck cancer and colon carcinomas.[10] In this Chapter, we will focus on melanoma and RCC which are generally considered to be immunogenic tumors in man.

Immunogenicity of Melanoma and Renal Cell Carcinoma

Beside the occurrence of rare spontaneous tumor regressions, the immunogenicity of melanoma and RCC is supported by the infiltration of a variety of lymphocytes and mononuclear phagocytes. T cells, macrophages, dendritic cells (DC), B cells and natural killer (NK) cells can be found in these tumors.[1,2,10,11] Especially the analysis of tumor infiltrating lymphocytes (TIL) provided more direct evidence for the existence of tumor reactive cytotoxic T lymphocytes (CTL). Melanoma-derived TIL were shown to react specifically with tumor cells but not with normal cells derived from the same individual.[1,2,12,13] In addition, the concept of common melanoma antigens, tumor-reactive CTL isolated from one individual that cross react with tumor cells from other patients expressing the same presenting HLA molecule, was based on studies involving melanoma-derived TIL.[1,2,14] The observation that some melanoma-derived TIL not only lysed melanoma-derived tumor cells but also their normal counterpart the melanocyte, suggested that melanocyte differentiation antigens could be targets for such CTL.[13,15-18] Moreover, these autoreactive CTL could possibly explain the occurrence of vitiligo (skin depigmentation) observed in some patients in which an anti-tumor immune response was ongoing.[19-21] Although most information has been obtained in melanoma, a similar picture is emerging for RCC.[22-25] Schendel et al characterized a CD8+ CTL line obtained from a RCC-derived TIL, which kills the autologous RCC but not normal kidney cells.[26] Interestingly, T cell receptor (TCR) analysis has demonstrated that both RCC and melanoma-derived CTL can persist in the periphery for many years after removal of the primary tumor.[27-30] This finding suggests an active role of these T cells in the immune response against the tumor.

Beside T cell mediated reactivity against the tumor, tumor-specific antibodies have been isolated from melanoma patients as well as other cancer patients.[31-33] These findings indicate that a humoral immune response, which requires CD4+ T cells and B cells, against tumors can be generated as well.

Based on the aforementioned observations, immunotherapy aiming at enhancing the anti-tumor reactivity of T cells or NK cells by providing cytokines such as IFNs, TNF-α and IL-2, has been and still is used in clinical trials. Especially high dose IL-2 treatment has been demonstrated to induce anti-tumor reactivity in melanoma and RCC with response rates of 17-25%, further emphasizing the potential of the immune system to fight cancer.[34] An alternative approach involves the ex vivo expansion of autologous TIL with IL-2 and subsequent re-infusion of these TIL into patients. Although melanoma-derived TIL are more easy to expand with IL-2 in vitro as compared to TIL from RCC, this approach has been followed in clinical trials in both melanoma and RCC patients.[34-37] The most effective results have been obtained in patients with melanoma (15-30%).[38] As will be discussed below, both TIL and peripheral blood-derived CTL have been instrumental to the identification of tumor-associated antigens expressed by melanoma and RCC.

Identification of Tumor-Associated Antigens in Melanoma and Renal Cell Carcinoma

Approaches to Identify Antigens Recognized by Tumor-Reactive T Cells

The availability of TIL and peripheral blood-derived CTL with anti-tumor reactivity boosted research aiming at the identification of the antigens recognized. Different approaches have been applied in search for these antigens: 1) expression cloning, 2) peptide elution, and 3) reversed immunology.[39] In the first approach, cells transfected with a tumor cell-derived cDNA library together with the cDNA encoding the presenting HLA molecule are screened as targets for tumor-reactive T cells.[40] In the peptide elution approach, peptides are eluted from tumor cell's MHC class I molecules, separated by high-performance liquid chromatography (HPLC) and subsequently used to screen CTL lines. The amino acid sequence of the peptides is subsequently determined by tandem mass spectroscopy.[41] In the third method, potential MHC class I binding epitopes of antigens are selected using computer scoring programs based on MHC class I binding motifs. The predicted peptides are synthesized and tested for their capacity to bind MHC class I and to induce CTLs in vitro and in vivo.[42-44] For this latter approach, a novel source of potentially interesting antigens comes from the serological analysis of tumor cell-derived cDNA expression libraries (SEREX).[32,33,45] Using reversed immunology, it was recently shown that existing or newly generated T cell clones react with SEREX-defined antigens, like NY-ESO-1 and HOM-MEL-40/SSX2 (de Vries et al, submitted).[46] As the SEREX methodology has recently been successfully applied for other tumor types as well, this approach is potentially valuable to define target antigens in those tumors types for which no CTL are available.

Categories of Tumor-Associated Antigens

Especially in melanoma, the aforementioned approaches resulted in the identification of a still growing number of tumor antigens. The characterized tumor antigens can be divided into three groups (Table 13.1). The first group, the melanocyte differentiation antigens, are expressed in tumors as well as in their normal counterpart, the melanocyte. Members of this group, like gp100, Melan A/MART-1 and tyrosinase, are expressed in a high percentage of melanoma tumors, but not in tumors of other origins.[13,16,17,47] The second group comprises so-called shared tumor antigens expressed in a wide variety of tumors of different origins. To illustrate this, tumor antigens BAGE, GAGE and MAGE are expressed in several distinct tumor types but not in normal tissue, except testis. Therefore, this particular group of antigens is often called cancer testis antigens.[40,48,49] The third group consists of mutated antigens, unique for each individual tumor such as a tumor-specific point mutation found in cyclin dependent kinase-4 (CDK-4) giving rise to a novel antigenic epitope.[50]

In case of RCC, only a few of the antigens have been elucidated. Strikingly, the shared tumor antigens identified in melanoma and expressed in multiple other tumor types are not expressed in RCC. The first RCC-specific antigen, called RAGE-1, was defined by expression cloning.[53] The RAGE-1 gene is expressed in less than 2% of primary RCC and is silent in normal tissue, except retina. A second RCC antigen identified belongs to the group of mutated antigens and consists of a mutated HLA-A2 protein.[54] Recently, the RCC-associated antigen recognized by a monoclonal antibody (mAb) called G250 was identified. This novel antigen, termed G250, is expressed in RCC but not in normal kidney tissue and may therefore be a potential target for RCC directed immunotherapy (Oosterwijk et al, submitted).[55] In the remainder of this Chapter, we will focus on the melanoma-associated antigen gp100 and the newly identified RCC antigen G250.

The Melanoma-Associated Antigen gp100

The melanoma-associated antigen gp100 is recognized by mAb NKI-beteb, HMB-45 and HMB-50, and is most commonly used to diagnose malignant melanoma.[56,57] Molecular cloning and subsequent analysis of the gp100 cDNA, demonstrated that gp100 is an intracellular type I transmembrane molecule highly homologous to Pmel17.[14,58] Analysis of the gp100 gene, which is localized on chromosome 12, indicated that both gp100 and Pmel17 transcripts are encoded by one single gene and arise by alternative splicing.[58,59] Pmel17 and gp100 are also simultaneously expressed in malignant and normal melanocytic cells and are located in melanosomes. They most likely play a role in the biosynthesis of melanin, either as enzymes or as structural components of melanosomes, but their exact biological function remains to be elucidated.[57,60]

Table 13.1. Three groups of tumor-associated antigens and some of their members as identified in melanoma

	Antigen	Melanocyte	Melanoma	References
Differentiation antigens	gp100	yes	yes	12,41
	Melan A/MART-1 1	yes	yes	17,18
	tyrosinase	yes	yes	16
Shared tumor antigens	BAGE	no	yes	48
	MAGE-1	no	yes	40
	MAGE-3	no	yes	51,52
Mutated antigens	CDK-4	no	yes	50

Screening of T cells from melanoma patients, revealed that several T cells do recognize peptides from the gp100 antigen in an HLA-A2.1 restricted manner.[13,41,61,62] As demonstrated for the melanoma-derived TIL 1200, these TIL lyse melanoma cells as well as normal melanocytes expressing gp100.[13] Using transfectants expressing truncated gp100 cDNAs, a peptide comprising amino acids 154-162 of the gp100 protein was identified as the dominant epitope recognized by TIL1200.[63] Based on the analysis of many TILs and blood-derived CTL by multiple laboratories, it has been shown that peptides derived from gp100 are among the most frequent epitopes recognized by anti-melanoma CTLs (Table 13.2). Importantly, the observation that infusion of TIL with anti-gp100 reactivity in combination with IL-2 can result in tumor regression in melanoma patients makes gp100 a promising target for anti-melanoma immunotherapy.[47,64]

The Renal Cell Carcinoma-Associated Antigen G250

The RCC-associated antigen G250 is a membrane-associated antigen defined by monoclonal antibody G250 (mAb G250).[55] mAb G250 reacts with approximately 85% of RCC, but does not cross-react with normal kidney tissue. Furthermore, G250 expression can be detected in sarcomas and colon-, ovarian- and cervical carcinomas. Expression of G250 in normal tissue is limited to gastric mucosal cells and cells of the larger bile ducts.[55,67] Clinical studies in RCC patients demonstrated exclusive targeting of the mAb G250 to RCC.[68] Therefore, the G250 antigen is the first broadly expressed RCC-associated antigen, and may constitute an interesting novel target for specific immunotherapy in RCC patients.

The recent cloning of the cDNA encoding the G250 antigen, indicate that the G250 protein is homologous to the recently characterized tumor-associated antigen MN (Oosterwijk et al, submitted).[67] We have recently defined a number of peptides derived from the G250 antigen that bind to HLA-A2.1 and are currently testing their immunogenicity by inducing CTL responses using DC in vitro as well as by vaccination of HLA-A2.1/Kb transgenic mice in vivo. Preliminary data suggest that at least one G250-derived peptide gives rise to CTL in HLA-A2.1/Kb transgenic mice and in man that lyse G250 expressing target cells indicating that this peptide is endogenously processed and presented.

Table 13.2. Gp100-derived epitopes recognized by T cells

Position	Epitope	MHC	References
17-25	ALLAVGATK	HLA-A3.1	65
154-162	KTWGQYWQV	HLA-A2.1	63
177-186	AMLGTHTMEV	HLA-A2	66
178-186	MLGTHTMEV	HLA-A2	66
209-217	ITDQVPGSV	HLA-A2.1	64
280-288	YLEPGPVTA	HLA-A2.1	41,62
457-466	LLDGTATLRL	HLA-A2.1	61
476-486	VLYRYGSFSV	HLA-A2.1	64
570-579	SLADTNSLAV	HLA-A2	66

Dendritic Cells

The availability of well characterized tumor-associated antigens allows specific targeting of the immune system towards the tumor. In addition, it allows proper monitoring of the immune response induced in vivo, which is essential to validate and optimize novel vaccination protocols. Multiple vaccination strategies can be envisaged, e.g., vaccination with immunogenic peptides or proteins with or without adjuvant, naked DNA/RNA or viruses encoding tumor-associated antigens, as well as vaccination with tumor antigen-loaded DC.[2,69] As the latter approach is explored in our department, we will focus on the use of tumor antigen-loaded DC.

Dendritic Cells are Professional Antigen Presenting Cells

DC constitute a family of antigen presenting cells (APC) defined by their morphology and their unique capacity to attract and interact with naive T cells to initiate a primary immune response.[70-73] Multiple studies have indicated that DC are superior in presenting antigen to unprimed resting T lymphocytes when compared with other APC, such as B cells, monocytes and macrophages.[70,72,74] DC are bone marrow-derived cells and DC progenitors are seeded through the blood into non-lymphoid tissues, where they develop into immature DC.[70,72] Such immature DC are very efficient in antigen uptake, mediated by high endocytotic activity and expression of a variety of cell surface receptors capable of capturing antigens.[75-77] Rerouting of DC from the non-lymphoid tissues into the secondary lymphoid organs is promoted by inflammatory mediators. In the secondary lymphoid tissues, DC mature and become fully stimulatory DC, having lost their capacity to capture and process antigen but well equipped to initiate a primary immune response (Figure 13.1). Importantly, depending on the type of DC, their maturation stage and environmental factors it has been suggested that DC can also have tolerizing effects. In addition, there is also evidence for the existence of a distinct lymphoid DC subset, which might play a role in tolerance induction.[78-82] Clearly, the application of DC-based vaccines in cancer therapy requires detailed analysis of these distinct DC-subsets and their biological function.

In Vitro Generation of Dendritic Cells

Research involving DC has been hampered for a long time by their low abundance in peripheral blood and the absence of specific markers. However, besides the conventional methods,[83]

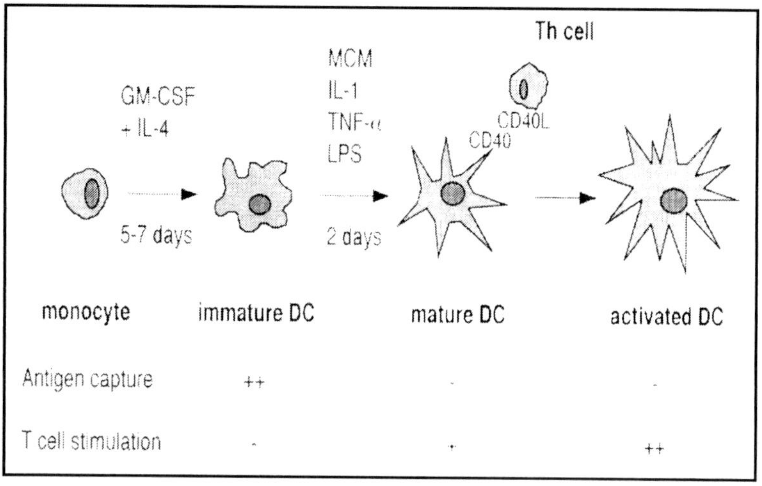

Fig. 13.1. Different maturation stages of monocyte-derived DC cultured in vitro.

several methods have now been described to generate significant numbers of DC in vitro. Starting with either bone marrow or blood cells, murine DC can be generated following culture in the presence of GM-CSF.[84-86] In humans, a common progenitor for DC and myeloid cells also exists which is dependent on GM-CSF.[86-88] Human cord blood-derived $CD34^+$ progenitor cells cultured with GM- CSF and TNF-α develop into two distinct precursor DC subpopulations.[88] DC can also develop from $CD14^+$ peripheral blood monocytes, but this process requires the presence of IL-4 in addition to GM-CSF and occurs in the absence of proliferation.[89-91] Monocyte-derived DC can be induced to mature further by inflammatory stimuli like TNF-α, IL-1, LPS or monocyte conditioned medium.[76,92,93] The subsequent engagement of mature DC with $CD4^+$ T helper cells via the CD40 pathway leads to increased expression of co-stimulatory molecules resulting in further activation of the DC (Figure 13.1).[94-96] The possibility to generate sufficient numbers of DC has boosted the use of these otherwise rare cells in clinical studies. Currently, several DC-based vaccination trials have been initiated in which ex vivo generated DC loaded with tumor-associated antigens are re-infused in the autologous tumor-bearing patient (Figure 13.2).

Towards Dendritic Cell-Based Vaccines

The goal of DC-based vaccination is the induction of an effective anti-tumor immune response, which is long lasting and generates long-lived memory. In murine tumor models DC loaded with tumor antigens are capable of inducing protective immunity as well as regression of established tumors in some models.[97-99] However, critical questions regarding the effect of the source of DC and their maturation stage on DC trafficking and their potency to induce anti-tumor immune response in vivo should be addressed to fully exploit the potential of DC-based vaccines.[100]

Dendritic Cell Subset and Maturation Stage

It is now well recognized that multiple DC subsets exist and that there are multiple ways to generate and mature DC in vitro. Two predominant sources of DC exploited in clinical studies are CD34+ progenitor-derived DC and monocyte-derived DC as they are most readily generated in vitro. We and others have demonstrated that monocyte-derived DC and bone

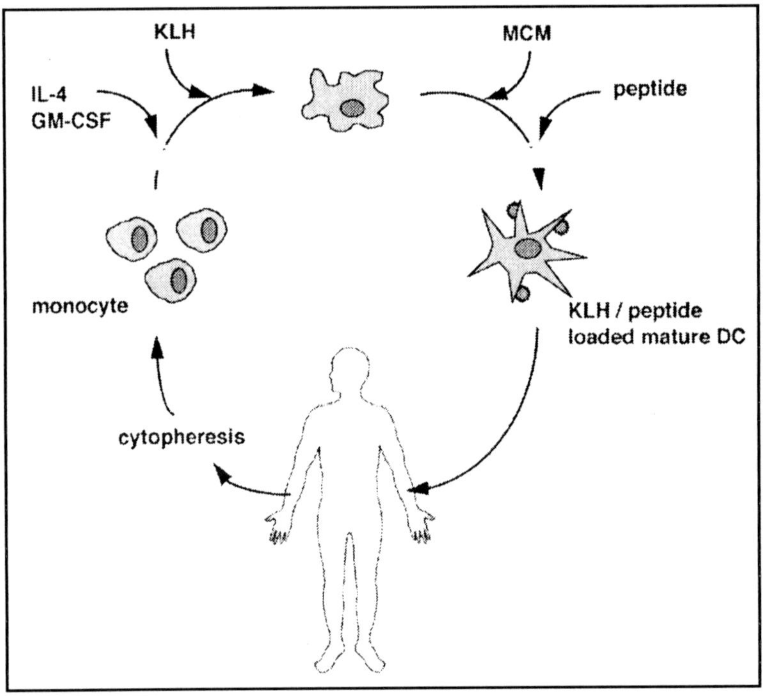

Fig. 13.2. Schedule of a DC vaccination procedure. The ex vivo generation, maturation and loading of monocyte-derived DC is exemplified by our currently used DC vaccination protocol.

marrow-derived DC loaded with the HLA binding peptides derived from e.g., melanocyte differentiation antigens are able to induce anti-tumor CTL in vitro (Figure 13.3).[44,46,66,101,102] However, which DC source is most effective in the therapeutical setting in vivo as well as the effect of their maturation stage has to be taken into account. It has clearly been documented that monocyte-derived DC cultured with GM-CSF and IL-4 for five to seven days in vitro require further stimulation to become stable mature DC, expressing increased levels of MHC and co-stimulatory molecules as well as the DC maturation marker CD83 (Figure 13.4).[92,93] Without these stimuli, immature DC revert back to macrophages in vitro, but whether maturation is essential for the in vivo application remains to be determined. In our ongoing clinical studies we have chosen to utilize monocyte-derived DC that have been matured with monocyte-conditioned medium (MCM) to vaccinate melanoma patients.

Antigen Delivery to Dendritic Cell

The availability of tumor-associated antigens allows loading of DC with these antigens or the immunogenic peptides derived thereof. A clear advantage of knowing the target antigen, is that immune responses generated can be readily monitored. We have chosen two HLA-A2.1 restricted peptides derived from gp100 (amino acid 1540-162 and 280-288) and one from tyrosinase (amino acid 369-377).These peptides are loaded separately onto the DC after maturation with MCM just prior to infusion into the patient. In addition to the MHC class I peptides, the DC are also loaded with a foreign protein antigen, keyhole limpet hemocyanin (KLH). KLH loading of DC is performed at day 3 to optimally make advantage of the uptake capacity of DC (Figure 2). The presence of KLH epitopes presented by DC in combination with the tumor antigen-derived peptides is twofold: 1) it will allow activation of the DC via

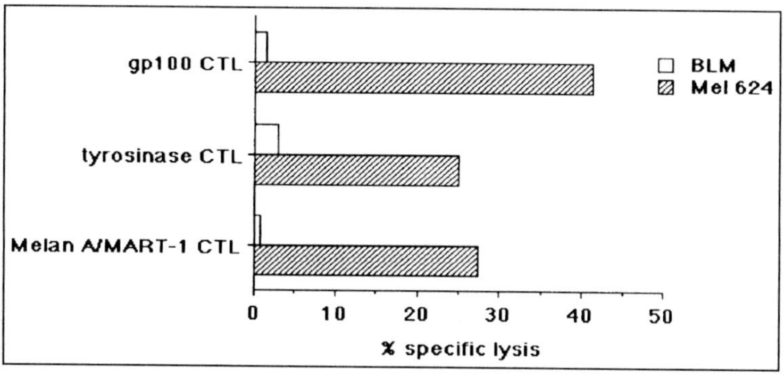

Fig. 13.3. Induction of CTL against the melanocyte differentiation antigens gp100, tyrosinase and Melan A/MART-1 using peptide-loaded DC. The HLA-A2.1 positive melanoma cells BLM (negative for all three antigens) and Mel 624 (endogenously expressing all three antigens) were used as target cells in a cytotoxicity assay.

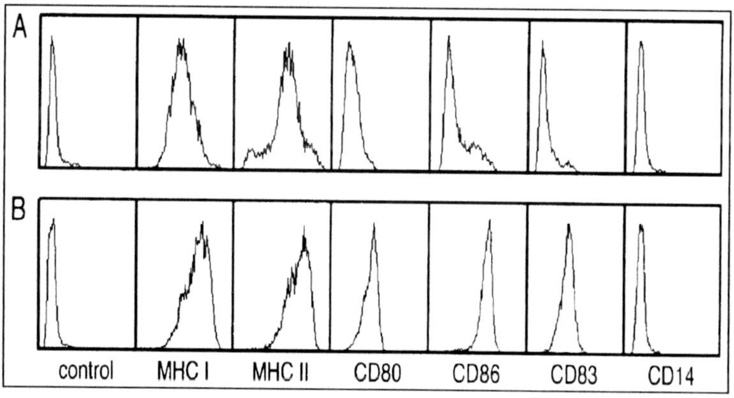

Fig. 13.4. Phenotypic changes in the expression level of MHC, co-stimulatory molecules (CD80 and CD86) and the DC marker CD83 from immature DC (A) to mature DC (B) as induced by monocyte-conditioned medium.

CD40-CD40 ligand interaction with KLH reactive helper T cells. and 2) the KLH-specific helper T cells provide essential cytokines for the development of cytotoxic T cells.[94-96,103-105] In time, MHC class II epitopes derived from antigens expressed by the tumor itself may be used for this purpose as this has recently been shown to improve the ongoing immune response in mouse models.[106]

A crucial aspect of peptide-loaded DC as a vaccine is that the life-span of the MHC class I-peptide complex is sufficiently long to allow full activation of responding T cells. Next to the binding characteristics of the peptide, the life span of the MHC-peptide complex will also be affected by the capacity of the DC to stabilize the MHC molecule on its cell surface. For MHC class II molecules it has recently been shown that the half life of MHC class II molecules is dramatically increased upon DC maturation, but optimal conditions for the stabilization of MHC class I still have to be defined.[107,108] A strategy we are currently exploring, is the use of DC loaded with modified peptide epitopes. Starting with melanocyte differentiation antigen-derived epitopes, we identified peptides analogues with increased MHC binding affinity as well as increased capacity to sensitize target cells. Importantly, CTL raised by DC loaded with these modified peptides specifically lysed melanoma cells which endogenously expressed the wild type epitope.[109-111] Recently, it was reported in mouse models as well as in a clinical study using modified peptides in adjuvant, that peptide analogues can have a beneficial effect in the induction of an immune response.[112-114] Next to the increased MHC binding affinity, the observed effects could potentially be explained by the induction of a different spectrum of the T cell repertoire by the peptide analogues and thereby increasing the possibility to mount an effective immune response against the tumor.

Application of Dendritic Cell-Based Vaccines

Another aspect regarding DC-based vaccines that should be considered is how to deliver a DC vaccine. In mouse models, several different administration routes (i.p., i.v. and s.c.) have been used, but no proper comparison of the different routes has been performed as yet. Likely, the route of administration is dependent on DC maturation stage as it has recently been reported that immature and mature DC express a different set of chemokine receptors and hence respond differently to chemo-attractants.[115-117] Using 111-Indium labeled mouse bone marrow-derived DC, we have recently demonstrated that after 48 hours, i.v. injected DC mainly accumulate in spleen whereas s.c. injected DC preferentially home to the T cell areas of the draining lymph nodes. In addition, we were able to demonstrate that these mouse DC, when loaded with a moderate MHC binding peptide, are more effective as a vaccine when given s.c. as compared to i.v. (Eggert et al, submitted). Interestingly, Nestlé and colleagues reported on the delivery of mature monocyte-derived DC directly into lymph nodes under ultrasound guidance and observed impressive responses in some patients.[118] Another application route that is being explored by several laboratories is the direct injection of the DC into the tumor.

Clinical Studies Using Dendritic Cell-Based Vaccines

The first DC-based vaccination trial was reported for B cell lymphoma patients in 1996. The DC utilized were obtained directly from peripheral blood mononuclear cells by a brief culture step and purification using gradient density centrifugation. Isolated DC were loaded with idiotype antigen plus KLH and administered intravenously. All vaccinated patients developed an immune response and one patient underwent a complete tumor regression.[119] Other reports using in vitro generated monocyte-derived DC include a clinical trial in prostate cancer using the prostate specific antigen (PSA) or PMSA as target antigen and studies in melanoma.[118,120,121] Interestingly, in the melanoma study a response rate of 42% was noted in patients receiving matured DC loaded with either MHC class I presented synthetic peptides from known melanoma tumor antigens or autologous tumor lysate. The DC, 1-3 x 10^6 per injection, were administered directly into the lymph node by ultrasound guidance.[118] Our first results using matured monocyte-derived DC pulsed with melanocyte differentiation antigen-derived peptides plus KLH, indicate that up to 6 x 10^6 DC s.c. and 30 x 10^6 i.v. can be safely administered and are well tolerated. Currently, the first three patients are analyzed for the presence of T cell reactivity against the applied peptides using the ELI-spot assay. A positive delayed type hypersensitivity reaction against KLH has been observed in one patient. No clear clinical responses has been observed so far.

Future Prospects

Ongoing studies investigating the optimal source and maturation stage of DC as well as their potency to stabilize MHC class I molecules and migratory capacity will provide insight in how to optimally apply DC-based vaccines clinically. In these studies, monitoring of the immune response will be crucial and will be facilitated by the use of well defined antigenic peptide epitopes. In addition, loading of DC with tumor-associated proteins or transfection of DC with the cDNA encoding such antigens will greatly enhance the applicability of DC-based vaccines as they become independent on HLA restriction. Ultimately, DC-based vaccines will have to be developed in which no prior knowledge regarding the target antigen is required. Several approaches, like transfection of DC with tumor cell-derived RNA,[122] loading DC with apoptotic tumor cells,[123,124] generation of DC-tumor cell hybridomas [125] as well as the expansion of DC in vivo using cytokines like Flt-3 ligand [126] are currently explored to create novel DC-based cancer vaccines for the future.

Acknowledgments

We would like to thank Drs. A. Eggert and M. Schreurs for critically reading the manuscript. We apologize to our colleagues in the field whose work we could not cite here due to space limitations. This work was supported by the Dutch Cancer Foundation.

References

1. Boon T, Coulie PG, van den Eynde B. Tumor antigens recognized by T cells. Immunol Today 1997; 18:267-268.
2. Rosenberg SA. Cancer vaccines based on the identification of genes encoding cancer regression antigens. Immunol Today 1997; 18:175-182.
3. Freed SZ, Halperin JP, Gordon M. Idiopathic regression of metastases from renal cell carcinoma. J Urol 1977; 118:538-542.
4. McGovern VJ. Spontaneous regression of melanoma. Pathology 1975; 7:91-99.
5. Bodurtha AJ, Berkelhammer J, Kim YH et al. A clinical histologic, and immunologic study of a case of metastatic malignant melanoma undergoing spontaneous remission. Cancer 1976; 37:735-742.
6. Marcus SG, Choyke PL, Reiter R et al. Regression of metastatic renal cell carcinoma after cytoreductive nephrectomy. J Urol 1993; 150:463-466.
7. Parker SL, Tong T, Bolden S et al. Cancer statistics, 1997 [published erratum appears in CA Cancer J Clin 1997 Mar-Apr; 47(2):68. CA Cancer J Clin 1997; 47:5-27.
8. Thoenes W, Storkel S, Rumpelt HJ. Histopathology and classification of renal cell tumors (adenomas, oncocytomas and carcinomas). The basic cytological and histopathological elements and their use for diagnostics. Pathol Res Pract 1986; 181:125-143.
9. Ritchie AW, deKernion JB. The natural history and clinical features of renal carcinoma. Semin Nephrol 1987; 7:131-139.
10. Lotze MT, Jaffe R. Cancer In: Lotze MT, Thomson AW, eds. Dendritic cells. San Diego: Academic press, 1999:326-338.
11. Kelly JW, Sagebiel RW, Blois MS. Regression in malignant melanoma. A histologic feature without independent prognostic significance. Cancer 1985; 56:2287-2291.
12. Muul LM, Spiess PJ, Director EP et al. Identification of specific cytolytic immune responses against autologous tumor in humans bearing malignant melanoma. J Immunol 1987; 138:989-995.
13. Bakker ABH, Schreurs MWJ, De Boer AJ et al. Melanocyte Lineage-Specific Antigen gp100 Is Rec ognized by Melanoma-Derived Tumor-Infiltrating Lymphocytes. J Exp Med 1994; 179:1005-1009.
14. Kawakami Y, Zakut R, Topalian SL et al. Shared human melanoma antigens. Recognition by tumor-infiltrating lymphocytes in HLA-A2.1-transfected melanomas. J Immunol 1992; 148:638-643.
15. Anichini A, Maccalli C, Mortarini R, et al. Melanoma cells and normal melanocytes share antigens recognized by HLA-A2-restricted cytotoxic T cell clones from melanoma patients. J Exp Med 1993; 177:989-998.
16. Brichard V, Van Pel A, Wolfel T et al. The tyrosinase gene codes for an antigen recognized by autologous cytolytic T lymphocytes on HLA-A2 melanomas. J Exp Med 1993; 178:489-495.
17. Coulie PG, Brichard V, Vanpel A et al. A new gene coding for a differentiation antigen recognized by autologous cytolytic T lymphocytes on HLA-A2 melanomas. J Exp Med 1994; 180:35-42.
18. Kawakami Y, Eliyahu S, Delgado CH et al. Cloning of the Gene Coding for a Shared Human Melanoma Antigen Recognized by Autologous T Cells Infiltrating into Tumor. Proc Natl Acad Sci USA 1994; 91:3515-3519.

19. Bystryn JC, Rigel D, Friedman RJ et al. Prognostic significance of hypopigmentation in malignant melanoma. Arch Dermatol 1987; 123:1053-1055.
20. Richards JM, Mehta N, Ramming K et al. Sequential chemoimmunotherapy in the treatment of metastatic melanoma. J Clin Oncol 1992; 10:1338-1343.
21. Nordlund JJ, Kirkwood JM, Forget BM et al. Vitiligo in patients with metastatic melanoma: A good prognostic sign. J Am Acad Dermatol 1983; 9:689-696.
22. Belldegrun A, Muul LM, Rosenberg SA. Interleukin 2 expanded tumor-infiltrating lymphocytes in human renal cell cancer: isolation, characterization, and antitumor activity. Cancer Res 1988; 48:206-214.
23. Finke JH, Rayman P, Alexander J et al. Characterization of the cytolytic activity of $CD4^+$ and $CD8^+$ tumor-infiltrating lymphocytes in human renal cell carcinoma. Cancer Res 1990; 50:2363-2370.
24. Koo AS, Tso CL, Shimabukuro T et al. Autologous tumor-specific cytotoxicity of tumor- infiltrating lymphocytes derived from human renal cell carcinoma. J Immunother 1991; 10:347-354.
25. Finke JH, Rayman P, Edinger M et al. Characterization of a human renal cell carcinoma specific cytotoxic $CD8^+$ T cell line. J Immunother 1992; 11:1-11.
26. Schendel DJ, Gansbacher B, Oberneder R et al. Tumor-specific lysis of human renal cell carcinomas by tumor-infiltrating lymphocytes. I. HLA-A2-restricted recognition of autologous and allogeneic tumor lines. J Immunol 1993; 151:4209-4220.
27. Ferradini L, Roman Roman S, Azocar J et al. Analysis of T-cell receptor alpha/beta variability in lymphocytes infiltrating a melanoma metastasis. Cancer Res 1992; 52:4649-4654.
28. Thor Straten P, Scholler J, Hou Jensen K et al. Preferential usage of T-cell receptor alpha beta variable regions among tumor-infiltrating lymphocytes in primary human malignant melanomas. Int J Cancer 1994; 57:138.
29. Sensi M, Parmiani G. Analysis of TCR usage in human tumors: a new tool for assessing tumor-specific immune responses. Immunol Today 1995; 16:588-595.
30. Jantzer P, Schendel DJ. Human renal cell carcinoma antigen-specific CTLs: Antigen-driven selection and long-term persistence in vivo. Cancer Res 1998; 58:3078-3086.
31. Vijayasaradhi S, Bouchard B, Houghton AN. The melanoma antigen gp75 is the human homologue of the mouse b (brown) locus gene product. J Exp Med 1990; 171:1375-1380.
32. Sahin U, Tureci O, Schmitt H et al. Human neoplasms elicit multiple specific immune responses in the autologous host. Proc Natl Acad Sci USA 1995; 92:11810-11813.
33. Chen YT, Scanlan MJ, Sahin U et al. A testicular antigen aberrantly expressed in human cancers detected by autologous antibody screening. Proc Natl Acad Sci USA 1997; 94:1914-1918.
34. Rosenberg SA, Lotze MT, Yang JC et al. Experience with the use of high-dose interleukin-2 in the treatment of 652 cancer patients. Ann Surg 1989; 210:474-484.
35. Bukowski RM, Sharfman W, Murthy S, et al. Clinical results and characterization of tumor- infiltrating lymphocytes with or without recombinant interleukin 2 in human metastatic renal cell carcinoma. Cancer Res 1991; 51:4199-4205.
36. Belldegrun A, Pierce W, Kaboo R et al. Interferon-alpha primed tumor-infiltrating lymphocytes combined with interleukin-2 and interferon-alpha as therapy for metastatic renal cell carcinoma. J Urol 1993; 150:1384-1390.
37. Rosenberg SA, Lotze MT, Muul LM et al. A progress report on the treatment of 157 patients with advanced cancer using lymphokine-activated killer cells and interleukin-2 or high-dose interleukin-2 alone. N Engl J Med 1987; 316:889-897.
38. Rosenberg SA, Yannelli JR, Yang JC, et al. Treatment of patients with metastatic melanoma with autologous tumor-infiltrating lymphocytes and interleukin 2. J Nat Cancer Inst 1994; 86:1159-1166.
39. Van den Eynde BJ, van der Bruggen P. T cell defined tumor antigens. Curr Opin Immunol 1997; 9:684-693.
40. van der Bruggen P, Traversari C, Chomez P et al. A gene encoding an antigen recognized by cytolytic T lymphocytes on a human melanoma. Science 1991; 254:1643-1647.
41. Cox AL, Skipper J, Chen Y et al. Identification of a peptide recognized by five melanoma-specific human cytotoxic T cell lines. Science 1994; 264:716-719.
42. Jung S, Schluesener HJ. Human T lymphocytes recognize a peptide of single point-mutated, oncogenic ras proteins. J Exp Med 1991; 173:273-276.
43. Celis E, Tsai V, Crimi C et al. Induction of anti-tumor cytotoxic T lymphocytes in normal humans using primary cultures and synthetic peptide epitopes. Proc Natl Acad Sci USA 1994; 91:2105-2109.

44. Salazar Onfray F, Nakazawa T, Chhajlani V et al. Synthetic peptides derived from the melanocyte-stimulating hormone receptor MC1R can stimulate HLA-A2-restricted cytotoxic T lymphocytes that recognize naturally processed peptides on human melanoma cells. Cancer Res 1997; 57:4348-4355.
45. Stockert E, Jager E, Chen YT et al. A survey of the humoral immune response of cancer patients to a panel of human tumor antigens. J Exp Med 1998; 187:1349-1354.
46. Jager E, Chen YT, Drijfhout JW et al. Simultaneous humoral and cellular immune response against cancer-testis antigen NY-ESO-1: Definition of human histocompatibility leukocyte antigen (HLA)-A2-binding peptide epitopes. J Exp Med 1998; 187:1349-1354.
47. Kawakami Y, Eliyahu S, Delgado CH et al. Cloning of the gene coding for a shared human melanoma antigen recognized by autologous T cells infiltrating into tumor. Proc Natl Acad Sci USA 1994; 91:3515-3519.
48. Boel P, Wildmann C, Sensi ML et al. BAGE: A new gene encoding an antigen recognized on human melanomas by cytolytic T lymphocytes. Immunity 1995; 2:167-175.
49. van den Eynde B, Peeters O, De Backer O et al. A new family of genes coding for an antigen recognized by autologous cytolytic T lymphocytes on a human melanoma. J Exp Med 1995; 182:689-698.
50. Wolfel T, Hauer M, Schneider J et al. A p16INK4a-insensitive CDK4 mutant targeted by cytolytic T lymphocytes in a human melanoma. Science 1995; 269:1281-1284.
51. Gaugler B, Vandeneynde B, Vanderbruggen P et al. Human Gene MAGE-3 Codes for an Antigen Recognized on a Melanoma by Autologous Cytolytic T Lymphocytes. J Exp Med 1994; 179:921-930.
52. Vanderbruggen P, Bastin J, Gajewski T et al. A peptide encoded by human gene MAGE-3 and presented by HLA-A2 induces cytolytic T lymphocytes that recognize tumor cells expressing MAGE-3. Eur J Immunol 1994; 24:3038-3043.
53. Gaugler B, Brouwenstijn N, Vantomme V et al. A new gene coding for an antigen recognized by autologous cytolytic T lymphocytes on a human renal carcinoma. Immunogenetics 1996; 44:323-330.
54. Brandle D, Brasseur F, Weynants P et al. A mutated HLA-A2 molecule recognized by autologous cytotoxic T lymphocytes on a human renal cell carcinoma. J Exp Med 1996; 183:2501-2508.
55. Oosterwijk E, Ruiter DJ, Hoedemaeker PJ et al. Monoclonal antibody G250 recognizes a determinant present in renal-cell carcinoma and absent from normal kidney. Int J Cancer 1986;38:489-494.
56. Vennegoor C, Hageman P, Van Nouhuijs H et al. A monoclonal antibody specific for cells of the melanocyte lineage. Am J Pathol 1988; 130:179-192.
57. Adema GJ, De Boer AJ, van 't Hullenaar R et al. Melanocyte lineage-specific antigens recognized by monoclonal antibodies NKI-beteb, HMB-50, and HMB-45 are encoded by a single cDNA. Am J Pathol 1993; 143:1579-1585.
58. Adema GJ, De Boer AJ, Vogel AM, et al. Molecular characterization of the melanocyte lineage-specific antigen gp100. J Biol Chem 1994; 269:20126-20133.
59. Kwon BS, Chintamaneni C, Kozak CA, et al. A melanocyte-specific gene, Pmel 17, maps near the silver coat color locus on mouse chromosome 10 and is in a syntenic region on human chromosome 12. Proc Natl Acad Sci USA 1991; 88:9228-9232.
60. Kobayashi T, Urabe K, Orlow SJ et al. The Pmel 17/silver locus protein—Characterization and investigation of its melanogenic function. J Biol Chem 1994; 269:29198-29205.
61. Kawakami Y, Eliyahu S, Delgado CH et al. Identification of a human melanoma antigen recognized by tumor-infiltrating lymphocytes associated with in vivo tumor rejection. Proc Natl Acad Sci USA 1994; 91:6458-6462.
62. Zarour H, De Smet C, Lehman F et al. The majority of autologous cytolytic T-lymphocyte clones derived from peripheral blood lymphocytes of a melanoma patient recognize an antigenic peptide derived from gene Pmel17/gp100. J Invest Dermatol 1996; 107:63-67.
63. Bakker ABH, Schreurs MWJ, Tafazzul G et al. Identification of a novel peptide derived from the melanocyte-specific gp100 antigen as the dominant epitope recognized by an HLA-A2.1-restricted anti-melanoma CTL line. Int J Cancer 1995; 62:97-102.
64. Kawakami Y, Eliyahu S, Jennings C et al. Recognition of multiple epitopes in the human melanoma antigen gp100 by tumor-infiltrating T lymphocytes associated with in vivo tumor regression. J Immunol 1995; 154:3961-3968.

65. Skipper JC, Kittlesen DJ, Hendrickson RC et al. Shared epitopes for HLA-A3-restricted melanoma-reactive human CTL include a naturally processed epitope from Pmel-17/gp100. J Immunol 1996; 157:5027-5033.
66. Tsai V, Southwood S, Sidney J et al. Identification of subdominant CTL epitopes of the gp100 melanoma-associated tumor antigen by primary in vitro immunization with peptide- pulsed dendritic cells. J Immunol 1997;158:1796-1802.
67. Pastorek J, Pastorekova S, Callebaut I et al. Cloning and characterization of MN, a human tumor-associated protein with a domain homologous to carbonic anhydrase and a putative helix-loop-helix DNA binding segment. Oncogene 1994; 9:2877-2888.
68. Oosterwijk E, Debruyne FM, Schalken JA. The use of monoclonal antibody G250 in the therapy of renal-cell carcinoma. Semin Oncol 1995; 22:34-41.
69. Lotze MT, Farhood H, Wilson CC et . Dendritic cell therapy of cancer and HIV infection. In: Lotze MT, Thomson AW, eds. Dendritic cells. San Diego: Academic press, 1999:459-485.
70. Steinman RM. The dendritic cell system and its role in immunogenicity. Annu Rev Immunol 1991; 9:271-296.
71. Marland G, Hartgers FC, Veltkamp R et al. Analysis of dendritic cells at the genetic level. Adv Exp Med Biol 1997; 417:443-448.
72. Hart DN. Dendritic cells: Unique leukocyte populations which control the primary immune response. Blood 1997; 90:3245-3287.
73. Adema GJ, Hartgers F, Verstraten R et al. A dendritic-cell-derived C-C chemokine that preferentially attracts naive T cells. Nature 1997; 387:713-717.
74. Levin D, Constant S, Pasqualini T et al. Role of dendritic cells in the priming of $CD4^+$ T lymphocytes to peptide antigen in vivo. J Immunol 1993; 151:6742-6750.
75. Austyn JM. Antigen uptake and presentation by dendritic leukocytes. Semin Immunol 1992; 4:227-236.
76. Sallusto F, Cella M, Danieli C et al. Dendritic cells use macropinocytosis and the mannose receptor to concentrate macromolecules in the major histocompatibility complex class II compartment: downregulation by cytokines and bacterial products. J Exp Med 1995; 182:389-400.
77. Jiang W, Swiggard WJ, Heufler C et al. The receptor DEC-205 expressed by dendritic cells and thymic epithelial cells is involved in antigen processing. Nature 1995; 375:151-155.
78. Ardavin C, Wu L, Li CL et al. Thymic dendritic cells and T cells develop simultaneously in the thymus from a common precursor population. Nature 1993; 362:761-763.
79. Marquez C, Trigueros C, Fernandez E, et al. The development of T and non-T cell lineages from CD34+ human thymic precursors can be traced by the differential expression of CD44. J Exp Med 1995;181:475-483.
80. Saunders D, Lucas K, Ismaili J, et al. Dendritic cell development in culture from thymic precursor cells in the absence of granulocyte/macrophage colony-stimulating factor. J Exp Med 1996;184:2185-2196.
81. Suss G, Shortman K. A subclass of dendritic cells kills CD4 T cells via Fas/Fas-ligand- induced apoptosis. J Exp Med 1996; 183:1789-1796.
82. Res P, Martinez Caceres E, Cristina Jaleco A et al. CD34+CD38dim cells in the human thymus can differentiate into T, natural killer, and dendritic cells but are distinct from pluripotent stem cells. Blood 1996; 87:5196-5206.
83. Freudenthal PS, Steinman RM. The distinct surface of human blood dendritic cells, as observed after an improved isolation method. Proc Natl Acad Sci USA 1990; 87:7698-7702.
84. Inaba K, Steinman RM, Pack MW et al. Identification of proliferating dendritic cell precursors in mouse blood. J Exp Med 1992; 175:1157-1167.
85. Inaba K, Inaba M, Romani N et al. Generation of large numbers of dendritic cells from mouse bone marrow cultures supplemented with granulocyte/macrophage colony-stimulating factor. J Exp Med 1992; 176:1693-1702.
86. Peters JH, Gieseler R, Thiele B et al. Dendritic cells: From ontogenetic orphans to myelomonocytic descendants. Immunol Today 1996; 17:273-278.
87. Caux C, Dezutter Dambuyant C, Schmitt D et al. GM-CSF and TNF-alpha cooperate in the generation of dendritic Langerhans cells. Nature 1992; 360:258-261.

88. Caux C, Vanbervliet B, Massacrier C et al. CD34+ hematopoietic progenitors from human cord blood differentiate along two independent dendritic cell pathways in response to GM-CSF+TNF alpha. J Exp Med 1996; 184:695-706.
89. Romani N, Gruner S, Brang D et al. Proliferating dendritic cell progenitors in human blood. J Exp Med 1994; 180:83-93.
90. Sallusto F, Lanzavecchia A. Efficient presentation of soluble antigen by cultured human dendritic cells is maintained by granulocyte/macrophage colony-stimulating factor plus interleukin 4 and downregulated by tumor necrosis factor alpha. J Exp Med 1994; 179:1109-1118.
91. Zhou LJ, Tedder TF. CD14+ blood monocytes can differentiate into functionally mature CD83+ dendritic cells. Proc Natl Acad Sci USA 1996; 93:2588-2592.
92. Bender A, SAPP M, Schuler G et al. Improved methods for the generation of dendritic cells from nonproliferating progenitors in human blood. J Immunol Methods 1996; 196:121-135.
93. Romani N, Reider D, Heuer M et al. Generation of mature dendritic cells from human blood. An improved method with special regard to clinical applicability. J Immunol Methods 1996; 196:137-151.
94. Caux C, Massacrier C, Vanbervliet B et al. Activation of human dendritic cells through CD40 cross-linking. J Exp Med 1994; 180:1263-1272.
95. Ludewig B, Graf D, Gelderblom HR et al. Spontaneous apoptosis of dendritic cells is efficiently inhibited by TRAP (CD40-ligand) and TNF-alpha, but strongly enhanced by interleukin-10. Eur J Immunol 1995; 25:1943-1950.
96. Cella M, Scheidegger D, Palmer Lehmann K et al. Ligation of CD40 on dendritic cells triggers production of high levels of interleukin-12 and enhances T cell stimulatory capacity: T-T help via APC activation. J Exp Med 1996; 184:747-752.
97. Mayordomo JI, Zorina T, Storkus WJ et al. Bone marrow-derived dendritic cells pulsed with synthetic tumor peptides elicit protective and therapeutic antitumour immunity. Nat Med 1995; 1:1297-1302.
98. Zitvogel L, Mayordomo JI, Tjandrawan T et al. Therapy of murine tumors with tumor peptide-pulsed dendritic cells: dependence on T cells, B7 costimulation, and T helper cell 1-associated cytokines. J Exp Med 1996; 183:87-97.
99. Celluzzi CM, Mayordomo JI, Storkus WJ et al. Peptide-pulsed dendritic cells induce antigen-specific CTL-mediated protective tumor immunity. J Exp Med 1996; 183:283-287.
100. Gilboa E, Nair SK, Lyerly HK. Immunotherapy of cancer with dendritic-cell-based vaccines. Cancer Immunol Immunother 1998; 46(2):82-87.
101. Bakker ABH, Marland G, De Boer AJ et al. Generation of antimelanoma cytotoxic T lymphocytes from healthy donors after presentation of melanoma-associated antigen-derived epitopes by dendritic cells in vitro. Cancer Res 1995; 55:5330-5334.
102. Mortarini R, Anichini A, Di Nicola M et al. Autologous dendritic cells derived from CD34+ progenitors and from monocytes are not functionally equivalent antigen-presenting cells in the induction of melan-A/MART-1(27-35)-specific CTLs from peripheral blood lymphocytes of melanoma patients with low frequency of CTL precursors. Cancer Res 1997; 57:5534-5541.
103. Schoenberger SP, Toes RE, van der Voort EI et al. T-cell help for cytotoxic T lymphocytes is mediated by CD40-CD40L interactions. Nature 1998; 393:480-483.
104. Ridge JP, Di Rosa F, Matzinger P. A conditioned dendritic cell can be a temporal bridge between a CD4+ T-helper and a T-killer cell. Nature 1998; 393:474-478.
105. Bennett SR, Carbone FR, Karamalis F et al. Help for cytotoxic-T-cell responses is mediated by CD40 signalling. Nature 1998; 393:478-480.
106. Ossendorp F, Mengede E, Camps M et al. Specific T helper cell requirement for optimal induction of cytotoxic T lymphocytes against major histocompatibility complex class II negative tumors. J Exp Med 1998; 187:693-702.
107. Pierre P, Turley SJ, Gatti E, et al. Developmental regulation of MHC class II transport in mouse dendritic cells. Nature 1997; 388:787-792.
108. Cella M, Engering A, Pinet V, et al. Inflammatory stimuli induce accumulation of MHC class II complexes on dendritic cells. Nature 1997; 388:782-787.
109. Parkhurst MR, Salgaller ML, Southwood S et al. Improved induction of melanoma-reactive CTL with peptides from the melanoma antigen gp100 modified at HLA-A*0201-binding residues. J Immunol 1996; 157:2539-2548.

110. Bakker ABH, van der Burg SH, Huijbens RJ et al. Analogues of CTL epitopes with improved MHC class-I binding capacity elicit anti-melanoma CTL recognizing the wild-type epitope. Int J Cancer 1997; 70:302-309.
111. Valmori D, Fonteneau JF, Lizana CM et al. Enhanced generation of specific tumor-reactive CTL in vitro by selected Melan-A/MART-1 immunodominant peptide analogues. J Immunol 1998; 160:1750-1758.
112. Overwijk WW, Tsung A, Irvine KR et al. gp100/pmel 17 is a murine tumor rejection antigen: induction of "self"-reactive, tumoricidal T cells using high-affinity, altered peptide ligand. J Exp Med 1998;188:277-286.
113. Rosenberg SA, Yang JC, Schwartzentruber DJ et al. Immunologic and therapeutic evaluation of a synthetic peptide vaccine for the treatment of patients with metastatic melanoma. Nat Med 1998; 4:269-270.
114. Dyall R, Bowne WB, Weber LW et al. Heteroclitic immunization induces tumor immunity. J Exp Med 1998; 188(9):1553-1561.
115. Dieu MC, Vanbervliet B, Vicari A et al. Selective recruitment of immature and mature dendritic cells by distinct chemokines expressed in different anatomic sites. J Exp Med 1998; 188:373-386.
116. Sallusto F, Schaerli P, Loetscher P et al. Rapid and coordinated switch in chemokine receptor expression during dendritic cell maturation. Eur J Immunol 1998; 28:2760-2769.
117. Lindhout E, Figdor CG, Adema GJ. Dendritic cells: migratory cells that are attractive. Cell Adhes Commun 1998; 6:117-123.
118. Nestle FO, Alijagic S, Gilliet M et al. Vaccination of melanoma patients with peptide- or tumor lysate-pulsed dendritic cells. Nat Med 1998; 4:328-332.
119. Hsu FJ, Benike C, Fagnoni F et al. Vaccination of patients with B-cell lymphoma using autologous antigen-pulsed dendritic cells. Nat Med 1996; 2:52-58.
120. Tjoa BA, Erickson SJ, Bowes VA et al. Follow-up evaluation of prostate cancer patients infused with autologous dendritic cells pulsed with PSMA peptides. Prostate 1997;32:272-278.
121. Murphy G, Tjoa B, Ragde H et al. Phase I clinical trial: T-cell therapy for prostate cancer using autologous dendritic cells pulsed with HLA-A0201-specific peptides from prostate-specific membrane antigen. Prostate 1996; 29:371-380.
122. Boczkowski D, Nair SK, Snyder D et al. Dendritic cells pulsed with RNA are potent antigen-presenting cells in vitro and in vivo. J Exp Med 1996; 184:465-472.
123. Celluzzi CM, Falo LDJ. Physical interaction between dendritic cells and tumor cells results in an immunogen that induces protective and therapeutic tumor rejection. J Immunol 1998; 160:3081-3085.
124. Albert ML, Sauter B, Bhardwaj N. Dendritic cells acquire antigen from apoptotic cells and induce class I-restricted CTLs. Nature 1998; 392:86-89.
125. Hart I, Colaco C. Immunotherapy. Fusion induces tumor rejection. Nature 1997; 388:626- 627.
126. Lynch DH, Andreasen A, Maraskovsky E, et al. Flt3 ligand induces tumor regression and antitumor immune responses in vivo. Nat Med 1997; 3:625-631.

Chapter 14

Melanoma Peptide Clinical Trials

Ian D. Davis and Michael T. Lotze

Introduction

Although various immunologic approaches to the treatment of cancer have been used for over a century,[1] it is only relatively recent that specific human cancer targets have been defined allowing specific therapy against them. Burnet[2] proposed a theory of immunosurveillance almost 30 years ago, postulating that the immune system is constantly surveying the internal milieu for cells with changes characteristic of malignancy. The next decade provided good evidence that this might be the case for malignancies related to viral infections, but for the majority of human tumors viruses are not believed to be implicated and the theory fell into disfavor in subsequent years.

However, there is now increasing evidence that immune surveillance for malignant cells does occur. Data from transplant registries indicate that in the setting of immunosuppression, there is an increase not only in virus-related malignancies and malignancies related to UV exposure, but also in some but not all other types of cancer. For example, there is a greater than five-fold increase in the risk of esophageal, ureteric, vulval and vaginal carcinomas and also melanoma in transplant recipients, but no increased risk of carcinomas of the breast.[3,4] Other types of cancer, such as prostate cancer, show a trend towards a reduction of risk in immunosuppressed transplant patients, although the 95% confidence limits for risk overlap unity and the numbers of patients studied are small.[4] Overall, immunosuppressed renal transplant recipients are 3.5-fold more likely to develop cancer than the age-matched population.[4] These findings suggest that the immune system plays an important role in detecting and preventing the growth of malignant tumors, and that this process can be manipulated.

Tumor Rejection Antigens

Early studies using methylcholanthrene-induced tumors in mice allowed identification of several tumor-specific rejection antigens in these models.[5-7] For many years it was believed that this might be a purely experimental phenomenon although early cancer immunotherapists hoped otherwise. It was not until the identification of the MAGE genes[8] that it was shown formally that in naturally occurring human cancers, antigens were expressed that could elicit a cellular immune response. Many more such antigens have since been described (Table 14.1), and subsequently, HLA class I- and II-restricted peptides have been defined for several HLA serotypes for many of these antigens.

Shared Antigens

Tumor antigens fall into several broad classifications. Antigens may be shared, i.e., found in cancers of the same or different types in different patients. The first of these antigens were defined using tumor-infiltrating lymphocytes (TIL) that were cloned and

Peptide-Based Cancer Vaccines, edited by W. Martin Kast. ©2000 Eurekah.com.

examined to determine their antigenic specificity. Lymphocytes showing specificity for defined shared tumor antigens may also be found in the peripheral blood or in regional lymph nodes.[9] Since these antigens have been shown to be relevant in other patients with the same disease, they are obvious candidates for vaccination strategies.

More recently, a technique known as SEREX (serological analysis of recombinant cDNA expression libraries)[10] has been used to identify additional tumor antigens, many of which had also been defined with T-cell reagents. This technique involves the construction of cDNA libraries from tumor or normal tissue samples, expression of the library in a bacterial expression vector, and screening patient serum for the presence of high titer IgG antibodies against tumor-derived antigens. SEREX strategies have identified many antigens that have not yet been shown to elicit cellular responses. However, since the technique relies upon the detection of IgG antibodies, the presence of these antibodies implies that the antigen of interest has been processed and presented by antigen presenting cells (APC) to B cells with appropriate T cell help, allowing immunoglobulin isotype switching from IgM to IgG. Thus, an antigen detected by SEREX must therefore contain epitopes that are recognized by both B and T cells. The number of antigens described using SEREX is now more than 800 (Jongeneel V, personal communication). Some shared antigens of interest are listed in Table 14.1. One antigen of particular interest is the NY-ESO-1 antigen, which can elicit a strong humoral and cellular response concurrently.[11] The first study using peptides derived from NY-ESO-1 commenced in Melbourne in early 1999.

Unique Antigens

Antigens may also be unique, i.e., expressed only in that particular cancer and not necessarily in similar cancers occurring in other patients. Although these antigens may be important potential immunologic targets for that particular patient, the sheer diversity of the outbred human population makes it very difficult to detect these antigens and to determine their significance. Since they may be of great importance for a given patient, some groups have developed techniques to allow the immune system to sample the antigenic repertoire of autologous tumors and to express them on autologous APC such as dendritic cells (DC). One interesting approach is to use peptides stripped from the surface of the tumor cell by means of acid elution. This crude preparation of class I peptides will contain mostly self epitopes, possibly including a small fraction of peptides derived from clinically relevant antigens. This technique has been successful in animal models, predominantly where the tumor model is immunogenic and the nature of the relevant peptide is known. It has also been shown to elicit relevant antitumor responses in less immunogenic tumor models in which the relevant epitope is not known.[9] This approach has now entered clinical trials (Thomas R, personal communication). Another approach is to purify and amplify mRNA from tumors and to transfect this into APC, allowing them to translate the relevant genes and to process and present appropriate peptides in the context of MHC class I or II molecules.[12] This approach has now been used in both animal models[12] and in preclinical human studies,[13] with some success. Since RNA may be able to be produced from small amounts of tumor tissue, this technique may eventually be applicable on a larger scale.

Tissue Specificity Of Tumor Antigens

Antigens may also be subclassified according to their tissue or tumor specificity. Some antigens are expressed in cells of a particular lineage and represent differentiation_antigens, such as the melanocyte antigens Melan-A/MART-1, tyrosinase, and gp100. These gene products are expressed in both normal and malignant melanocytes, and immune responses against these targets have been associated with the development of autoimmune reactions against nor-

Table 14.1. Human melanoma particles used in clinical trials

Antigen	Peptide sequence	Amino acids	HLA restriction	Comments
Differentiation antigens:				
Melan-A/MART-1	AAGIGILTV	27-35	A2	Naturally processed but has less than optimal anchor residue at position 2.
	EAAGIGILTV	26-35	A2	
	ELAGIGILTV	26-35	A2	Modified at position 2.
	ILTVILGVL	32-40	A2	
gp100	YLEPGPVTA	280-288	A2	
	KTWGQYWQV	154-162	A2	
	ITDQVPFSV	209-217	A2	
	IMDQVPFSV	209-217	A2	210M substitution
	YLEPGPVTA	280-288	A2	
	YLEPGPVTV	280-288	A2	288V substitution
	ALLAVGATK	17-25	A3	
Tyrosinase	MLLAVLYCL	1-9	A2	Leader sequence
	YMDGTMSQV	369-377	A2	Internal sequence
	DAEKCDICTDEY	240-251	A1	Cysteines may crosslink
	DAEKSDICTDEY	240-251	A1	244S substitution
	AFLPWHRLF	206-214	A24	
TRP-1 (ORF3)	MSLQRQFLR	1-9	A31/A33	Translated from alternative open reading frame

Table 14.1, continued.

Antigen	Peptide sequence	Amino acids	HLA restriction	Comments
CT antigens:				
MAGE-1	EADPTGHSY	27-35	A1	
MAGE-3	EVDPIGHLY	168-176	A1	A2 peptide may not be naturally processed by proteasome
	FLWGPRALV	271-279	A2	
NY-ESO-1	SLLMWITQCFL	157-167	A2	Cysteines may crosslink
	SLLMWITQC	157-165		
	QLSLLMWIT	155-163		

Table 14.2. Mechanisms of immunologic escape by non-viral-induced tumors

Mechanism	Potential solution	Reference
Downregulation of antigen expression	Induction of expression, e.g., by demethylation in the case of MAGE genes	16
Downregulation of MHC class I complexes expression	Induction of expression: Systemic or local treatment with interferons	20
Proteasomal cleavage to generate peptides	Cytokine treatment may allow use of alternate proteasomal subunits with production of different cleavage peptides	18
TAP-mediated peptide transport	Target HLA types and antigens not dependent on TAP	
β2-microglobulin expression	Cytokine treatment, e.g., interferons	20
MHC class I gene expression Total loss Haplotype loss Locus loss eg HLA-A Allelic loss eg HLA-A2	Cytokine treatment, e.g., interferons	20

Table 14.3. Types of melanoma peptide/protein clinical trials

In vivo peptide vaccination:	
Peptide alone	
Chemical adjuvants	QS21, MF59, monophosphoryl lipid A
Cytokine adjuvants	GM-CSF, IL-2, IL-12, FLT-3L
Cellular adjuvants	Dendritic cell
Ex vivo expansion and adoptive transfer of T cells reactive with individual melanoma peptides:	
CTL	
Following selection with specific tetramers	
Protein vaccination:	
Other:	
Genetic vaccination	cDNA, cRNA, polytope vaccines

mal melanocytes.[14] However, these genes are not expressed in tumors derived from other cell types. Therefore vaccination against these targets should allow a relatively tissue-specific response to be elicited.

Another broad group of antigens is known as the "cancer-testis" (CT antigen) group.[15] These antigens are found in cancers of different histologic types, but the only normal tissues in which they are found are testis and occasionally ovary. The cells in the testis expressing these antigens are immunologically privileged by dint of their anatomical localization and because they do not express HLA class I molecules, so the CT antigens are essentially specific for malignant cells (Table 14.1). Many of these antigens have also been detected using SEREX, implying productive crosstalk has occurred between APC, T and B cells. Interestingly, expression of some of these antigens appears to be acquired with dedifferentiation of tumors and expression is often higher in metastatic disease when compared with primary lesions (Gibbs et al, submitted). Induction of MAGE gene family expression appears to be related to demethylation of the promoter.[16]

Issues Related to Cancer Immunotherapy with Peptides

Translating such basic research findings into the clinic has been problematic. This may in part be due to the fact that most tumors are derived from autologous cells, i.e., of "self" origin, with very few exceptions in immunosuppressed transplant patients where lymphoid or other inadvertently transplanted tumors are sometimes of donor origin. Antigens that are expressed in normal tissues may contain potential T cell epitopes, however deletion in the thymus will remove many of these clones. In other cases, the tumor antigen of interest may be an embryonic antigen or neoantigen, but reactive T cells have been depleted through peripheral tolerance. The challenge is to break both central and peripheral tolerance, and allow development of an effective antitumor immune response specific for that malignancy. In some cases, peptides have been predicted to have good binding affinity for particular HLA class I molecules and can elicit specific T cell reactivity in vitro. However, many of these T cells are not capable of recognizing peptides derived from natural proteasomal cleavage expressed on HLA-matched melanoma cells.[17,18] Many of the processes involved in antigen processing are not clearly understood.[19] Therefore, it is important to ensure that peptides that are predicted to be useful targets are tested against relevant targets, i.e., HLA-matched tumor cells expressing such antigens naturally (and not solely after pulsing with synthetic peptide).

Tumors are under selection pressures ("tumor Darwinism") not to express particular tumor antigens or their restricting molecules. Various mechanisms for evading the immune response have been identified (Table 14.2). A majority of melanomas (and probably many other tumor types) exhibit deficiencies in HLA class I expression and resultant antigen presentation. It is important that there be at least a small amount of class I expression in order to protect against natural killer (NK) cell activity.[21] By judicious selection of specific class I alleles the tumor cell may avoid T cell killing through class I[22] while inhibiting NK activity by expressing alleles that have killer cell inhibitory receptor (KIR) activity.[23, 24] This may not be detected unless specifically looked for, since HLA class I immunohistochemistry is usually performed using a pan-class-I antigen which will be positive even if five of six alleles are missing. These points highlight one of the problems inherent in vaccination using specific peptides: there are obvious and easy ways for tumors to escape peptide-specific cytotoxic T lymphocytes (CTL), particularly if a vaccine is targeting one epitope only.[25, 26] It may be possible to overcome this problem by means of various therapeutic means (Table 14.2).

Types of Clinical Trials

The clinical use of peptides derived from melanoma antigens has followed a logical evolution. The initial studies were with peptides alone followed by studies using various synthetic, cytokine or cellular (DC) adjuvants. The various approaches used to date are shown in Table #.3.

Peptides Alone

Initial studies using melanoma peptides were performed using peptides alone. One of the first studies involved the use of the MAGE-3 HLA-A1-restricted peptide EVDPIGHLY[27] in the absence of any adjuvant. In this study, three of 12 patients with melanoma displayed tumor regression and toxicity was minimal. In one patient, tumor regression occurred several months after completing peptide vaccination although the tumor had progressed while on treatment, leading to the recommencement of vaccination. Eventually four of five lung metastases in this patient disappeared entirely. This illustrates an important issue: clinical responses to immunotherapy may be delayed and may occur despite initial tumor progression on treatment.[28] This initial study demonstrated that synthetic peptides derived from self tumor antigens were safe and that meaningful clinical responses could be obtained in the absence of any other adjuvant, somewhat surprisingly. The same group later reported that CTL responses against melanoma differentiation antigens correlate inversely with expression of the antigen in melanoma tissues: patients with progressing disease often displayed antigen-loss variants, implying that in vivo immunoselection was occurring under the selection pressure of peptide vaccination.[26,29] The optimal dose and timing of peptide injections is not known: in another study from the same group, weekly rather than monthly injections of peptides led to immunologic responses but no clinical responses.[30]

A series of studies was performed at the NCI involving peptides derived from the melanoma differentiation antigen gp100.[31-33] Initially, HLA-A2+ patients were treated with the immunodominant nonamer peptide ITDQVPFSV (residues 209-217), and subsequent patients were treated with a modified version of this peptide with a methionine substituted for threonine at position 210.[33] This substituted peptide has a higher binding affinity for HLA-A*0201 than the native peptide, with a correspondingly greater ability to induce CTL.[33] Patients treated with the substituted peptide developed CTL capable of responding to either native or modified peptide-pulsed target cells or to HLA-A2+ tumor cell lines expressing gp100 in almost all cases. Later studies examined the role of systemic cytokines in peptide responses[31] (see below).

Chemical Adjuvants

Work quickly moved to the use of chemical adjuvants for peptide vaccination, since these have been known for many decades to enhance antibody responses to protein vaccines. Since there was only a little preclinical data [34, 35] and virtually no clinical information to support the notion that cellular responses could also be stimulated using chemical adjuvants, this was somewhat of a leap of faith. The QS21 formulation had been used for several years and was known to be effective in inducing antibody and cellular responses in animals.[35, 36] It has been used successfully in humans and is being developed in combination with monophosphoryl lipid A.[37] The oil-in-water adjuvant MF59 was shown to be safe and effective in humans in pediatric studies as well as those involving HIV and influenza, [38,39] and was a logical choice for these studies.

Study 94-95 at the University of Pittsburgh Cancer Institute involved treating HLA-A2+ melanoma patients with peptides derived from Melan-A/MART-1, gp100 or tyrosinase, formulated in MF59 and administered intramuscularly on four occasions one week apart. 28 patients entered this study. Most patients showed progressive disease, although one patient remained on study for 16 months before progressing. Although there were no formal clinical responses, several patients exhibited mixed responses with regression of some disease sites (predominantly cutaneous disease) while progressing at other sites (predominantly visceral or soft tissue). No CTL responses to peptides could be detected consistently in the peripheral blood or in lymphocytes isolated from sites where peptide had been injected to examine for delayed-type hypersensitivity responses.

Some investigators have used other chemical adjuvants, such as incomplete Freund's adjuvant (IFA).[40] Patients treated with the Melan-A/MART-1 9-mer AAGIGILTV in incomplete Freund's adjuvant showed enhanced CTL responses, although no clinical responses were seen in 18 patients in this study. Another study recently reported in abstract form involved vaccination of HLA-A1+ melanoma patients with a MAGE-3 peptide plus the class II consensus peptide PADRE.[41] Toxicity was modest in this study, and six/16 patients developed MAGE-3 peptide skin reactions following vaccination. Five/15 patients showed peripheral blood lymphocyte reactivity against the MAGE-3 peptide but not against PADRE. No clinical response data was reported in this abstract. It is unlikely that IFA will be used for future studies given the availability of other more suitable or potent adjuvants. Other chemical manipulations, such as linking peptides to lipids, have been used in hepatitis B vaccination and will probably be useful in humans.[42]

Cytokine Adjuvants

Many cytokines have potent effects on cellular responses to peptides in vitro and in vivo. However, for most cytokines the optimal dose and even route of administration has not been defined. It is unlikely that the maximum tolerated dose will be the most biologically effective dose. A striking example of this was seen in a model of p53 peptide vaccination together with IL-12 in a murine BALB/c Meth A sarcoma model.[35] In this study, adjuvant IL-12 together with vaccination with the 234CM relevant mutated p53 peptide led to an antitumor effect, and a dose response was seen for IL-12. However, the highest doses of IL-12 actually suppressed CTL generation.[35] It is therefore important when designing studies using adjuvant cytokines to use a biologically optimal dose rather than maximum tolerated dose wherever possible.

One of the first cytokines studied as an adjuvant to peptide vaccination was GM-CSF. GM-CSF is an important cytokine in the biology of DC.[43-45] In gene transfer models, GM-CSF gene transfer into poorly immunogenic tumors leads to rejection of those tumors in animal models,[46] or to clinical and immunologic responses in humans.[47,48] Direct injection of GM-CSF into subcutaneous melanoma metastases may also lead to regression of injected and non-injected tumor deposits.[49] GM-CSF was therefore an obvious cytokine to study in the context of peptide vaccination. A study using GM-CSF together with peptides derived from Melan-A/MART-1, tyrosinase and gp100 in HLA-A2 melanoma patients was reported in 1996.[50] In this study, the peptides were injected alone on three occasions at monthly intervals. Patients still on study at the end of this treatment then received GM-CSF at a dose of 75 g/d. Immunologic responses were observed, consisting of development of skin reactions to intradermal injections of peptide, infiltration of injection site by $CD4^+$ and $CD8^+$ lymphocytes, and detectable CTL in peripheral blood. The treatment was virtually free of toxicity. Three patients were reported in whom tumor regressions were seen despite having progressive melanoma at the time of entry into the study: one patient showed complete regression of tumor and the others displayed partial regression. One difficulty in the interpretation of this study is to determine to what degree the GM-CSF was involved in the tumor regression. Temporally, the regressions predominantly occurred after the institution of the GM-CSF injections, however in one patient tumor shrinkage was seen prior to GM-CSF. Further studies have been performed using GM-CSF as an adjuvant but are yet to be reported in the literature.

An interesting approach involving the use of GM-CSF plus melanoma peptides has been tested in animal models but is yet to be used in humans. This approach involves the use of a polysaccharide matrix containing GM-CSF and the peptide of interest, allowing slow release of both the cytokine and the peptide.[51] Such depot preparations may enable a reduction in the number of painful intradermal injections required by current peptide vaccination approaches. Alternative strategies under development include the use of pegylated forms of GM-CSF as well as liposomal delivery.

IL-12 is a heterodimeric cytokine with important effects on T helper cells, as well as activity on NK cells and anti-angiogenic effects (for review, see ref. 52). Three studies have used IL-12 as an adjuvant to peptide vaccination. Two of these studies involve vaccination with the HLA-A2 Melan-A/MART-1 10-mer peptide EAAGIGILTV together with IL-12 given either subcutaneously or intravenously (Knuth A, personal communication; Cebon J, personal communication). These studies are still ongoing and no results are yet available. A third study at the University of Pittsburgh involving 25 patients has recently been completed without any meaningful clinical responses being seen.

IL-2 is an important cytokine for the expansion and maintenance of both helper and cytotoxic T cells. IL-2 alone has clinical activity in melanoma,[53] and has been used as an adjuvant in peptide vaccination. In a study using modified gp100 HLA-A2 peptides, 13 of 31 patients receiving the peptide plus systemic IL-2 showed regression of metastatic melanoma, with an additional 4 patients showing minor or mixed responses.[31] Immunologic responses were able to be generated in 91% of patients. IL-2 may also have important effects on the modulation of effector responses to dendritic cell (DC) vaccination (see below).[54]

Cellular Adjuvants

Peptides have been used with cellular adjuvants in two different ways. Peptides may be pulsed onto APC such as autologous DC; or they may be used to extract from the blood CTL specific for the peptide of interest, which can then be expanded ex vivo and reinfused into the patient.

DC are specialized APC that have the unique ability to be able to stimulate naïve T cells.[55,56] They are also capable of presenting exogenous antigens either by the MHC class I or II pathways·[57-65] Following transmigration across the endothelium DC will encounter tumor, becoming tumor associated DC, about whose function little is known. Much is now known about some of the signaling pathways responsible for maturation of DC and the delivery of apoptotic tumor to the DC plays an important part in this.[64] However, the survival of these DC is regulated by different signaling pathways.[66] As such, it may involve T-cell counterreceptors such as RANKL[67] or DC-derived autocrine signals such as IL-12.[68] In addition to the known function of DC as antigen presenting cells, which presumably occurs after migration to regional nodal sites, DC may also play a critical role at the site of the tumor.[69] DC can regulate angiogenesis (for example through elaboration of the antiangiogenic factors alpha interferon, IL-12, and IL-18; Odoux C, Wong M, and Lotze MT, in preparation) or protect T-cells from premature programmed cell death at the site of tumor. The available literature suggests improved prognosis for patients with infiltration of tumor by DC. This finding applies to a variety of tumor types and suggests that the therapeutic outcome could be improved if DC can be delivered to tumor sites.

It has been known for many years that DC per se can mediate antitumor effects,[70] and recent animal studies using synthetic[71] or stripped[9] peptides pulsed onto DC have shown potent inhibition of tumor growth. Other preclinical murine studies [72-74] have also shown promising results. In humans, DC can be generated easily from peripheral blood monocytes,[44] CD34+ progenitors,[45,75] or by direct purification from blood,[76] and have entered clinical trials in a variety of malignant diseases including lymphoma[77] and melanoma.[78-81]

Nestle et al have performed a clinical trial involving 30 patients with malignant melanoma using DC vaccination.[79,82] Monocyte-derived DC were generated from peripheral blood by culture in IL-4 and GM-CSF, and in later patients DC were matured by the use of monocyte-conditioned medium or by adding the cytokines IL-1, IL-6 and TNF-α. Patients of HLA serotype HLA-A1 were treated with appropriate peptides derived from the MAGE-1 and MAGE-3 antigens; those who were HLA-A2+ were treated with peptides from Melan-A/MART-1, gp100 and tyrosinase; and those who expressed HLA-B44 received treatment with peptides derived from MAGE-3 and tyrosinase. Keyhole limpet hemocyanin (KLH) was used to provide non-specific T cell helper epitopes during priming. Patients were treated with four weekly vaccinations, a fifth injection in week six, and thereafter monthly for up to 10 vaccinations, depending on clinical response. Peptide-pulsed DC were injected in most cases into an inguinal lymph node under ultrasound control, or close to regional lymph nodes. The treatment was well tolerated. All patients developed DTH reactivity to KLH and reactivity to the relevant peptides was seen in six patients. 27% (8/30) patients showed clinical responses, including 3 complete remissions and five partial remissions. This study is the most mature of all the current DC/peptide studies in melanoma and has indicated that this treatment is feasible, safe and well tolerated, and that meaningful clinical responses can be induced in a group of patients with poor prognosis disease.

Study 95-060 at the University of Pittsburgh Cancer Institute has been completed[83] In this study, DC were derived from peripheral blood monocytes and cultured in IL-4 and GM-CSF without additional maturation factors. The resulting immature DC were then pulsed with HLA-A2 peptides from Melan-A/MART-1, gp100 and tyrosinase, and injected back into patients. 90% of the peptide-pulsed DC were injected intravenously and the remaining 10% subcutaneously, in order to allow later biopsy of the injection site to determine the nature of residual or infiltrating cells at the site. Four weekly vaccinations were given in each cycle of treatment. After culture, peptide pulsing, washing and pooling of DC, a mean of approximately 10^6 DC were available for injection in each vaccination. 28 patients were treated. The first patient on study had a relapsing/remitting course of her melanoma over several years. She has very high levels of CTL in her peripheral blood and has taken part in several previous immunotherapy strategies. After DC vaccination, this patient developed a complete remission and remained free of disease for 20 months. DC vaccination was tolerated very well, however about one year after commencing therapy she developed overt rheumatoid arthritis, with high C-reactive protein and rheumatoid factor. Retrospective review of banked serum in this patient indicated that rheumatoid factor had been detectable in the past and had risen slightly during a previous protocol involving IL-12 treatment, but had risen sharply and remained high during DC vaccination. This is the first report of autoimmune disease in DC vaccination and warrants a cautionary note that DC pulsed with peptide and activated either in vitro or in vivo will present other self peptides to autoreactive T and B cells also. Another patient on this study has attained a complete remission and a third has had a partial remission. Other patients did not respond during therapy but have responded during subsequent treatment with high-dose systemic IL-2 off study, suggesting that subclinical priming of an immune response against relevant antigens may have occurred during DC therapy but required IL-2 therapy to become overt. A further study using DC derived from peripheral blood mononuclear cells or CD34+ progenitors following G-CSF mobilization has now commenced at the University of Pittsburgh and another study using CD34+ progenitor-derived DC [45,75] is due to commence soon at the Ludwig Institute in Melbourne, Australia.

Others have used peripheral blood monocytes cultured in GM-CSF as APC.[78, 84] In this study, 17 HLA-A1+ patients with metastatic melanoma were treated with APC pulsed with the synthetic MAGE-1 nonapeptide EADPTGHSY. Pulsed APC were injected intradermally once a month and the dose of cells was escalated over four monthly injections. Thirteen patients completed the study, three coming off study due to progressive disease and one due to unrelated cardiac disease. Toxicity was minimal and no autoimmunity was seen. One patient showed partial regression of a skin nodule. The median survival for the surviving 13 patients was 17 months at the time of reporting. CTL derived from peripheral blood showed no differences in specific reactivity against peptides post vaccination compared to pre vaccination. However, in nine patients in whom vaccine-infiltrating lymphocytes could be grown, three patients showed antigen-specific cytotoxicity or TNF production, indicating that specific CTL could be induced and that these cells could be found preferentially at the vaccine site rather than in the peripheral blood. This has implications for the immunologic monitoring of patients on tumor vaccine studies, where most investigators are trying to detect CTL responses in the peripheral blood. The precursor frequency of such CTL is low and often difficult to detect without many rounds of in vitro stimulation to expand the CTL to detectable levels. In addition, some peptides may elicit different clonal CTL responses in the same individual, and these clones may be present at different frequencies in the peripheral blood compared to the population of CTL infiltrating tumor.[85] Therefore, it may be that the peripheral blood is the wrong compartment in which to search for relevant antigen-specific cells.

Cellular Effectors

Ultimately, a successful antitumor vaccine strategy will result in the expansion of effector cells capable of attacking tumor cells with some specificity. This specificity may occur by dint of targeting HLA/peptide complexes on the surface of tumor cells by specific CTL. It may also occur by means of the production of antibodies with tumor specificity (implying T cell help in order to allow antibody isotype switching and affinity maturation). Finally, it may occur by activation of natural killer cells, which although lacking specificity in terms of peptide targets, will nevertheless target cells attempting to hide from the specific immune system by means of downregulating class I molecules. Direct in vivo peptide vaccination is one means of expanding cellular effectors, however peptides may also be used to facilitate ex vivo expansion of these cells.

Recently, a technique has been developed that allows isolation and expansion of low-frequency peptide-specific CTL.[86, 87] This technique consists of fluorogenic complexes of HLA class I molecules plus specific peptide. Tetrameric complexes are assembled which bind with high affinity to T cell receptors on T cells that have specificity for the peptide/HLA complex of interest. Fluorescent cell sorting is then used to isolate this low frequency population, which can then be expanded ex vivo. This technique is currently being used in a clinical trial, and immunologic responses have been observed.[88]

Future Directions

There has probably been many more tumor antigen peptides described since the preparation of this Chapter. As more peptides become available for a larger number of HLA serotypes, more patients will be potentially suitable for peptide-based immunotherapy. Several databases are now available for cataloging described peptides, including the MHCPEP database at the Walter and Eliza Hall Institute for Medical Research (http://wehih.wehi.edu.au/mhcpep/).[89] Algorithms are available for the prediction of potential HLA-binding peptide epitopes, such as the web-based version at the NIH (http://bimas.dcrt.nih.gov/molbio/hla_bind/).[90-93] However, most peptides to date have been defined because they are dominant epitopes, i.e., most

CTL induced against that protein antigen have specificity for a particular peptide sequence. There is some evidence that CTL with specificity for dominant epitopes may be preferentially deleted or inactivated in cancer patients,[94] and it may be that subdominant epitopes are more clinically useful.[95-97] A potential criticism of strategies based on whole protein or gene vaccination is that by definition CTL recognizing dominant epitopes will be induced preferentially, and these may not be the most useful CTL for an antitumor effect. Interestingly, most epitopes defined to date have been discovered because they are dominant epitopes recognized by tumor-infiltrating lymphocytes or induced in vitro by gene transfer. Other techniques such as SEREX may now allow the definition of other subdominant but clinically important epitopes.

Some investigators have devised ingenious methods to improve potential immune responses. There is some evidence that the binding affinity of the peptide to the class I molecule is important in determining immunogenicity, since peptides with very low binding affinities often do not elicit significant CTL responses. There appears to be a lower limit of affinity of 500 nM: peptides with binding affinities weaker than this do not elicit CTL responses.[98] The converse is not always true: some immunodominant peptides bind to class I with only moderate affinity and peptides that bind with high affinity may not necessarily lead to useful CTL responses. The important interaction is probably complex and involves the peptide/class I/β2 microglobulin complex as a whole, binding to the T cell receptor. The kinetics of this interaction are very difficult to measure. Considerable work has been done for several HLA class I subtypes to map binding pockets and to determine preferred anchor residues for binding peptides.[99] In an attempt to improve the stability of the class I complex, some investigators have produced modified class I peptides with amino acid variations designed to introduce residues that improve the binding to the class I molecule. This has resulted in enhanced generation of CTL with specificity not only for the modified peptide pulsed onto target cells, but also for unmodified peptide and for naturally-processed peptide.[33,100] Interestingly, modified peptides may differ in their ability to induce cytokine release, and these differences do not always correlate with the ability of the modified peptide to induce lysis of peptide-pulsed target cells by specific CTL.[101,102] The approach of vaccination using modified peptides has now entered clinical trials and CTL have been generated in humans in vivo.[31] Confirmatory studies using these peptides in conjunction with FLT-3L or IL-2 are to be initiated soon.

Other characteristics of tumor cells may also lend themselves to peptide-based strategies allowing specific targeting of tumors rather than normal cells. For example, mutations in particular proteins in some cases lead to expression of alternate open reading frames, capable of eliciting CTL responses.[96, 103] Clearly, such CTL will not recognize normal tissues in which the gene is either not expressed or not mutated.

There is growing interest in the possibility that some peptides may be applicable for patients of different HLA class I serotypes. Recent work has shown that HLA class I peptides often can be subdivided into "supertypes" characterized by conserved anchor residues and binding affinities. [104-108] Thus far, supertypes have been defined for HLA- A2-like, A3-like and B7-like peptide families. Within each supertype, peptides bind with good affinity to all members of that group. This phenomenon has probably evolved to ensure that most individuals express class I molecules capable of binding an epitope from most antigens, allowing for serotype variations among different ethnic groups. A given supertype is expressed in about 40-50% of the population.[106] Within particular ethnic groups, total coverage by the three supertype families occurs in about 85-95% of patients, although the frequency of expression of specific class I serotypes varies.[106] It is likely that peptides can be defined that will bind to most or all members of a given supertype, making peptide vaccination more widely applicable than it currently is. Such peptides have already been defined for antigens such as the hepatitis B core antigen (amino acids 18-27, binding to members of the A2 supertype).[105]

It is unlikely that targeting one peptide or even one antigen only will allow a meaningful antitumor response to be generated, since it is almost a trivial matter for tumor cells to escape such a limited attack. The hope of immunotherapists is that the initial antitumor response directed against a single or a few antigens will lead to "epitope spreading," or a second tier immune response in which antigens exposed in the first wave of immune attack then elicit responses against that wider range of antigens. There is some preclinical evidence to support this notion.[72] Another approach is to immunize with multiple peptide targets at once. This can be achieved by using whole protein, which will contain all the relevant epitopes or by transfer of cDNA or RNA encoding the protein of interest. Interestingly, for at least one melanoma antigen (Melan A/MART-1), there is evidence that the only peptide epitope able to be generated for any class I allele is the A*0201 27-35 9-mer AAGIGILTV.[109] This suggests that for at least that antigen, there may be little advantage in using recombinant protein over peptide vaccination, and that patients who are HLA-A*0201 negative may not be able to generate a CTL response. It is not clear whether this phenomenon is unique to the Melan A/MART-1 antigen or whether other antigens may also show such limited class I immunodominance.

It is dangerous to inject or express oncogenic or transforming proteins in patients, but other non-oncogenic tumor-specific antigens may be appropriate for use in this way.[110-114] Alternatively, multiple peptide epitopes can be linked to form a "polytope" vaccine,[115,116] which allows expression of each peptide and permits immunologic recognition by peptide-specific CTL.[117] CTL induction using a polytope vaccine may also overcome issues of domi-

Table 14.4. Immunologic assays used in monitoring results from immunotherapy clinical trials

Skin testing against the appropriate antigens
- Skin
- Protein
- Cells (autologous or allogeneic)

Determination of lymphocyte number and phenotype
- Peripheral blood
- Within a tumor biopsy

Assays of lytic function of T cells against specific antigen or whole tumor cells
- Peripheral blood lymphocytes
- Tumor-infiltrating lymphocytes
- Draining lymph nodes
- Lymphocytes infiltrating vaccination sites (VILs)

Proliferative responses of T cells from the above sites against relevant antigens

Cytokine expression
- Release by bulk culture of lymphocytes (directly isolated or cloned)
- PCR analysis of cytokine mRNA
- Intracellular cytokine detection by flow cytometry
- ELISpot assays for antigen-specific cytokine release by individual cells

Tetramer assays for specific T cells

T cell receptor characterization and enumeration

Antitumor response

Survival

nance and subdominance of epitopes [117] This technique is still under investigation for both cancer and infectious disease applications.

Other investigators are using clever chemical methods for linking multiple peptides.[118] Another particularly favored approach at this point is using so called autologous tumor apoptotic bodies and cells fed to DC and then administered in vivo. We have treated three patients to date with such an approach.

New adjuvants are also becoming available. One cytokine of great potential is FLT-3L, a hemopoietic cytokine similar to SCF which preferentially expands DC and DC progenitors in the peripheral blood and many tissues, both in mice and in humans. [119-121] Natural killer cell progenitors may also be expanded after FLT-3L therapy,[122] and these cells may be activated with IL-2. Thus, FLT-3L may provide a mechanism for targeting tumors that have escaped CTL killing by downregulation of HLA class I/peptide complexes. ISCOMs ("immunostimulatory complexes") are also being investigated as potential adjuvants to be used in cancer vaccination. These lipid/saponin constructs are able to incorporate peptide or protein antigens and preliminary studies using influenza viral antigens[123] and other infective agents indicate that potent antibody and cellular responses can be induced with ISCOMs. Clinical protocols involving ISCOMs and melanoma antigens are under development by the Ludwig Institute for Cancer Research and its collaborators.

It is possible that non-classical pathways of antigen presentation may also be relevant in generating an antitumor response. CD1 is structurally related to MHC class I and is probably important in presenting carbohydrate or related epitopes.[57, 124-126] Recently the resident of the nominal antigen binding groove in CD1d expressed on DC has been identified as alpha galactosyl ceramide, capable of stimulating the so called monomorphic NKT cells with characteristic T-cell receptor rearrangements.[127-129] Although no melanoma targets have yet been identified using this antigen presentation pathway, this should be an interesting area of exploration.

A major difficulty in conducting trials of cancer immunotherapy is the selection of appropriate endpoints. Most patients on such studies are heavily pretreated and immunosuppressed, either non-specifically by dint of tumor burden or prior therapy or specifically immunosuppressed due to deletion or inactivation of CTL clones.[94] It is logical to apply these treatments to patients with a much lower burden of disease such as in the adjuvant setting, where all obvious disease is removed but there is a high chance of relapse due to microscopic residual or metastatic disease. Performing such studies using the clinical endpoints of time to tumor progression or survival involves studying very large numbers of patients over a long period of time. This is justifiable if phase II studies in patients with advanced disease indicate that there is significant clinical activity, but for most strategies this has not yet been shown to be the case. In addition, some types of immunotherapy, such as DC therapy, may theoretically also put patients at risk of long term autoimmune disease. Therefore cancer immunotherapists are caught in a classic "Catch 22" situation: the very population likely to benefit most from these interventions cannot be studied, because the population least likely to benefit has shown no response.

A surrogate marker capable of predicting clinical response is required urgently. Most studies so far have used immunologic assays as endpoints to demonstrate an immune response to the therapeutic intervention. These assays are listed in Table 14.4. To date, no such test has been shown to correlate with the ultimate clinical outcome of the patient. A criticism of most of the in vitro assays listed in Table 14.4 is that they rely on expansion of responder T cells prior to assay. This means that T cells of low affinity, low proliferative potential or simply in low numbers will not be detected, and the cells measured in the final assay may bear little resemblance to the true situation in vivo. In an attempt to overcome these limitations more modern assays have been developed allowing assays on lymphocytes without prior expansion. The ELISpot assay, initially developed to measure single cell antibody production,[130] has been modified to allow measurement of cytokine production by individual cells.[131] Measurement of these spots can be automated and quantitative assays are possible.[132] Another recently described technique involves the use of tetramerized complexes of HLA class I molecules, 2-microglobulin and appropriate peptides, labeled with a fluorescent tag. [86,88] Such tetramers can be used to detect T cells in the peripheral blood that have specificity for the class I complex of interest. The sensitivity of this assay is about 1 in 10^4 cells, depending on the sensitivity and settings of the flow cytometer. It remains to be seen whether these assays will correlate with clinical outcome.

Peptides may also be used in cancer immunotherapy in ways not involving T cell recognition. Conjugated muramyl dipeptides have been used as adjuvants in vaccination studies involving HIV[39] and influenza,[38] and elicit good antibody responses. Toxicity may be problematic especially if other adjuvants such as MF59 are used.[39] These peptides have also been used in animal studies to augment NK responses against melanoma.[133] Muramyl dipeptides may eventually find a supporting role in the immunotherapy of melanoma.

Other peptides have direct effects on the biology of tumor cells per se, rather than through immune effector cells. Some peptides block adhesion between cells and the extracellular matrix, and may be relevant in blocking metastasis of cells such as renal cell or gastric carcinoma.[134,135] These peptides include the RGDV motif and the blockade of adhesion and spreading in vitro may also be mediated by related peptides such as RGDS, LRGDV and ELRGDV.[134] It is likely that these effects are mediated through interactions with various integrin molecules.[136]

Class II epitopes have not been studied thoroughly so far in melanoma immunotherapy, because of the intrinsic greater complexity of the molecules constituting the Class II MHC molecules. Presentation of class II epitopes is important in initiating an immune response, and also for maintaining CTL and protecting them from apoptotic stimuli in the periphery. Cognate help, i.e., involving helper epitopes not derived from the immunizing antigen, is not required for the initiation of an immune response: such responses can be enhanced by using non-cognate antigens such as KLH,[79] consensus sequences such as the PADRE peptide[41] or others such as tetanus toxoid. Cognate help probably is required for the survival of CTL in the periphery, and arrives in the form of IL-2 secreted by $CD4^+$ T cells responding to cognate epitopes presented by local APC at the site of the tumor or other danger signal.[137] Recently, class II epitopes have been identified for several melanoma antigens and are under investigation to determine whether they will prove to be useful adjuncts for peptide vaccination (Storkus WJ, personal communication).

Conclusions

Clinical trials of peptide vaccination using epitopes derived from melanoma antigens are now well advanced. Clinical responses were seen somewhat surprisingly with class I peptides in the absence of any adjuvant, and this has led to an explosion of activity in the field. Many more antigens are now available and potent adjuvants designed to enhance peptide-specific cellular immune responses are also in use or under development. Response rates remain low, probably reflecting the myriad ways in which tumors may evade a CTL attack directed against one or a few antigens, particularly if only peptides from one antigen or binding to one HLA class I serotype only are being used. Future studies will involve peptides derived from multiple antigens, using several HLA class I serotypes and potent adjuvants. At the same time measures will be used to facilitate the induction of memory T cells specific for those peptides and to maintain the expanded CTL pool. A surrogate marker will need to be developed allowing prediction of patients destined to respond to immune situations, permitting the study of patients with low burdens of disease without the need for large numbers of patients and long term follow-up. Finally, strategies will need to be developed to enhance non-specific antitumor activity such as that mediated by NK cells. There is much work yet to do and sufficient reason to justify a cautious optimism for cancer immunotherapists.

Acknowledgments

IDD is supported by the Ludwig Institute for Cancer Research. MTL is supported by NCI grant 1UO1 CA 74329-01 (Clinical Trials of Biological Response Modifiers).

References

1. Coley WB. The treatment of malignant tumors by repeated inoculations of erysipelas: with a report of ten original cases. Am J Med Sci 1893; 105:487-511.
2. Burnet FM. The concept of immunological surveillance. Prog Exp Tumor Res 1970; 13:1-27.
3. Sheil AGR, Flavel S. Cancer analysis in dialysis and transplant patients. In: Disney APS, ed. Ninth report of the Australian and New Zealand combined dialysis and transplant registry. Woodville, SA: Queen Elizabeth Hospital, 1986:
4. Disney APS, Russ GR, Walker R, Sheil AGR, eds. ANZDATA registry report 1997. Adelaide, South Australia: Australia and New Zealand Dialysis and Transplant Registry, 1997.
5. Gross L. Intradermal immunization of C3H mice against a sarcoma that originated in an animal of the same line. Cancer Res 1943; 3:326-333.
6. Prehn RT, Main JM. Immunity to methylcholanthrene-induced sarcomas. J Natl Cancer Inst 1957; 18:769-778.
7. Old LJ, Boyse EA, Clarke DA, Carswell EA. Antigenic properties of chemically- induced tumors. Ann N Y Acad Sci 1962; 101:80-106.
8. Knuth A, Wölfel T, Klehmann E, Boon T, Meyer zum Buschenfelde KH. Cytolytic T-cell clones against an autologous human melanoma: specificity study and definition of three antigens by immunoselection. Proc Natl Acad Sci U S A 1989; 86:2804-2808.
9. Zitvogel L, Mayordomo J, Tjandrawan T, et al. Therapy of murine tumors with tumor peptide-pulsed dendritic cells: dependence on T cells, B7 costimulation, and T helper cell 1-associated cytokines. J Exp Med 1996; 183(1):87-97.
10. Tureci O, Sahin U, Schobert I, et al. The SSX-2 gene, which is involved in the t(X;18) translocation of synovial sarcomas, codes for the human tumor antigen HOM-MEL-40. Cancer Res 1996; 56(20):4766-4772.
11. Jäger E, Chen Y-T, Drijfhout JW, et al. Simultaneous humoral and cellular immune response against cancer-testis antigen NY-ESO-1: definition of human histocompatibility leukocyte antigen (HLA)-A2-binding epitopes. J Exp Med 1998; 187(2):265-270.
12. Boczowski D, Nair SK, Snyder D, Gilboa E. Dendritic cells pulsed with RNA are potent antigen-presenting cells in vitro and in vivo. J Exp Med 1996; 184(2):465-472.
13. Morse MA, Lyerly HK, Gilboa E, Thomas E, Nair SK. Optimization of the sequence of antigen loading and CD40-ligand-induced maturation of dendritic cells. Cancer Res 1998; 58(14):2965-2968.
14. Kawakami Y, Rosenberg SA. Immunobiology of human melanoma antigens MART-1 and gp100 and their use for immuno-gene therapy. Int Rev Immunol 1997; 14(2-3):173-192.
15. Gure AO, Türeci Ö, Sahin U, et al. SSX: a multigene family with several members transcribed in normal testis and human cancer. Int J Cancer 1997; 72:965-971.
16. De Smet C, De Backer O, Faraoni I, Lurquin C, Brasseur F, Boon T. The activation of human gene MAGE-1 in tumor cells is correlated with genome-wide demethylation. Proc Natl Acad Sci U S A 1996; 93(14):7149-7153.
17. Valmori D, Liénard D, Waanders G, Rimoldi D, Cerottini J-C, Romero P. Analysis of MAGE-3-specific cytolytic T lymphocytes in human leukocyte antigen-A2 melanoma patients. Cancer Res 1997; 57:735-741.
18. Levy F, Servis C, Cerottini J-C, Romero P, Valmori D. Modulation of proteasomal activity in vitro induces the generation of an HLA-A*0201 specific CTL-defined epitope derived from the melanoma-associated antigen MAGE-3. Cancer Vaccine Week 1998. New York: Cancer Research Institute, 1998:p1-12.
19. Viner NJ, Nelson CA, Deck B, Unanue ER. Complexes generated by the binding of free peptides to class II MHC molecules are antigenically diverse compared with those generated by intracellular processing. J Immunol 1996; 156:2365-2368.
20. Restifo NP, Esquivel F, Kawakami Y, et al. Identification of human cancers deficient in antigen processing. J Exp Med 1993; 177(2):265-272.

21. Storkus WJ, Howell DN, Salter RD, Dawson JR, Cresswell P. NK susceptibility varies inversely with target cell class I HLA antigen expression. J Immunol 1987; 138(6):1657-1659.
22. Jager E, Ringhoffer M, Altmannsberger M, et al. Immunoselection in vivo: independent loss of MHC class I and melanocyte differentiation antigen expression in metastatic melanoma. Int J Cancer 1997; 71(2):142-147.
23. Ikeda H, Lethe B, Lehmann F, et al. Characterization of an antigen that is recognized on a melanoma showing partial HLA loss by CTL expressing an NK inhibitory receptor. Immunity 1997; 6(2):199-208.
24. Lehmann F, Marchand M, Hainaut P, et al. Differences in the antigens recognized by cytolytic T cells on two successive metastases of a melanoma patient are consistent with immune selection. Eur J Immunol 1995; 25(2):240-247.
25. Toes REM, Blom RJJ, Offringa R, Kast WM, Melief CJM. Enhanced tumor outgrowth after peptide vaccination. Functional deletion of tumor-specific CTL induced by peptide vaccination can lead to the inability to reject tumors. J Immunol 1996; 156:3911-3918.
26. Jäger E, Ringhoffer M, Karbach J, Arand M, Oesch F, Knuth A. Inverse relationship of melanocyte differentiation antigen expression in melanoma tissues and CD8+ cytotoxic-T-cell responses: evidence for immunoselection of antigen-loss variants in vivo. Int J Cancer 1996; 66(4):470-476.
27. Marchand M, Weynants P, Rankin E, et al. Tumor regression responses in melanoma patients treated with a peptide encoded by gene MAGE-3 (letter). Int J Cancer 1995; 63:883-885.
28. Davis ID. Cytokine therapy in metastatic renal cancer (letter). N Engl J Med 1998; 339(12):199.
29. Jäger E, Ringhoffer M, Altmannsberger M, et al. Immunoselection in vivo: independent loss of MHC class I and melanocyte differentiation antigen expression in metastatic melanoma. Int J Cancer 1997; 71(2):142-147.
30. Jäeger E, Bernhard H, Romero P, et al. Generation of cytotoxic T cell responses with synthetic melanoma associated peptides in vivo: implications for tumor vaccines with melanoma associated antigens. Int J Cancer 1996; 66:162-169.
31. Rosenberg SA, Yang JC, Schwartzentruber DJ, et al. Immunologic and therapeutic evaluation of a synthetic peptide vaccine for the treatment of patients with metastatic melanoma. Nature Med 1998; 4(3):321-327.
32. Kawakami Y, Eliyahu S, Jennings C, et al. Recognition of multiple epitopes in the human melanoma antigen gp100 by tumor-infiltrating T lymphocytes associated with in vivo tumor regression. J Immunol 1995; 154(8):3961-3968.
33. Parkhurst MR, Salgaller ML, Southwood S, et al. Improved induction of melanoma-reactive CTL with peptides from the melanoma antigen gp100 modified at HLA-A*0201-binding residues. J Immunol 1996; 157(6):2539-2548.
34. Zhu X, Tommasino M, Vousden K, et al. Both immunization with protein and recombinant vaccinia virus can stimulate CTL specific for the E7 protein of human papilloma virus 16 in H-2d mice. Scand J Immunol 1995; 42(5):557-563.
35. Noguchi Y, Richards EC, Chen YT, Old LJ. Influence of interleukin 12 on p53 peptide vaccination against established Meth A sarcoma. Proc Natl Acad Sci U S A 1995; 92(6):2219-2223.
36. Fattom A, Li X, Cho YH, et al. Effect of conjugation methodology, carrier protein, and adjuvants on the immune response to Staphylococcus aureus capsular polysaccharides. Vaccine 1995; 13(14):1288-1293.
37. Lewis JJ, Janetzki S, Wang S, et al. Phase I trial of vaccination with tyrosinase peptide plus QS-21 in melanoma. Proceedings of the American Society of Clinical Oncology. Los Angeles: American Society of Clinical Oncology, 1998:428a.
38. Keitel W, Couch R, Bond N, Adair S, Van Nest G, Dekker C. Pilot evaluation of influenza virus vaccine (IVV) combined with adjuvant. Vaccine 1993; 11(9):909-913.
39. Kahn JO, Sinangil F, Baenziger J, et al. Clinical and immunologic responses to human immunodeficiency virus (HIV) type 1SF2 gp120 subunit vaccine combined with MF59 adjuvant with or without muramyl tripeptide dipalmitoyl phosphatidylethanolamine in non-HIV-infected human volunteers. J Infect Dis 1994; 170(5):1288-1291.
40. Cormier JN, Salgaller ML, Prevette T, et al. Enhancement of cellular immunity in melanoma patients immunized with a peptide from MART-1/Melan A. Cancer J Sci Am 1997; 3(1):37-44.

41. Weber J, Spears L, Marty V, Hua F, Kuniyoshi C, Celis E. Phase I trial of a MAGE-3 peptide vaccine in patients with resected stages IIb, III and IV melanoma. Proceedings of the American Society of Clinical Oncology. Los Angeles: American Society of Clinical Oncology, 1998:489a.
42. Vitiello A, Ishioka G, Gray HM, et al. Development of a lipopeptide-based therapeutic vaccine to treat chronic HBV infection. I. Induction of a primary cytotoxic T lymphocyte response in humans. J Clin Invest 1995; 95:341-349.
43. Caux C, Dezutter-Dambuyant C, Schmitt D, Banchereau J. GM-CSF and TNF-alpha cooperate in the generation of dendritic Langerhans cells. Nature 1992; 360:258-261.
44. Sallusto F, Lanzavecchia A. Efficient presentation of soluble antigen by cultured human dendritic cells is maintained by granulocyte/macrophage colony-stimulating factor plus interleukin 4 and down-regulated by tumor necrosis factor . J Exp Med 1994; 179:1109-1118.
45. Luft T, Pang KC, Thomas E, et al. A serum-free culture model for studying the differentiation of human dendritic cells from adult CD34+ progenitor cells. Exp Hematol 1998; 26:489-500.
46. Dranoff G, Jaffee E, Lazenby A, et al. Vaccination with irradiated tumor cells engineered to secrete murine granulocyte-macrophage colony-stimulating factor stimulates potent, specific, and long-lasting anti-tumor immunity. Proc Natl Acad Sci U S A 1993; 90:3539-3543.
47. Ellem KA, O'Rourke MG, Johnson GR, et al. A case report: immune responses and clinical course of the first human use of granulocyte/macrophage-colony-stimulating-factor-transduced autologous melanoma cells for immunotherapy. Cancer Immunol Immunother 1997; 44:10-20.
48. Simons JW, Jaffee EM, Weber CE, et al. Bioactivity of autologous irradiated renal cell carcinoma vaccines generated by ex vivo granulocyte-macrophage colony-stimulating factor gene transfer. Cancer Res 1997; 57:1537-1546.
49. Si Z, Hersey P, Coates AS. Clinical responses and lymphoid infiltrates in metastatic melanoma following treatment with intralesional GM-CSF. Melanoma Res 1996; 6:247-255.
50. Jäger E, Ringhoffer M, Karbach J, Arand M, Oesch F, Knuth A. Granulocyte-macrophage-colony-stimulating factor enhances immune responses to melanoma-associated peptides in vivo. Int J Cancer 1996; 67:54-62.
51. Cole DJ, Gattonicelli S, Mcclay EF, et al. Characterization of a sustained-release delivery system for combined cytokine/peptide vaccination using a poly-N-acetyl glucosamine-based polymer matrix. Clin Cancer Res 1997; 3(6):867-873.
52. Storkus WJ, Tahara H, Lotze MT. Interleukin-12. In: Thomson A, ed. The cytokine handbook. San Diego: Academic Press, 1998:391-425.
53. Atkins MB, Lotze M, Wiernik P, et al. High-dose IL-2 therapy alone results in long-term durable complete remissions in patients with metastatic melanoma. Proceedings of the American Society of Clinical Oncology. Denver: American Society of Clinical Oncology, 1997:1780.
54. Shimizu K, Fields RC, Giedlin M, Mulé JJ. Systemic administration of interleukin 2 enhances the therapeutic efficacy of dendritic cell-based tumor vaccines. Proc Natl Acad Sci U S A 1999; 96(5):2268-2273.
55. Inaba K, Metlay JP, Crowley MT, Witmer-Pack M, Steinman RM. Dendritic cells as antigen presenting cells in vivo. Int Rev Immunol 1990; 6(2-3):197-206.
56. Takamizawa M, Rivas A, Fagnoni F, et al. Dendritic cells that process and present nominal antigens to naive T lymphocytes are derived from CD2+ precursors. J Immunol 1997; 158:2134-2142.
57. Reimann J, Kaufmann SH. Alternative antigen processing pathways in anti-infective immunity. Curr Opin Immunol 1997; 9(4):462-469.
58. Brossart P, Bevan MJ. Presentation of exogenous protein antigens on major histocompatibility complex class I molecules by dendritic cells: pathway of presentation and regulation by cytokines. Blood 1997; 90(4):1594-1599.
59. Shen Z, Reznikoff G, Dranoff G, Rock KL. Cloned dendritic cells can present exogenous antigens on both MHC class I and class II molecules. J Immunol 1997; 158(6):2723-2730.
60. Norbury CC, Chambers BJ, Prescott AR, Ljunggren HG, Watts C. Constitutive macropinocytosis allows TAP-dependent major histocompatibility complex class I presentation of exogenous soluble antigen by bone marrow-derived dendritic cells. Eur J Immunol 1997; 27(1):280-288.
61. Harding CV. Class I MHC presentation of exogenous antigens. J Clin Immunol 1996; 16(2):90-96.

62. Bohm W, Schirmbeck R, Elbe A, et al. Exogenous hepatitis B surface antigen particles processed by dendritic cells or macrophages prime murine MHC class I-restricted cytotoxic T lymphocytes in vivo. J Immunol 1995; 155(7):3313-3321.
63. Rock KL, Rothstein L, Gamble S, Fleischacker C. Characterization of antigen-presenting cells that present exogenous antigens in association with class I MHC molecules. J Immunol 1993; 150(2):438-446.
64. Albert ML, Sauter B, Bhardwaj N. Dendritic cells acquire antigen from apoptotic cells and induce class I-restricted CTLs. Nature 1998; 392(6671):86-89.
65. Ingulli E, Mondino A, Khoruts A, Jenkins MK. In vivo detection of dendritic cell antigen presentation to CD4(+) T cells. J Exp Med 1997; 185(12):2133-2141.
66. Rescigno M, Martino M, Sutherland CL, Gold MR, Ricciardi-Castagnoli P. Dendritic cell survival and maturation are regulated by different signaling pathways. J Exp Med 1998; 188(11):2175-2180.
67. Galibert L, Tometsko ME, Anderson DM, Cosman D, Dougall WC. The involvement of multiple tumor necrosis factor receptor (TNFR)-associated factors in the signaling mechanisms of receptor activator of NF-kappaB, a member of the TNFR superfamily. J Biol Chem 1998; 273(51):34120-34127.
68. Grohmann U, Belladonna ML, Bianchi R, et al. IL-12 acts directly on DC to promote nuclear localization of NF-kappaB and primes DC for IL-12 production. Immunity 1998; 9(3):315-323.
69. Celluzzi CM, Falo LD, Jr. Physical interaction between dendritic cells and tumor cells results in an immunogen that induces protective and therapeutic tumor rejection. J Immunol 1998; 160:3081-3085.
70. Knight SC, Hunt R, Dore C, Medawar PB. Influence of dendritic cells on tumor growth. Proc Am Soc Clin Oncol 1985; 82:4495-4497.
71. Mayordomo JI, Zorina T, Storkus WJ, et al. Bone marrow-derived dendritic cells pulsed with synthetic tumour peptides elicit protective and therapeutic antitumour immunity. Nature Med 1995; 1(12):1297-1302.
72. Celluzzi CM, Mayordomo JI, Storkus WJ, Lotze MT, Falo LD, Jr. Peptide-pulsed dendritic cells induce antigen-specific, CTL-mediated protective tumor immunity. J Exp Med 1996; 183:283-287.
73. Paglia P, Chiodoni C, Rodolfo M, Colombo MP. Murine dendritic cells loaded in vitro with soluble protein prime cytotoxic T lymphocytes against tumor antigen in vivo. J Exp Med 1996; 183:317-322.
74. Young JW, Inaba K. Dendritic cells as adjuvants for class I major histocompatibility complex-restricted antitumor immunity. J Exp Med 1996; 183(1):7-11.
75. Luft T, Pang KC, Thomas E, et al. Type I IFNs enhance the terminal differentiation of dendritic cells. J Immunol 1998; 161:1947-1953.
76. McLellan AD, Sorg RV, Williams LA, Hart DNJ. Human dendritic cells activate T lymphocytes via a CD40:CD40 ligand-dependent pathway. Eur J Immunol 1996; 26:1204-1210.
77. Hsu FJ, Benike C, Fagnoni F, Liles TM, Czerwinski D, Taidi B. Vaccination of patients with B-cell lymphoma using autologous antigen-pulsed dendritic cells. Nature Med 1996; 2:52-58.
78. Mukherji B, Chakraborty NG, Yamasaki S, et al. Induction of antigen-specific cytolytic T cells in situ in human melanoma by immunization with synthetic peptide-pulsed autologous antigen presenting cells. Proc Natl Acad Sci U S A 1995; 92:8078-8082.
79. Nestle FO, Alijagic S, Gilliet M, et al. Vaccination of melanoma patients with peptide- or tumor lysate-pulsed dendritic cells. Nature Med 1998; 4(3):328-332.
80. Lotze MT, Shurin M, Davis I, Amoscato A, Storkus WJ. Dendritic cell based therapy of cancer. Adv Exp Med Biol 1997; 417:551-569.
81. Lotze MT, Hellerstedt B, Stolinski L, et al. The role of interleukin-2, interleukin-12, and dendritic cells in cancer therapy. Cancer J Sci Am 1997; 3(Supplement 1):S109-S114.
82. Nestle FO. DC vaccination: clinical trial in melanoma. In: Lotze MT, Steinman R, Banchereau J, eds. 5th International Symposium on Dendritic Cells in Fundamental and Clinical Immunology. Pittsburgh, PA: Society for Leukocyte Biology, 1998.
83. Lotze M, Elder E, Whiteside T, et al. Dendritic cell therapy of cancer - not just an antigen presenting cell. In: Lotze MT, Steinman R, Banchereau J, eds. 5th International Symposium on Dendritic Cells in Fundamental and Clinical Immunology. Pittsburgh, PA: Society for Leukocyte Biology, 1998.

84. Chakraborty NG, Sporn JR, Tortora AF, et al. Immunization with a tumor-cell-lysate-loaded autologous-antigen-presenting-cell-based vaccine in melanoma. Cancer Immunol Immunother 1998; 47(1):58-64.
85. Lee K-H, Panelli MC, Kim CJ, et al. Functional dissociation between local and systemic immune response during anti-melanoma peptide vaccination. J Immunol 1998; 161(8):4183-4194.
86. Dunbar PR, Ogg GS, Chen J, Rust N, van der Bruggen P, Cerundolo V. Direct isolation, phenotyping and cloning of low-frequency antigen-specific cytotoxic T lymphocytes from peripheral blood. Current Biol 1998; 8(7):413-416.
87. Yee C, Savage PA, Lee PP, Davis MM, Greenberg PD. Isolation of high avidity melanoma-reactive CTL from heterogeneous populations using peptide-MHC tetramers. J Immunol 1999; 162:2227-2234.
88. Greenberg P, Yee C, Öhlén C, et al. Therapy of human viral and malignant diseases with T-cell clones: lessons for vaccine design. Cancer Vaccine Week 1998. New York, NY: Cancer Research Institute, 1998:S26.
89. Brusic V, Rudy G, Harrison LC. MHCPEP, a database of MHC-binding peptides: update 1997. Nucl Acids Res 1998; 26(1):368-371.
90. Parker KC, Bednarek MA, Coligan JE. Scheme for ranking potential HLA-A2 binding peptides based on independent binding of individual peptide side-chains. J Immunol 1994; 152:163-175.
91. Brusic V, Rudy G, Harrison LC. Prediction of MHC binding peptides using artificial neural networks. In: Stonier RJ, Yu XS, eds. Complex Systems: Mechanism of Adaptation. Amsterdam/ OMSHA Tokyo: IOS Press, 1994:253-260.
92. Brusic V, Honeyman G, Hammer J, Harrison L. Prediction of MHC class II-binding peptides using an evolutionary algorithm and artificial neural network. Bioinformatics 1998; 14(2):121-130.
93. Honeyman MC, Brusic V, Stone N, Harrison LC. Neural network-based prediction of candidate T-cell epitopes. Nature Biotechnol 1998; 16(10):966-969.
94. Chen Q, Jackson H, Gibbs P, Davis ID, Trapani J, Cebon J. Spontaneous T cell responses to melanoma differentiation antigens from melanoma patients and healthy subjects. Cancer Immunol Immunother 1998; 47(4):191-197.
95. Tsai V, Southwood S, Sidney J, et al. Identification of subdominant CTL epitopes of the GP100 melanoma-associated tumor antigen by primary in vitro immunization with peptide-pulsed dendritic cells. J Immunol 1997; 158(4):1796-1802.
96. Wang R-F, Johnston L, Zeng G, Topalian SL, Schwartzentruber DJ, Rosenberg SA. A breast and melanoma-shared tumor antigen: T cell responses to antigenic peptides translated from different open reading frames. J Immunol 1998; 161(7):3598-3606.
97. Kawakami Y, Robbins PF, Wang RF, Parkhurst M, Kang X, Rosenberg SA. The use of melanosomal proteins in the immunotherapy of melanoma. J Immunother 1998; 21(4):237-246.
98. Sette A, Vitiello A, Reherman B, et al. The relationship between class I binding affinity and immunogenicity of potential cytotoxic T cell epitopes. J Immunol 1994; 153:5586-5592.
99. Ruppert J, Sidney J, Celis E, Kubo RT, Grey HM, Sette A. Prominent role of secondary anchor residues in peptide binding to HLA-A2.1 molecules. Cell 1993; 74:929-937.
100. Valmori D, Fonteneau J-F, Lizana CM, et al. Enhanced generation of specific tumor-reactive CTL in vitro by selected Melan-A/MART-1 immunodominant peptide analogues. J Immunol 1998; 160:1750-1758.
101. Gervois N, Guilloux Y, Diez E, Jotereau F. Suboptimal activation of melanoma infiltrating lymphocytes (TIL) due to low avidity of TCR/MHC-tumor peptide interactions. J Exp Med 1996; 183(5):2403-2407.
102. Maeurer MJ, Chan HW, Karbach J, et al. Amino acid substitutions at position 97 in HLA-A2 segregate cytolysis from cytokine release in MART-1/Melan-A peptide AAGIGILTV-specific cytotoxic T lymphocytes. Eur J Immunol 1996; 26(11):2613-2623.
103. Wang RF, Parkhurst MR, Kawakami Y, Robbins PF, Rosenberg SA. Utilization of an alternative open reading frame of a normal gene in generating a novel human cancer antigen. J Exp Med 1996; 183(3):1131-1140.
104. Sidney J, del Guercio M-F, Southwood S, et al. Several HLA alleles share overlapping peptide specificities. J Immunol 1995; 154:247-259.
105. del Guercio M-F, Sidney J, Hermanson G, et al. Binding of a peptide antigen to multiple HLA alleles allows definition of an A2-like supertype. J Immunol 1995; 154:685-693.

106. Sidney J, Grey HM, Southwood S, et al. Definition of an HLA-A3-like supermotif demonstrates the overlapping peptide-binding repertoires of common HLA molecules. Human Immunol 1996; 45:79-93.
107. Bertoni R, Sidney J, Fowler P, Chesnut RW, Chisari FV, Sette A. Human histocompatibility leukocyte antigen-binding supermotifs predict broadly cross-reactive cytotoxic T lymphocyte responses in patients with acute hepatitis. J Clin Invest 1997; 100(3):503-513.
108. Southwood S, Sidney J, Kondo A, et al. Several common HLA-DR types share largely overlapping peptide binding repertoires. J Immunol 1998; 160:3363-3373.
109. Bettinotti MP, Kim CJ, Lee K-H, et al. Stringent allele/epitope requirements for MART-1/Melan A immunodominance: implications for peptide-based immunotherapy. J Immunol 1998; 161:877-889.
110. Nair SK, Boczkowski D, Morse M, Cumming RI, Lyerly HK, Gilboa E. Induction of primary carcinoembryonic antigen (CEA)-specific cytotoxic T lymphocytes in vitro using human dendritic cells transfected with RNA. Nature Biotechnol 1998; 16(4):364-369.
111. Tuting T, DeLeo AB, Lotze MT, Storkus WJ. Genetically modified bone marrow-derived dendritic cells expressing tumor-associated viral or "self" antigens induce antitumor immunity in vivo. Eur J Immunol 1997; 27(10):2702-2707.
112. Fernandez N, Duffour MT, Perricaudet M, Lotze MT, Tursz T, Zitvogel L. Active specific T-cell-based immunotherapy for cancer: nucleic acids, peptides, whole native proteins, recombinant viruses, with dendritic cell adjuvants or whole tumor cell-based vaccines. Principles and future prospects. Cytokines Cell Mol Ther 1998; 4(1):53-65.
113. Tuting T, Zorina T, Ma DI, et al. Development of dendritic cell-based genetic vaccines for cancer. Adv Exp Med Biol 1997; 417:511-518.
114. Tuting T, Storkus WJ, Lotze MT. Gene-based strategies for the immunotherapy of cancer. J Mol Med 1997; 75(7):478-491.
115. Suhrbier A. Multi-epitope DNA vaccines. Immunol Cell Biol 1997; 75(4):402-408.
116. Thomson SA, Sherritt MA, Medveczky J, et al. Delivery of multiple CD8 cytotoxic T cell epitopes by DNA vaccination. J Immunol 1998; 160(4):1717-1723.
117. Mateo L, Gardner J, Chen Q, Schmidt C, Cebon J, Suhrbier A. Polytope vaccines for melanoma. Australian Society for Immunology 28th Annual Conference, Tumour Immunology Workshop. Melbourne, Australia: Australian Society for Immunology, 1998:S38-34.
118. Jackson DC, O'Brien-Simpson N, Ede NJ, Brown L. Free radical induced polymerization of synthetic peptides into polymeric immunogens. Vaccine 1997; 15(15):1697-1705.
119. Chen K, Braun S, Lyman S, et al. Antitumor activity and immunotherapeutic properties of Flt3-ligand in a murine breast cancer model. Cancer Res 1997; 57(16):3511-3516.
120. Fields RC, Osterholzer JJ, Fuller JA, Thomas EK, Geraghty PJ, Mule JJ. Comparative analysis of murine dendritic cells derived from spleen and bone marrow. J Immunother 1998; 215):323-329.
121. Rosenzwajg M, Camus S, Guigon M, Gluckman JC. The influence of interleukin (IL)-4, IL-13, and Flt3 ligand on human dendritic cell differentiation from cord blood CD34+ progenitor cells. Exp Hematol 1998; 26(1):63-72.
122. Shaw SG, Maung AA, Steptoe RJ, Thomson AW, Vujanovic NL. Expansion of functional NK cells in multiple tissue compartments of mice treated with Flt3-ligand: implications for anti-cancer and anti-viral therapy. J Immunol 1998; 161(6):2817-2824.
123. Coulter A, Wong TY, Drane D, Bates J, Macfarlan R, Cox J. Studies on experimental adjuvanted influenza vaccines: comparison of immune stimulating complexes (Iscoms) and oil-in-water vaccines. Vaccine 1998; 16(11-12):1243-1253.
124. Castaño AR, Tangri S, Miller JEW, et al. Peptide binding and presentation by mouse CD1. Science 1995; 269:223-226.
125. Park S-H, Roark JH, Bendelac A. Tissue-specific recognition of mouse CD1 molecules. J Immunol 1998; 160:3128-3134.
126. Sallusto F, Nicolò C, Maria RD, Corinti S, Testi R. Ceramide inhibits antigen uptake and presentation by dendritic cells. J Exp Med 1996; 184:2411-2416.
127. Kawano T, Cui J, Koezuka Y, et al. Natural killer-like nonspecific tumor cell lysis mediated by specific ligand-activated Valpha14 NKT cells. Proc Natl Acad Sci U S A 1998; 95(10):5690-5693.

128. Brossay L, Naidenko O, Burdin N, Matsuda J, Sakai T, Kronenberg M. Structural requirements for galactosylceramide recognition by CD1-restricted NK T cells. J Immunol 1998; 161(10):5124-5128.
129. Brossay L, Chioda M, Burdin N, et al. CD1d-mediated recognition of an alpha-galactosylceramide by natural killer T cells is highly conserved through mammalian evolution. J Exp Med 1998; 188(8):1521-1528.
130. Lycke N. A sensitive method for the detection of specific antibody production in different isotypes from single lamina propria plasma cells. Scand J Immunol 1986; 24(4):393-403.
131. Czerkinsky C, Andersson G, Ekre HP, Nilsson LA, Klareskog L, Ouchterlony O. Reverse ELISPOT assay for clonal analysis of cytokine production. I. Enumeration of gamma-interferon-secreting cells. J Immunol Methods 1988; 110(1):29-36.
132. Herr W, Linn B, Leister N, Wandel E, Meyer zum Buschenfelde KH, Wolfel T. The use of computer-assisted video image analysis for the quantification of CD8+ T lymphocytes producing tumor necrosis factor alpha spots in response to peptide antigens. J Immunol Methods 1997; 203(2).141-152.
133. Sosnowska D, Mysliwski A, Dzierzbicka K, Kolodziejczyk AM. The in vitro effect of new muramyl peptide derivatives on cytotoxic activity of NK (natural killer) cells from hamsters bearing Ab Bomirski melanoma. Biotherapy 1997; 10(2):161-168.
134. Takagaki M, Honke K, Tsukamoto T, et al. Zn-alpha 2-glycoprotein is a novel adhesive protein. Biochem Biophys Res Commun 1994; 201(3):1339-1347.
135. Matsuoka T, Hirakawa K, Chung YS, et al. Adhesion polypeptides are useful for the prevention of peritoneal dissemination of gastric cancer. Clin Exp Metastasis 1998; 16(4):381-388.
136. Lanza P, Felding Habermann B, Ruggeri ZM, Zanetti M, Billetta R. Selective interaction of a conformationally-constrained Arg-Gly-Asp (RGD) motif with the integrin receptor alphavbeta3 expressed on human tumor cells. Blood Cells Mol Dis 1997; 23(2):230-241.
137. Ossendorp F, Mengede E, Camps M, Filius R, Melief CJ. Specific T helper cell requirement for optimal induction of cytotoxic T lymphocytes against major histocompatibility complex class II negative tumors. J Exp Med 1998; 187(5):693-702.

Index